T0142804

Recent Results in Cancer Research

Volume 207

More information about this series at http://www.springer.com/series/392

Thorsten Cramer · Clemens A. Schmitt
Editors

Metabolism in Cancer

 Springer

Editors
Thorsten Cramer
Molecular Tumor Biology
University Hospital RWTH
Aachen
Germany

Clemens A. Schmitt
Department of Hematology and Oncology
Charité
Berlin
Germany

ISSN 0080-0015   ISSN 2197-6767   (electronic)
Recent Results in Cancer Research
ISBN 978-3-319-82501-4   ISBN 978-3-319-42118-6   (eBook)
DOI 10.1007/978-3-319-42118-6

Printed on acid-free paper

This Springer imprint is published by Springer Nature
The registered company is Springer International Publishing AG Switzerland

Contents

Contributors

Ivayla Apostolova Department of Radiology and Nuclear Medicine, Medical School, Otto-von-Guericke University, Magdeburg, Germany

Nikolaus Berndt Institute of Biochemistry—Computational Systems Biochemistry Group, Charité—Universitätsmedizin Berlin, Berlin, Germany

Sebastian Brandhorst Department of Biological Sciences, School of Gerontology, Longevity Institute, University of Southern California, Los Angeles, CA, USA

Charlene Brault Biocenter, Theodor-Boveri-Institute, Würzburg, Germany

Winfried Brenner Department of Nuclear Medicine, University Medicine Charité, Berlin, Germany

Jan Budczies Institute of Pathology, Charité—Universitätsmedizin Berlin, Berlin, Germany

Lewis C. Cantley Department of Medicine, The Cancer Center, Weill Cornell Medical College, New York, NY, USA

Chi V. Dang Division of Hematology-Oncology, Department of Medicine, Abramson Family Cancer Research Institute, Perelman School of Medicine, University of Pennsylvania, Philadelphia, PA, USA

Carsten Denkert Institute of Pathology, Charité—Universitätsmedizin Berlin, Berlin, Germany

Giulio F. Draetta Department of Genomic Medicine, The University of Texas MD Anderson Cancer Center, Houston, TX, USA; Department of Molecular and Cellular Oncology, The University of Texas MD Anderson Cancer Center, Houston, TX, USA

Adrian L. Harris Molecular Oncology Laboratories, Department of Medical Oncology, Weatherall Institute of Molecular Medicine, University of Oxford, Oxford, UK

Hermann-Georg Holzhütter Institute of Biochemistry—Computational Systems Biochemistry Group, Charité—Universitätsmedizin Berlin, Berlin, Germany

Annie L. Hsieh Abramson Family Cancer Research Institute, Perelman School of Medicine, University of Pennsylvania, Philadelphia, PA, USA; Department of Pathology, Johns Hopkins University School of Medicine, Baltimore, MD, USA

Stefan Kempa Integrative Proteomics and Metabolomics, Max-Delbrueck-Center of Molecular Medicine in the Helmholtz Association, Berlin, Germany

Evan C. Lien Department of Pathology, Beth Israel Deaconess Medical Center, Harvard Medical School, Boston, MA, USA

Valter D. Longo Department of Biological Sciences, School of Gerontology, Longevity Institute, University of Southern California, Los Angeles, CA, USA; IFOM, FIRC Institute of Molecular Oncology, Milan, Italy

Costas A. Lyssiotis Department of Molecular and Integrative Physiology, University of Michigan, Ann Arbor, MI, USA; Department of Internal Medicine, Division of Gastroenterology, University of Michigan, Ann Arbor, MI, USA

Alan McIntyre Molecular Oncology Laboratories, Department of Medical Oncology, Weatherall Institute of Molecular Medicine, University of Oxford, Oxford, UK

Wolfgang Mueller-Klieser Institute of Pathophysiology, University Medical Center, Johannes Gutenberg-University of Mainz, Mainz, Germany

Almut Schulze Biocenter, Theodor-Boveri-Institute, Würzburg, Germany; Comprehensive Cancer Center Mainfranken, Würzburg, Germany

Christopher Smyl Department of Hepatology and Gastroenterology Charité—Universitätsmedizin Berlin, Berlin, Germany

Andrea Viale Department of Genomic Medicine, The University of Texas MD Anderson Cancer Center, Houston, TX, USA

Nadine F. Voelxen Institute of Pathophysiology, University Medical Center, Johannes Gutenberg-University of Mainz, Mainz, Germany

Stefan Walenta Institute of Pathophysiology, University Medical Center, Johannes Gutenberg-University of Mainz, Mainz, Germany

Florian Wedel Department of Nuclear Medicine, University Medicine Charité, Berlin, Germany

Christin Zasada Integrative Proteomics and Metabolomics, Max-Delbrueck-Center of Molecular Medicine in the Helmholtz Association, Berlin, Germany

Description and Purpose of the Work

The notion that tumor cells display characteristic alterations of metabolic pathways has significantly changed our understanding of cancer. While the first description of tumor-specific changes in cellular energetics was published over 90 years ago, the causal significance of this observation for the pathogenesis of cancer was only discovered in the post-genome era. The first 10 years of the twenty-first century were characterized by a rapid gain in knowledge on the functional role of cancer-specific metabolism as well as the underlying molecular pathways. Various unanticipated interrelations of metabolic alterations with cancer-driving pathways were identified and are awaiting translation into diagnosis and therapy of cancer. Velocity, quantity, and complexity of these new discoveries render it difficult for researchers to keep up-to-date with the latest developments. This textbook provides concise chapters of internationally renowned experts on various important aspects of cancer-associated metabolism and hence a comprehensive platform for an overview of the central features of this exciting research field.

The Role of Glucose and Lipid Metabolism in Growth and Survival of Cancer Cells

Charlene Brault and Almut Schulze

Abstract

One of the prerequisites for cell growth and proliferation is the synthesis of macromolecules, including proteins, nucleic acids and lipids. Cells have to alter their metabolism to allow the production of metabolic intermediates that are the precursors for biomass production. It is now evident that oncogenic signalling pathways target metabolic processes on several levels and metabolic reprogramming has emerged as a hallmark of cancer. The increased metabolic demand of cancer cells also produces selective dependencies that could be targeted for therapeutic intervention. Understanding the role of glucose and lipid metabolism in supporting cancer cell growth and survival is crucial to identify essential processes that could provide therapeutic windows for cancer therapy.

Keywords

Glucose metabolism · Lipid metabolism · Cancer · Fatty acids · Lipid peroxidation · Oncogenic signalling pathways

1 Glucose Metabolism in Cancer Cells

The german biochemist Otto Warburg already established in the first half of the last century that cancer tissue consumes large amounts of glucose irrespective of the availability of oxygen (Warburg et al. 1924). This observation led to the definition

C. Brault · A. Schulze (✉)
Biocenter, Theodor-Boveri-Institute, Am Hubland, 97074 Würzburg, Germany
e-mail: almut.schulze@uni-wuerzburg.de

A. Schulze
Comprehensive Cancer Center Mainfranken, Josef-Schneider-Str. 6, 97080 Würzburg, Germany

© Springer International Publishing Switzerland 2016
T. Cramer and C.A. Schmitt (eds.), *Metabolism in Cancer*,
Recent Results in Cancer Research 207, DOI 10.1007/978-3-319-42118-6_1

of aerobic glycolysis in cancer cells, meaning that the relative use of oxidative and non-oxidative glucose metabolism is uncoupled from oxygen levels, which is now better known as the 'Warburg effect'. It was initially thought that increased glycolytic ATP production in cancer cells is the consequence of reduced mitochondrial function, potentially caused by the mutation of the mitochondrial genome. However, it is now clear that the increased glucose uptake and glycolytic activity in cancer cells allow the redirection of glucose-derived metabolites into biosynthetic pathways (Fig. 1).

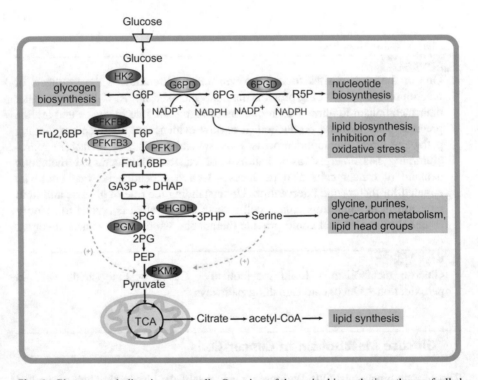

Fig. 1 Glucose metabolism in cancer cells. Overview of the major biosynthetic pathways fuelled by glucose in cancer cells. Phosphorylation of glucose by hexokinase 2 (HK2) retains glucose within the cell. Glucose-6-phosphate (G6P) can enter glycolysis, pentose phosphate pathway or glycogen biosynthesis. Regulation of the levels of the allosteric activator fructose-2,6-bisphosphate (FR2,6BP) by phosphofructokinase-2/fructose-2,6-bisphosphatase (PFKFB3 and PFKFB4) controls the activity of phosphofructokinase 1 (PFK1). Redirection of 3-phosphoglycerate (3PG) into serine biosynthesis by phosphoglycerate dehydrogenase (PHGDH) fuels the production of glycine, purines and lipid head groups as well as the one-carbon metabolism. The low activity of the M2 isoform of pyruvate kinase (PKM2) controls the last step of the glycolytic cascade, thereby allowing the use of glycolytic intermediates into biosynthetic reactions. Production of NADPH from NADP + by glucose-6-phosphate dehydrogenase (G6PDH) and 6-phosphogluconate dehydrogenase (6PGD), two enzymes within the oxidative arm of the pentose phosphate pathway, provides reducing equivalents for lipid biosynthesis and antioxidant production

Many oncogenic signalling pathways affect glycolytic activity in cancer cells. The expression of many glycolytic enzymes is induced by the oncogenic transcription factor c-Myc and the hypoxia-inducible factor (HIF) (Cairns et al. 2011). One of the best-studied pathways is the phosphatidylinositol 3-phosphate kinase (PI3K) pathway, which is frequently activated in human cancer. In normal cells, PI3K is activated in response to growth factor binding to receptor tyrosine kinases at the cell surface. This stimulates the lipid kinase activity of PI3K and leads to the generation of phosphatidylinositol 3,4,5-trisphosphate (PIP3), a lipid second messenger (Vanhaesebroeck et al. 2012). This activates several downstream effectors including the serine/threonine kinase Akt (c-Akt/PKB). Akt is one of the major mediators of insulin signalling and has a number of important metabolic target proteins in different tissues, thereby controlling the removal of glucose from the bloodstream (Manning and Cantley 2007). In cancer cells, Akt increases glucose uptake by enhancing the localisation of the glucose transporters 1 and 4 (GLUT1 and GLUT4) to the plasma membrane. Akt also leads to the phosphorylation of hexokinase 2 (HK2), the enzyme that catalyses the first and irreversible step of the glycolytic cascade, and enhances its localisation to the mitochondrial membrane (Gottlob et al. 2001; Majewski et al. 2004). This couples the conversion of glucose to glucose-6-phosphate to mitochondrial ATP production and protects cancer cells from apoptosis. Increased hexokinase activity in cancer cells is also exploited for diagnostic purposes as it causes the retention of ^{18}F-fluorodeoxyglucose (^{18}F-FDG) for positron emission tomography (PET) imaging. Targeting hexokinase activity to inhibit glycolysis in cancer cells has been discussed for some time. But it has only been shown quite recently that deletion of HK2 efficiently blocks tumour development in several genetically engineered mouse models of human cancer (Patra et al. 2013). Importantly, the same study showed that systemic deletion of HK2 in adult mice has no detrimental effects, making the inhibition of this enzyme a valid strategy for drug development.

Akt also modulates glycolytic activity by phosphorylating the heart isoform of phosphofructokinase-2/fructose-2,6-bisphosphatase (PFK-2/Fru-2,6BPase). This family of bifunctional enzymes regulates the interconversion of fructose 6-phosphate (F6-P) and fructose-2,6-bisphosphate (F2,6-BP), which is an allosteric activator of the enzyme that catalyses the second irreversible step of glycolysis, phosphofructokinase 1 (PFK1). The mammalian genome encodes several PFKFB isoforms (PFKFB1-4), which show differences in their tissue-specific expression and relative activity of their kinase and bisphosphatase domains (Rider et al. 2004). The high kinase activity of the brain isoform (PFKFB3) can promote glycolysis in tumour cells (Yalcin et al. 2009), and PFKFB3 has been shown to be required for Ras-dependent transformation (Telang et al. 2006). However, PFKFB3 inhibition also stimulates autophagy, thereby providing a survival mechanism in tumour cells (Klarer et al. 2014). Interestingly, PFKFB3 activity promotes glycolysis in endothelial cells and supports the induction of vessel sprouting during developmental angiogenesis (De Bock et al. 2013). This observation exemplifies that metabolic processes in both tumour and stromal cells could be targeted therapeutically.

Another PFKFB isoform, PFKFB4, was identified as an important gene for prostate cancer cell survival in a functional screen (Ros et al. 2012). This isoform seems to function mainly as a fructose-2,6-bisphosphatase, at least in prostate cancer cells, as depletion of PFKFB4 increased the levels of F2,6-BP leading to enhanced glycolytic flux and depletion of metabolites from the pentose phosphate pathway. This reduces the production of NADPH and, consequently, the reduced form of glutathione, an important antioxidant, leading to oxidative stress and cell death (Ros et al. 2012). Another screen, this time in glioma stemlike cells, also demonstrated a role for PFKFB4 in cancer cell survival (Goidts et al. 2012). While normal brain expresses high levels of PFKFB3, PFKFB4 expression was increased in primary high-grade glioma and was predictive of poor survival outcome (Goidts et al. 2012).

The regulation of glycolytic activity by PFKFB proteins is also tightly associated with the function of the AMP-activated protein kinase (AMPK). Several PFKFB isoforms, including PFKFB3, can be phosphorylated by AMPK on a conserved serine in the C-terminal regulatory domain. AMPK-dependent phosphorylation increases the activity of the kinase domain of PFKFB (Barford et al. 1991). As AMPK is activated in response to low ATP levels, phosphorylation of PFKFB by AMPK increases glycolytic ATP generation under conditions of energy deprivation. While the AMPK pathway is generally considered to have tumour suppressor functions, experimental evidence also suggests that AMPK could be important to limit biosynthetic processes in cancer cells when nutrients are scarce, thereby preserving NADPH for the regeneration of antioxidant molecules (Jeon et al. 2012). Interestingly, the regulatory domain, which contains the AMPK phosphorylation site in PFKFB3, is missing in PFKFB4, suggesting that the regulation of PFKFB4 is quite different from that of the other isoforms. As isoform-specific inhibitors of PFKFB3 have been developed and their efficacy is currently tested in different cancer cells (Clem et al. 2013), a more detailed understanding of the differential roles of PFKFB3 and PFKFB4 in cancer will emerge.

Two of the enzymes of the oxidative pentose phosphate pathway (PPP) have also been associated with metabolic reprogramming in cancer cells. Glucose-6-phosphate dehydrogenase (G6PD) is part of the metabolic transcriptional signature downstream of the mammalian target of rapamycin complex 1 (mTORC1) (Duvel et al. 2010). The activity of G6PD is inhibited by the p53 tumour suppressor through direct binding (Jiang et al. 2011). Moreover, knockdown of 6-phosphogluconate dehydrogenase (6PGD), the second NADPH-producing enzyme of the PPP, induces senescence in lung cancer cells (Sukhatme and Chan 2012). 6PGD is also inhibited by the accumulation of 3-phosphoglycerate (3PG) in response to the inhibition of phosphoglycerate mutase (PGM) by p53 (Hitosugi et al. 2012). TP53 also regulates the activity of PFK1 by inducing the expression of the TP53-induced glycolysis and apoptosis regulator (TIGAR), which has structural similarities to the fructose-2,6-bisphosphatase domain of PFKFB proteins. TIGAR reduces the amounts of F2,6-BP, resulting in the inhibition of glycolysis and redirection of metabolites into the PPP in response to DNA damage (Bensaad et al.

2006). Together, the modulation of glycolytic activity by TP53 allows cells to increase the production of nucleotides for DNA repair. However, loss of TP53 through deletion results in increased NADPH production to support biosynthetic reactions.

Another glycolytic enzyme that has been shown to be important for cancer cell survival is phosphoglycerate dehydrogenase (PHGDH) (Locasale et al. 2011; Possemato et al. 2011). PHGDH promotes the redirection of 3-phosphoglycerate into the biosynthesis of serine and glycine, two non-essential amino acids. As glycine is required for the production of glutathione, enhanced flux through the serine biosynthesis pathway could ensure sufficient production of this important antioxidant. Moreover, serine and glycine are closely linked to the one-carbon metabolism (also known as folate-mediated one-carbon metabolism or folate and methionine cycles), which provides intermediates for purine biosynthesis, head groups for lipid synthesis and methyl groups for the modification of DNA and histones (Locasale 2013). Several enzymes in the metabolism of glycine have been shown to be upregulated in non-small-cell lung cancer (NSCLC) stem cells, and overexpression of glycine decarboxylase, which converts glycine into CO_2, ammonia and methyl tetrahydrofolate (methyl-THF), is required for lung cancer development (Zhang et al. 2012). Glycine uptake has also been linked to the proliferation of cancer cells (Zhang et al. 2012), while serine is essential for the survival of p53 null cancer cells in vitro and in vivo (Maddocks et al. 2013). Moreover, a systems biology approach identified one-carbon metabolism as part of the unique metabolic phenotype of clear cell renal carcinoma (ccRCC) (Gatto et al. 2014). As one-carbon metabolism occupies such a central position and connects to many important metabolic processes, it will be challenging to unravel the exact contribution of this important metabolic node to cancer cell growth and tumour development.

Probably, the best-studied glycolytic enzyme in cancer cells is pyruvate kinase. The muscle isoform of this enzyme (PKM) comes in two splice variants, M1 and M2, which differ in a single exon. Exon 9 is specific to the M2 isoform (also known as the embryonic isoform, PKM2), while the M1 isoform (PKM1) contains exon 10. Most proliferating cells, including cancer cells, express mainly PKM2 (Christofk et al. 2008a). In contrast to PKM1, PKM2 can switch between a tetrameric state with high activity and a dimeric state with low activity (Mazurek et al. 2005). This allows cancer cells to fine-tune the flux of metabolites through the last glycolytic reaction, thereby allowing the use of glycolytic intermediates for biosynthetic reactions. The alternative splicing of exons 9 and 10 is controlled by heterogeneous nuclear ribonucleoproteins (hnRNPs), which are, in turn, regulated downstream of the oncogenic transcription factor c-Myc (David et al. 2010). The activity of PKM2 is also controlled by allosteric regulation. F1,6-BP, the product of PFK1, binds to and stabilises the high-activity tetrameric form of PKM2, providing a coupling between pyruvate production and metabolite levels of upper glycolysis. The association of F2,6-BP with the PKM2 tetramer is prevented by tyrosine-phosphorylated peptides (Christofk et al. 2008b) or ROS-dependent cysteine oxidation (Anastasiou et al. 2011). In addition, PKM2 activity is also regulated by serine (Chaneton et al. 2012). Serine starvation thus reduces PKM2 activity allowing the redirection of 3-phosphogluconate for serine biosynthesis. Interestingly, a study to address the requirement of PKM2 for cancer development using a genetic model revealed differential metabolic requirements of quiescent and proliferating cancer cells (Israelsen et al. 2013), which could have important therapeutic implications as PKM2 activating compounds are currently being developed (Anastasiou et al. 2012).

Research over the past decade has demonstrated the complexity of the connections between glycolytic reactions and other metabolic processes in cancer cells. It is now clear that the impact of glycolysis goes far beyond energy metabolism, but also affects macromolecule biosynthesis, antioxidant generation, methylation and epigenetic regulation.

2 Lipid Metabolism in Cancer Cells

Besides altered glucose metabolism, changes in the synthesis and utilisation of lipids are among the main metabolic features of cancer cells. Increased de novo lipid synthesis has been described in early studies investigating the metabolic alteration in cancer (Medes et al. 1953). This study found that slices of tissue from transplanted tumours show incorporation of labelled glucose and acetate carbons into lipids, at a rate that was lower compared to liver of kidney tissue, but substantially higher compared to other normal tissues (Medes et al. 1953). Despite being the first to demonstrate de novo lipogenesis in cancer tissue, this study also concluded that cancer cells must acquire a substantial amount of lipids from the host (Medes et al. 1953).

Since these early findings, numerous studies have addressed the lipogenic phenotype in cancer cells (Menendez and Lupu 2007). Fatty acid synthase (FASN), the enzyme that catalyses the condensation of acetyl groups to form palmitate (Fig. 2), has been identified as a tumour-specific antigen in breast cancer (Kuhajda et al. 1994). Interestingly, expression of FASN in cancer cells was found to be independent of the amount of circulating lipids, suggesting that oncogenic transformation must override the normal regulatory mechanisms that control the expression of lipogenic enzymes. Today, the induction of FASN and other enzymes within the lipid biosynthesis pathway is well documented. One of the underlying mechanisms for this induction involves the activation of the sterol regulatory element-binding proteins (SREBPs) by Akt and mTORC1 (Krycer et al. 2010; Shao and Espenshade 2012). SREBPs are basic helix–loop–helix transcription factors and regulate the expression of many enzymes within the fatty acid and cholesterol biosynthesis pathway (Nohturfft 2008). There are three SREBP isoforms in mammals: SREBP1a and SREBP1c, which only differ in their N-terminal transactivation domain and are expressed from the same gene through alternative splicing, and SREBP2, which is the product of a separate gene (Bengoechea-Alonso and Ericsson 2007). SREBPs are expressed as inactive precursors and require proteolytic processing by two Golgi-resident proteases to release the N-terminal part of the protein, which contains the DNA-binding

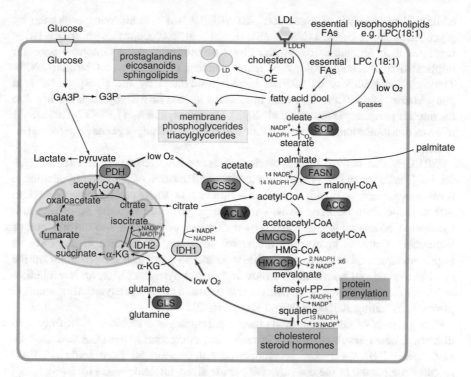

Fig. 2 Fatty acid and cholesterol biosynthesis in cancer cells. Overview of the pathways involved in the synthesis of saturated fatty acids and cholesterol. Elongation, desaturation and uptake of essential polyunsaturated fatty acids (i.e. ω-3 and ω-6 unsaturated fatty acids) produce the cellular fatty acid pool. Cancer cells can also take up exogenous fatty acids (e.g. palmitate) or lysophospholipids (e.g. lysophosphatidylcholine (LPC)) to supplement their demand of saturated and unsaturated fatty acids. Uptake of lipids via LDL particles also contributes to the pool of essential polyunsaturated fatty acids, while conversion of free cholesterol to cholesteryl esters (CE) and storage in lipid droplets (LD) maintains SREBP-dependent expression of the LDL receptor (LDLR) in aggressive prostate cancer cells. Hypoxia blocks the formation of acetyl-CoA from glucose-derived pyruvate, while increasing the reductive carboxylation of glutamine-derived α-ketoglutarate (α-KG) to citrate. Direct conversion of acetate to acetyl-CoA is also induced in hypoxic cells. Desaturation of fatty acids and the conversion of squalene to cholesterol is also oxygen dependent

and transactivation domain (Bengoechea-Alonso and Ericsson 2007). Processing of SREBP is regulated by cellular sterol concentrations through a mechanism termed regulated intramembrane processing (RIP) (Brown and Goldstein 1997). Moreover, an alternative pathway that regulates SREBP1 downstream of phosphoglyceride biosynthesis has been identified. Low levels of phosphatidylcholine (PC) induce the retrograde transport of the Golgi-resident proteases to the ER where they induce the processing of SREBP1 (Walker et al. 2011). SREBP processing and the accumulation of the mature form are induced following the activation of Akt and mTORC1 (Duvel et al. 2010; Porstmann et al. 2008). Mature SREBPs are also subject to additional mechanism of regulation. All SREBP isoforms are phosphorylated by glycogen synthase kinase 3 (GSK3) (Sundquist et al. 2005), and this phosphorylation regulates their ubiquitination by the Fbw7 ubiquitin ligase, a tumour suppressor implicated in the pathobiology of many cancers (Welcker and Clurman 2008). Transcriptional activity of SREBP is also modulated by the phosphatidic acid phosphatase lipin 1 (LPIN1), which inactivates mature SREBP by sequestering it to the nuclear periphery (Peterson et al. 2011). Phosphorylation of LPIN1 by mTORC1 relieves this inhibition, allowing the induction of SREBP target genes (Peterson et al. 2011).

SREBPs also control the expression of genes involved in cholesterol biosynthesis. This pathway involves the sequential condensation of 3 acetate molecules to form mevalonate, a 6-carbon unit. Mevalonate is subsequently converted into activated 5-carbon isoprene units, which are polymerised into the 30-carbon unit squalene. Cholesterol is then formed by cyclisation of squalene and introduction of a number of double bonds into the sterol structure. Cholesterol is not only an important structural component of cellular membranes but also the precursor for the synthesis of steroid hormones. Moreover, farnesyl pyrophosphate, an intermediate in the cholesterol biosynthesis pathway, is required for the prenylation of small G proteins, including K-Ras, Rab and Rho (Fig. 2).

The product of fatty acid biosynthesis, palmitate, is a saturated C18 fatty acid that is subjected to subsequent desaturation and elongation to generate various fatty acid species. However, the generation of polyunsaturated fatty acids (PUFAs) requires the uptake of the essential fatty acids alpha-linolenic acid (18:3n-3), a ω-3 polyunsaturated fatty acid, and linoleic acid (18:2n-6), a ω-6 polyunsaturated fatty acid. Together, these fatty acids provide the precursors for the synthesis of multiple lipid classes, including phosphoglycerides, sphingolipids and eicosanoids.

Structural lipids, such as phosphoglycerides, are the major components of biological membranes. Sphingolipids, eicosanoids and phosphoinositides are classes of lipids that include important signalling molecules, such as ceramide, sphingosine-1-phosphate (S1P), prostaglandins and phosphatidylinositol 3-phosphate, which control numerous cellular processes and are involved in cell–cell communication, inflammation and modulation of the immune response (Wang and Dubois 2010; Wymann and Schneiter 2008; Vanhaesebroeck et al. 2001). The synthesis of these lipids is highly regulated and controlled by complex signalling networks, including oncogenic signalling pathways, but it is also likely that alterations in fatty acid biosynthesis can affect the production of certain species of these signalling lipids.

In addition to transcriptional induction, other regulatory processes also regulate lipid biosynthesis in cancer cells. For example, FASN is a substrate for the USP2A deubiquitinating enzyme and androgen-dependent overexpression of USP2A leads to increased FASN protein stability in prostate cancer (Graner et al. 2004). Moreover, inactivation of AMP-activated protein kinase (AMPK) through the loss of the STK11/LKB-1 tumour suppressor decreases the inactivating phosphorylation of acetyl-CoA carboxylase (ACC), thereby removing an important regulatory mechanism that limits lipid biosynthesis (Carling et al. 2012). It has also been shown that AMPK can phosphorylate SREBP1c in liver cells, thereby inhibiting its processing and resulting in a reduced expression of SREBP target genes (Li et al. 2011).

Some of the processes involved in lipid biosynthesis are sensitive to oxygen deprivation. The conversion of squalene into cholesterol is dependent on molecular oxygen, and the cholesterol biosynthesis pathway has been identified as a major oxygen sensor in yeast (Hughes et al. 2005). Fatty acid desaturation (discussed in more detail below) requires NADPH and molecular oxygen and is consequently inhibited in hypoxia. One major consequence of reduced oxygen availability is the inhibition of pyruvate dehydrogenase (PDH) activity, which is inhibited through the phosphorylation by pyruvate dehydrogenase kinases (PDHKs), some of which are transcriptional targets of HIF (Kim et al. 2006) (Fig. 2). This inhibition of PDH reduces the availability of glucose-derived citrate for lipid biosynthesis. However, it was shown that hypoxic cancer cells, or cells with defective mitochondria, use reductive carboxylation of glutamine-derived α-ketoglutarate by isocitrate dehydrogenase (IDH) to replenish the citrate pool to maintain lipid biosynthesis (Wise et al. 2011; Metallo et al. 2011; Mullen et al. 2012). Moreover, direct conversion of cytosolic acetate to acetyl-CoA by acetyl-CoA carboxylase 2 (ACSS2) can contribute to fatty acid biosynthesis in hypoxia (Yoshii et al. 2009). As highly aggressive cancers, including triple-negative breast cancers or high-grade gliomas, frequently display pronounced tumour hypoxia, the exact regulation of the switch between different carbon sources for lipid biosynthesis is of substantial clinical and therapeutic interest.

3   The Cost of Increased Lipid Biosynthesis in Cancer Cells

In addition to providing important intermediates for the synthesis of both structural and signalling lipids, enhanced fatty acid biosynthesis in cancer cells also imposes liabilities that can render cancer cells highly sensitive to the disruption of selected metabolic processes. One important modification that determines the biophysical properties of the resulting lipids is the introduction of double bonds within the acyl chain. The conversion of stearate into oleate, a monounsaturated fatty acid, is catalysed by stearoyl-CoA desaturase (SCD), another target gene of SREBP. As already mentioned above, this reaction requires molecular oxygen and is inhibited under hypoxic conditions. It has been shown that mouse embryo fibroblasts deficient for the tuberous sclerosis complex proteins 1 or 2 (TSC1 or TSC2), two negative regulators of mTORC1, become highly dependent on monounsaturated fatty acids when they are grown in low serum in hypoxia, a condition representing tumour-like stress (Young et al. 2013). Depletion of SREBP also blocks cellular desaturase activity while maintaining the synthesis of long-chain saturated fatty acids, resulting in an imbalance in the ratio of saturated to monounsaturated fatty acids (Griffiths et al. 2013; Williams et al. 2013). This imbalance in lipid saturation causes mitochondrial dysfunction, induction of oxidative stress and engages the unfolded protein response pathway (UPR), resulting in inhibition of global protein synthesis. Importantly, inhibition of SREBP also reduced the growth of U87 glioblastoma cells in a xenograft model (Griffiths et al. 2013; Williams et al. 2013), suggesting that the tumour microenvironment does not provide a sufficient supply of monounsaturated fatty acids. Targeting SREBP, or indeed SCD, could therefore be a strategy to starve cancer cells of monounsaturated fatty acids while maintaining the lipogenic drive. This would lead to the accumulation of saturated fatty acid species leading to lipotoxicity and cell death.

Another liability imposed by the induction of de novo lipid biosynthesis is the enhanced demand for reducing cofactors in the form of NADPH. The synthesis of one molecule of palmitate from acetyl-CoA requires 14 molecules of NADPH, while the production of one molecule of cholesterol requires even 26 NADPH molecules (Lunt and Vander Heiden 2011). Thus, cancer cells need to simultaneously increase processes that convert $NADP^+$ to NADPH to meet this demand. Two recent studies have investigated the relative contribution of different metabolic processes to the production of NADPH in cancer cells using quantitative flux analysis. Surprisingly, these studies revealed a substantial contribution of the serine-driven 1-carbon metabolism (almost 50 % of total), while the reactions of the pentose phosphate pathway and malic enzymes accounted for approx. 25 % of the total each (Fan et al. 2014; Lewis et al. 2014). As already discussed, NADPH is a cofactor for the production of antioxidants and enhanced fatty acid biosynthesis is likely to put additional strain on the NADPH pool. Oncogenic signalling pathways that drive macromolecule biosynthesis must therefore also ensure sufficient production of NADPH. Nevertheless, increased fatty acid biosynthesis can sensitise cancer cells to inhibitors of NADPH or antioxidant synthesis resulting in oxidative stress. For example, inhibition of the allosteric regulation of glycolytic flux by phosphofructokinase 2/fructose-2,6-bisphosphatase 4 (PFKFB4) reduces NADPH production in prostate cancer cells, leading to enhanced oxidative stress and xenograft tumour regression (Ros et al. 2012). Moreover, malic enzymes 1 and 2 (ME1 and ME2), which generate NADPH from the conversion of malate into pyruvate, were shown to be involved in a reciprocal regulation with p53. Loss of p53 function increases expression of these enzymes, while their inhibition leads to the activation of p53 resulting in induction of senescence and reduced tumour growth (Jiang et al. 2013). The increased NADPH demand imposed by enhanced fatty acid and cholesterol biosynthesis in rapidly proliferating cancer cells could therefore produce a therapeutic window for the development of treatment strategies.

4   Lipid Uptake and Remodelling in Cancer

In addition to de novo synthesis, cancer cells can also take up endogenous lipids. Early studies already concluded that cancer cells must satisfy a substantial amount of their lipid demand from external sources (Medes et al. 1953). More recently, it has been shown that hypoxic cancer cells, or cells transformed by oncogenic Ras, scavenge unsaturated fatty acids from exogenous lysophospholipids, presumably because the SCD reaction is repressed by hypoxia (Kamphorst et al. 2013). In contrast, cells expressing activated Akt showed reduced lipid uptake but increased sensitivity towards chemical inhibitors of SCD (Kamphorst et al. 2013), most likely as they increase endogenous fatty acid synthesis and desaturation via activation of SREBP (Griffiths et al. 2013; Porstmann et al. 2005).

In addition, isotopic labelling of fatty acids showed that cancer cells can take up exogenous palmitate and remodel it to generate not only structural lipids but also lipids that have important signalling functions, including ceramide-1-phosphate (C1P), platelet-activating factor (PAF) and lysophosphatidic acid (LPA) (Louie et al. 2013). At the same time, usage of lipids for oxidative processes was reduced, suggesting that the increased lipid demand of cancer cells drives tumorigenic lipid signalling (Louie et al. 2013). Moreover, the release of free fatty acids from monoacylglycerols by monoacylglycerol lipase (MAGL) in aggressive cancer cells promotes a specific lipidomic signature that correlates with cell migration (Nomura et al. 2010).

Cancer cells can also take up lipids as low-density lipoprotein (LDL) particles (Fig. 2). These particles contain high levels of polyunsaturated fatty acids (PUFAs) as well as esterified and non-esterified cholesterol molecules and are imported to the cell by the LDL receptor (LDLR), a canonical SREBP target gene. It has been shown that aggressive prostate cancer cells convert the free cholesterol released from LDL particles into cholesteryl esters (Yue et al. 2014). These are then stored in lipid droplets, a specialised cellular structure that is separated from the cytosol by a single lipid membrane. Inhibition of cholesterol esterification by the depletion of acetyl-CoA acyltransferase 1 (ACAT1) resulted in the accumulation of free cholesterol in the cytoplasm leading to the inhibition of SREBP-dependent LDLR expression, inhibition of lipid uptake and reduced tumour growth (Yue et al. 2014).

Increased lipid content also favours the production of lipid peroxidation products. Indeed, PUFAs can be attacked directly by reactive oxygen species resulting in spontaneous or chain reaction lipid peroxidation processes (Fig. 3). They can also be actively oxidised by cellular oxidases (enzymatic lipid peroxidation), such as cyclooxygenases (COX-1 and COX-2) and lipooxygenases (12-LOX and 15-LOX). Direct products of both lipid peroxidation processes are short-lived and are quickly converted into advanced products of lipid peroxidation, such as malondialdehyde (MDA) and hydroxyalkenals, among them 4-hydroxy-2-nonenal (4-HNE), which is the most commonly measured lipid peroxide. Isoprostanes, such as $15\text{-}F_{2t}$-Isoprostane, are also lipid peroxidation products specifically generated by spontaneous lipid peroxidation processes (Fig. 3).

There is a constitutive basal level of lipid peroxidation within the cell, generating specific lipid peroxides such as prostaglandins (PG) and thromboxanes (through the action of COX enzymes) and hydroperoxyeicosatetraenoic acids (HPETEs) and leukotrienes (through the action of LOX enzymes) (Fig. 4). These enzymatically generated lipid peroxides are important signalling entities and are involved in inflammation, remodelling of the tumour microenvironment and modulation of the immune response (Wang and Dubois 2010). Prostaglandin E2 (PGE_2) is found in many tumours and is generally considered to promote neoplastic growth. For example, PGE_2 promotes tumour burden in APC^{min} mice (Wang et al. 2004). However, the exact role of some prostaglandins in cancer remains ambiguous and

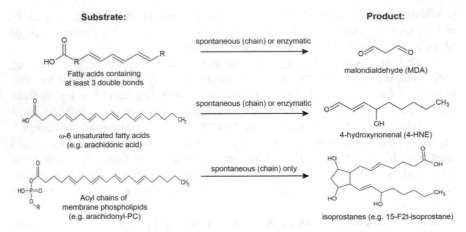

Fig. 3 Structure and origin of lipid peroxidation products. Malondialdehyde (MDA) and 4-hydroxynonenal (4-HNE) are generated from cytosolic polyunsaturated fatty acids either through spontaneous (chain) lipid peroxidation or by the action of cyclooxygenases (COX1 or COX2) or lipooxygenases (12- or 15-LOX). ω-6 fatty acids are used for the production of 4-HNE, while MDA is produced from fatty acids containing at least 3 double bonds. However, only the spontaneous chain peroxidation of acyl chains in membrane phospholipids can produce isoprostanes, including the most common form 15-F_{2t}-isoprostane. After peroxidation, these products have to be released from the membrane by action of a cellular phospholipase

may be highly context dependent. Genetic ablation of the prostaglandin D2 (PGD$_2$) receptor enhanced tumour progression and vascular expansion, suggesting that PGD$_2$ inhibits angiogenesis (Murata et al. 2008).

Lipid peroxides regulate gene expression by modulating the activity of peroxisome proliferator-activated receptors (PPAR) (Murata et al. 2008). For example, PGE$_2$ promotes survival in colon cancer cells by activating PPARδ (Wang et al. 2004), while PGD$_2$ acts via modulation of the activity of PPARγ (Sales et al. 2007). The products of non-enzymatic lipid oxidation by free oxygen radicals (i.e. 4-hydroxyalkenals and MDA) also represent ligands for these transcription factors and can induce changes in gene expression in response to oxidative stress (Niki et al. 2005; Riahi et al. 2010). In addition to fatty acids, cholesterol can also act as a substrate for lipid oxidation, again via enzymatic and non-enzymatic mechanisms. Different oxysterols function as intermediates in bile acid production but can also have functions as signalling molecules (Niki et al. 2005).

An increase in cellular oxidative stress or an overactivation of cellular oxygenases will generate larger and potentially cell-damaging amounts of lipid peroxides. Accumulation of these oxidation products can alter the activation state of the signalling pathways in which they participate. However, oxidised lipids can also react with cellular macromolecules and cause oxidative damage. For example, 4-HNE or MDA can react with thiol residues in proteins or purine bases in DNA, generating protein adducts and DNA mutations, respectively (Esterbauer 1993; Marnett 1999). Protein modification by lipid peroxides can lead to the induction of ER stress,

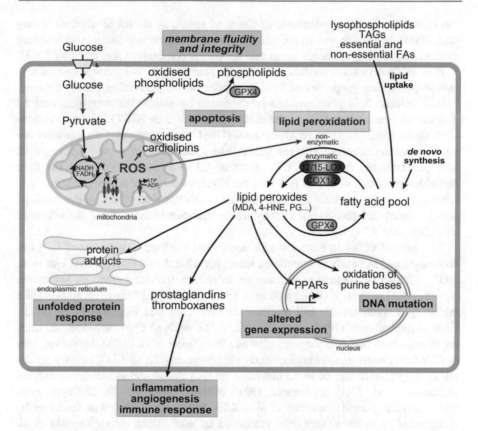

Fig. 4 Lipid peroxides in cell signalling and oxidative damage. Reactive oxygen species produced in the mitochondrial respiratory chain (complexes 1 and 3) can lead to the non-enzymatic peroxidation of unsaturated free fatty acids or the unsaturated acyl chains of membrane phospholipids and cardiolipins. Polyunsaturated free fatty acids (i.e. arachidonic acid) can also be subjected to enzymatic peroxidation by cyclooxygenases 1 and 2 (COX1/2) or lipooxygenases (12/15-LOX). Lipid peroxides, including malondialdehyde (MDA), 4-hydroxynonenal (4-HNE) and prostaglandins (PG), are involved in different cellular functions: Oxidation of purine bases can lead to DNA mutation; formation of protein adducts can induce the unfolded protein response (UPR); modulation of the activity of the peroxisome proliferator-activated receptors (PPAR) can later gene expression; some lipid peroxides are involved in cell–cell communication and modulate inflammation, angiogenesis and immune response. Peroxidation of membrane phospholipids can alter membrane fluidity and integrity, while peroxidation of cardiolipins affects mitochondrial function and can lead to apoptosis. Exogenous lipids, including lysophospholipids, triacylglycerides (TAGs) and free essential and non-essential fatty acids (FAs), can also contribute to the cellular fatty acid pool and be subjected to lipid peroxidation

resulting in the activation of autophagy in rat aortic smooth muscle cells (Haberzettl and Hill 2013) and activation of endothelial cells (Vladykovskaya et al. 2012). Cellular membranes, which are composed of phospholipids that contain saturated and unsaturated acyl chains as well as cholesterol, are also at risk of peroxidation. Indeed, peroxidation of membrane lipids induces changes in membrane fluidity and can contribute to the development of diseased states, as shown in diabetes (Hong et al. 2004). Moreover, excess peroxidation of cardiolipins will cause mitochondrial dysfunction leading to energy stress and apoptosis (Gonzalvez and Gottlieb 2007).

It is evident that the modulation of lipid peroxidation is essential to facilitate growth-promoting properties of lipid peroxides while preventing excess damage. Indeed, cellular lipid peroxides can be degraded by antioxidant enzymes, such as glutathione peroxidases (GPx), glutathione-S-transferases (GST), or small antioxidant molecules, like reduced glutathione (GSH) or vitamin E. GPx proteins are selenocysteine-containing enzyme that utilise glutathione to protect cells against lipid peroxidation. Among all GPx proteins, GPx4 is considered to be the lipid peroxide-specific isoform, as this enzyme displays a higher activity towards this substrate than the other GPxs and is the only GPx capable of degrading oxidised phospholipids and cholesterol directly inside the membrane without the action of phospholipase A2 (PLA2).

The role of GPx4 in cancer is still somewhat unclear. Expression of GPx4 is downregulated in poorly differentiated breast invasive ductal carcinoma (Cejas et al. 2007). GPx4 overexpression was shown to reduce tumour growth of pancreatic cancer cells (Liu et al. 2006) as well as in fibrosarcoma (Heirman et al. 2006), while inhibiting smooth muscle cell proliferation (Banning and Brigelius-Flohe 2005). Genetic depletion of GPx4 increases 12/15-LOX-derived lipid peroxidation, leading to cell death and neurodegeneration in mice (Seiler et al. 2008). Moreover, loss of GPx4 increased tumour angiogenesis via increasing 12/15-LOX activity resulting in the upregulation of basic fibroblast growth factor and vessel normalisation (Schneider et al. 2010). In contrast, GPx4 overexpression has no effect on melanoma tumour growth (Heirman et al. 2006). Moreover, GPx4 was found to be upregulated in colon cancer cells compared to non-cancer cells (Yagublu et al. 2011). This could be explained by the fact that GPx4 has antiapoptotic properties that could be hijacked by cancer cells to protect them from lipid peroxide-mediated damage. Indeed, GPx4 has been described to inhibit mitochondrial apoptosis by inhibiting cardiolipin peroxidation, thereby preventing the release of cytochrome C to the cytosol (Nomura et al. 2000; Nakagawa 2004). In hepatocellular carcinoma (HCC) patients, lipid peroxide levels are significantly increased compared to healthy controls and correlate with the levels of VEGF, which is known to be an indicator of poor prognosis (Rohr-Udilova et al. 2012). In HCC cell lines, treatment of cells with linoleic acid peroxides increases VEGF and IL8. This phenotype can be rescued in vitro and in vivo by selenium supplementation that specifically activates GPx4 activity towards lipid peroxides, reducing VEGF and c-fos mRNA expression and significantly decreasing preneoplastic liver nodule growth in a rat hepatocarcinogenesis model (Rohr-Udilova et al. 2012). However, the potential influence of selenium supplementation on cancer risk is still unclear, as different studies suggest either a protective role or no effect on prostate cancer risk (Duffield-Lillico et al. 2003; Lippman et al. 2009). Moreover, selenium supplementation seemed to have no effect on basal cell carcinoma risk, while potentially increasing the risk of developing squamous cell carcinoma and non-melanoma skin cancers (Duffield-Lillico et al. 2003).

GPx4 was also identified as a mediator of ferroptosis (Yang et al. 2014). This iron-dependent form of cell death does not involve mitochondria but is associated with metabolic dysfunction, oxidative stress and accumulation of lethal levels of lipid peroxidation products (Dixon et al. 2012). Agents that induce this form of cell death were shown to inhibit either glutathione synthesis or the activity of GPx4, resulting in the accumulation of lipid peroxides (Yang et al. 2014). Interestingly, large B cell lymphoma and clear cell renal cell carcinoma cell lines were highly susceptible to ferroptosis-inducing compounds (Yang et al. 2014), suggesting that differential regulation of lipid peroxidation and antioxidant production could determine the sensitivity of cancer cells to particular therapeutic agents.

5 Concluding Remarks

Enhanced glucose uptake and reactivation of de novo lipid biosynthesis are hallmarks of cancer cells. This altered metabolic phenotype distinguishes cancer cells from normal non-proliferative tissue in the body. Enhanced glucose uptake and lipid biosynthesis form the molecular basis for the visualisation of tumours by PET using ^{18}F-fluorodeoxyglucose (^{18}F-FDG) or ^{11}C-choline tracers. Glucose not only contributes to energy-generating processes but also provides important precursors for biosynthetic processes, most importantly ribosomes for the production of nucleotides for the synthesis of DNA and RNA molecules. Lipids have important structural functions but also constitute an important component of cellular signalling pathways and cell–cell communication. Studying the regulation of these processes in cancer cells is likely to yield insight into the intimate connection between oncogenic transformation and metabolic reprogramming and highlight novel targets for therapeutic development.

References

Anastasiou D, Poulogiannis G, Asara JM, Boxer MB, Jiang JK, Shen M, Bellinger G, Sasaki AT, Locasale JW, Auld DS, Thomas CJ, Vander Heiden MG, Cantley LC (2011) Inhibition of pyruvate kinase M2 by reactive oxygen species contributes to antioxidant responses. Science. doi:10.1126/science.1211485 science.1211485 [pii]

Anastasiou D, Yu Y, Israelsen WJ, Jiang JK, Boxer MB, Hong BS, Tempel W, Dimov S, Shen M, Jha A, Yang H, Mattaini KR, Metallo CM, Fiske BP, Courtney KD, Malstrom S, Khan TM, Kung C, Skoumbourdis AP, Veith H, Southall N, Walsh MJ, Brimacombe KR, Leister W, Lunt SY, Johnson ZR, Yen KE, Kunii K, Davidson SM, Christofk HR, Austin CP, Inglese J, Harris MH, Asara JM, Stephanopoulos G, Salituro FG, Jin S, Dang L, Auld DS, Park HW, Cantley LC, Thomas CJ, Vander Heiden MG (2012) Pyruvate kinase M2 activators promote tetramer formation and suppress tumorigenesis. Nat Chem Biol 8(10):839–847. doi:10.1038/nchembio.1060

Banning A, Brigelius-Flohe R (2005) NF-kappaB, Nrf2, and HO-1 interplay in redox-regulated VCAM-1 expression. Antioxid Redox Signal 7(7–8):889–899. doi:10.1089/ars.2005.7.889

Barford D, Hu SH, Johnson LN (1991) Structural mechanism for glycogen phosphorylase control by phosphorylation and AMP. J Mol Biol 218(1):233–260

Bengoechea-Alonso MT, Ericsson J (2007) SREBP in signal transduction: cholesterol metabolism and beyond. Curr Opin Cell Biol 19(2):215–222

Bensaad K, Tsuruta A, Selak MA, Vidal MN, Nakano K, Bartrons R, Gottlieb E, Vousden KH (2006) TIGAR, a p53-inducible regulator of glycolysis and apoptosis. Cell 126(1):107–120

Brown MS, Goldstein JL (1997) The SREBP pathway: regulation of cholesterol metabolism by proteolysis of a membrane-bound transcription factor. Cell 89(3):331–340 S0092-8674(00) 80213-5 [pii]

Cairns RA, Harris IS, Mak TW (2011) Regulation of cancer cell metabolism. Nat Rev Cancer 11 (2):85–95. doi:10.1038/nrc2981 nrc2981 [pii]

Carling D, Thornton C, Woods A, Sanders MJ (2012) AMP-activated protein kinase: new regulation, new roles? Biochem J 445(1):11–27. doi:10.1042/BJ20120546

Cejas P, Garcia-Cabezas MA, Casado E, Belda-Iniesta C, De Castro J, Fresno JA, Sereno M, Barriuso J, Espinosa E, Zamora P, Feliu J, Redondo A, Hardisson DA, Renart J, Gonzalez-Baron M (2007) Phospholipid hydroperoxide glutathione peroxidase (PHGPx) expression is downregulated in poorly differentiated breast invasive ductal carcinoma. Free Radical Res 41(6):681–687. doi:10.1080/10715760701286167

Chaneton B, Hillmann P, Zheng L, Martin AC, Maddocks OD, Chokkathukalam A, Coyle JE, Jankevics A, Holding FP, Vousden KH, Frezza C, O'Reilly M, Gottlieb E (2012) Serine is a natural ligand and allosteric activator of pyruvate kinase M2. Nature 491(7424):458–462. doi:10.1038/nature11540

Christofk HR, Vander Heiden MG, Harris MH, Ramanathan A, Gerszten RE, Wei R, Fleming MD, Schreiber SL, Cantley LC (2008a) The M2 splice isoform of pyruvate kinase is important for cancer metabolism and tumour growth. Nature 452(7184):230–233

Christofk HR, Vander Heiden MG, Wu N, Asara JM, Cantley LC (2008b) Pyruvate kinase M2 is a phosphotyrosine-binding protein. Nature 452(7184):181–186

Clem BF, O'Neal J, Tapolsky G, Clem AL, Imbert-Fernandez Y, Kerr DA 2nd, Klarer AC, Redman R, Miller DM, Trent JO, Telang S, Chesney J (2013) Targeting 6-phosphofructo-2-kinase (PFKFB3) as a therapeutic strategy against cancer. Mol Cancer Ther 12(8):1461–1470. doi:10.1158/1535-7163.MCT-13-0097

David CJ, Chen M, Assanah M, Canoll P, Manley JL (2010) HnRNP proteins controlled by c-Myc deregulate pyruvate kinase mRNA splicing in cancer. Nature 463(7279):364–368. doi:10.1038/ nature08697 nature08697 [pii]

De Bock K, Georgiadou M, Schoors S, Kuchnio A, Wong BW, Cantelmo AR, Quaegebeur A, Ghesquiere B, Cauwenberghs S, Eelen G, Phng LK, Betz I, Tembuyser B, Brepoels K, Welti J, Geudens I, Segura I, Cruys B, Bifari F, Decimo I, Blanco R, Wyns S, Vangindertael J, Rocha S, Collins RT, Munck S, Daelemans D, Imamura H, Devlieger R, Rider M, Van Veldhoven PP, Schuit F, Bartrons R, Hofkens J, Fraisl P, Telang S, Deberardinis RJ, Schoonjans L, Vinckier S, Chesney J, Gerhardt H, Dewerchin M, Carmeliet P (2013) Role of PFKFB3-driven glycolysis in vessel sprouting. Cell 154(3):651–663. doi:10.1016/j.cell.2013. 06.037

Dixon SJ, Lemberg KM, Lamprecht MR, Skouta R, Zaitsev EM, Gleason CE, Patel DN, Bauer AJ, Cantley AM, Yang WS, Morrison B 3rd, Stockwell BR (2012) Ferroptosis: an iron-dependent form of nonapoptotic cell death. Cell 149(5):1060–1072. doi:10.1016/j.cell.2012.03.042

Duffield-Lillico AJ, Slate EH, Reid ME, Turnbull BW, Wilkins PA, Combs GF, Jr., Park HK, Gross EG, Graham GF, Stratton MS, Marshall JR, Clark LC, Nutritional Prevention of Cancer Study G (2003) Selenium supplementation and secondary prevention of nonmelanoma skin cancer in a randomized trial. J Nat Cancer Inst 95(19):1477–1481

Duvel K, Yecies JL, Menon S, Raman P, Lipovsky AI, Souza AL, Triantafellow E, Ma Q, Gorski R, Cleaver S, Vander Heiden MG, Mackeigan JP, Finan PM, Clish CB, Murphy LO, Manning BD (2010) Activation of a metabolic gene regulatory network downstream of mTOR complex 1. Mol Cell 39(2):171–183. doi:10.1016/j.molcel.2010.06.022 S1097-2765(10) 00463-6 [pii]

Esterbauer H (1993) Cytotoxicity and genotoxicity of lipid-oxidation products. Am J Clin Nutr 57 (5 Suppl):779S–785S (discussion 785S–786S)

Fan J, Ye J, Kamphorst JJ, Shlomi T, Thompson CB, Rabinowitz JD (2014) Quantitative flux analysis reveals folate-dependent NADPH production. Nature 510(7504):298–302. doi:10. 1038/nature13236

Gatto F, Nookaew I, Nielsen J (2014) Chromosome 3p loss of heterozygosity is associated with a unique metabolic network in clear cell renal carcinoma. Proc Natl Acad Sci USA 111(9):E866–E875. doi:10.1073/pnas.1319196111

Goidts V, Bageritz J, Puccio L, Nakata S, Zapatka M, Barbus S, Toedt G, Campos B, Korshunov A, Momma S, Van Schaftingen E, Reifenberger G, Herold-Mende C, Lichter P, Radlwimmer B (2012) RNAi screening in glioma stem-like cells identifies PFKFB4 as a key molecule important for cancer cell survival. Oncogene 31(27):3235–3243. doi:10.1038/onc. 2011.490

Gonzalvez F, Gottlieb E (2007) Cardiolipin: setting the beat of apoptosis. Apoptosis Int J Programmed Cell Death 12(5):877–885. doi:10.1007/s10495-007-0718-8

Gottlob K, Majewski N, Kennedy S, Kandel E, Robey RB, Hay N (2001) Inhibition of early apoptotic events by Akt/PKB is dependent on the first committed step of glycolysis and mitochondrial hexokinase. Genes Dev 15(11):1406–1418

Graner E, Tang D, Rossi S, Baron A, Migita T, Weinstein LJ, Lechpammer M, Huesken D, Zimmermann J, Signoretti S, Loda M (2004) The isopeptidase USP2a regulates the stability of fatty acid synthase in prostate cancer. Cancer Cell 5(3):253–261

Griffiths B, Lewis CA, Bensaad K, Ros S, Zhang X, Ferber EC, Konisti S, Peck B, Miess H, East P, Wakelam M, Harris AL, Schulze A (2013) Sterol regulatory element binding protein-dependent regulation of lipid synthesis supports cell survival and tumour growth. Cancer Metabol 1(3)

Haberzettl P, Hill BG (2013) Oxidized lipids activate autophagy in a JNK-dependent manner by stimulating the endoplasmic reticulum stress response. Redox Biol 1(1):56–64. doi:10.1016/j. redox.2012.10.003

Heirman I, Ginneberge D, Brigelius-Flohe R, Hendrickx N, Agostinis P, Brouckaert P, Rottiers P, Grooten J (2006) Blocking tumor cell eicosanoid synthesis by GP x 4 impedes tumor growth and malignancy. Free Radical Biol Med 40(2):285–294. doi:10.1016/j.freeradbiomed.2005.08.033

Hitosugi T, Zhou L, Elf S, Fan J, Kang HB, Seo JH, Shan C, Dai Q, Zhang L, Xie J, Gu TL, Jin P, Aleckovic M, LeRoy G, Kang Y, Sudderth JA, DeBerardinis RJ, Luan CH, Chen GZ, Muller S, Shin DM, Owonikoko TK, Lonial S, Arellano ML, Khoury HJ, Khuri FR, Lee BH, Ye K, Boggon TJ, Kang S, He C, Chen J (2012) Phosphoglycerate mutase 1 coordinates glycolysis and biosynthesis to promote tumor growth. Cancer Cell 22(5):585–600. doi:10. 1016/j.ccr.2012.09.020

Hong JH, Kim MJ, Park MR, Kwag OG, Lee IS, Byun BH, Lee SC, Lee KB, Rhee SJ (2004) Effects of vitamin E on oxidative stress and membrane fluidity in brain of streptozotocin-induced diabetic rats. Clin chimica acta; Int J Clin Chem 340(1–2):107–115

Hughes AL, Todd BL, Espenshade PJ (2005) SREBP pathway responds to sterols and functions as an oxygen sensor in fission yeast. Cell 120(6):831–842

Israelsen WJ, Dayton TL, Davidson SM, Fiske BP, Hosios AM, Bellinger G, Li J, Yu Y, Sasaki M, Horner JW, Burga LN, Xie J, Jurczak MJ, DePinhó RA, Clish CB, Jacks T, Kibbey RG, Wulf GM, Di Vizio D, Mills GB, Cantley LC, Vander Heiden MG (2013) PKM2 isoform-specific deletion reveals a differential requirement for pyruvate kinase in tumor cells. Cell 155(2):397–409. doi:10.1016/j.cell.2013.09.025

Jeon SM, Chandel NS, Hay N (2012) AMPK regulates NADPH homeostasis to promote tumour cell survival during energy stress. Nature 485(7400):661–665. doi:10.1038/ nature11066nature11066 nature11066 [pii]

Jiang P, Du W, Wang X, Mancuso A, Gao X, Wu M, Yang X (2011) p53 regulates biosynthesis through direct inactivation of glucose-6-phosphate dehydrogenase. Nat Cell Biol 13(3):310–316. doi:10.1038/ncb2172 ncb2172 [pii]

Jiang P, Du W, Mancuso A, Wellen KE, Yang X (2013) Reciprocal regulation of p53 and malic enzymes modulates metabolism and senescence. Nature 493(7434):689–693. doi:10.1038/nature11776

Kamphorst JJ, Cross JR, Fan J, de Stanchina E, Mathew R, White EP, Thompson CB, Rabinowitz JD (2013) Hypoxic and Ras-transformed cells support growth by scavenging unsaturated fatty acids from lysophospholipids. Proc Nat Acad Sci U.S.A. 110(22):8882–8887. doi:10.1073/pnas.1307237110

Kim JW, Tchernyshyov I, Semenza GL, Dang CV (2006) HIF-1-mediated expression of pyruvate dehydrogenase kinase: a metabolic switch required for cellular adaptation to hypoxia. Cell Metab 3(3):177–185

Klarer AC, O'Neal J, Imbert-Fernandez Y, Clem A, Ellis SR, Clark J, Clem B, Chesney J, Telang S (2014) Inhibition of 6-phosphofructo-2-kinase (PFKFB3) induces autophagy as a survival mechanism. Cancer Metab 2(1):2. doi:10.1186/2049-3002-2-2

Krycer JR, Sharpe LJ, Luu W, Brown AJ (2010) The Akt-SREBP nexus: cell signaling meets lipid metabolism. Trends Endocrinol Metab. doi:10.1016/j.tem.2010.01.001 S1043-2760(10) 00003-2 [pii]

Kuhajda FP, Jenner K, Wood FD, Hennigar RA, Jacobs LB, Dick JD, Pasternack GR (1994) Fatty acid synthesis: a potential selective target for antineoplastic therapy. Proc Natl Acad Sci USA 91(14):6379–6383

Lewis CA, Parker SJ, Fiske BP, McCloskey D, Gui DY, Green CR, Vokes NI, Feist AM, Vander Heiden MG, Metallo CM (2014) Tracing compartmentalized NADPH metabolism in the cytosol and mitochondria of mammalian cells. Mol Cell 55(2):253–263. doi:10.1016/j.molcel.2014.05.008

Li Y, Xu S, Mihaylova MM, Zheng B, Hou X, Jiang B, Park O, Luo Z, Lefai E, Shyy JY, Gao B, Wierzbicki M, Verbeuren TJ, Shaw RJ, Cohen RA, Zang M (2011) AMPK Phosphorylates and inhibits SREBP activity to attenuate hepatic steatosis and atherosclerosis in diet-induced insulin-resistant mice. Cell Metab 13(4):376–388. doi:10.1016/j.cmet.2011.03.009 S1550-4131(11)00096-9 [pii]

Lippman SM, Klein EA, Goodman PJ, Lucia MS, Thompson IM, Ford LG, Parnes HL, Minasian LM, Gaziano JM, Hartline JA, Parsons JK, Bearden JD 3rd, Crawford ED, Goodman GE, Claudio J, Winquist E, Cook ED, Karp DD, Walther P, Lieber MM, Kristal AR, Darke AK, Arnold KB, Ganz PA, Santella RM, Albanes D, Taylor PR, Probstfield JL, Jagpal TJ, Crowley JJ, Meyskens FL Jr, Baker LH, Coltman CA Jr (2009) Effect of selenium and vitamin E on risk of prostate cancer and other cancers: the Selenium and Vitamin E Cancer Prevention Trial (SELECT). JAMA, J Am Med Assoc 301(1):39–51. doi:10.1001/jama.2008.864

Liu J, Du J, Zhang Y, Sun W, Smith BJ, Oberley LW, Cullen JJ (2006) Suppression of the malignant phenotype in pancreatic cancer by overexpression of phospholipid hydroperoxide glutathione peroxidase. Hum Gene Ther 17(1):105–116. doi:10.1089/hum.2006.17.105

Locasale JW (2013) Serine, glycine and one-carbon units: cancer metabolism in full circle. Nat Rev Cancer 13(8):572–583. doi:10.1038/nrc3557

Locasale JW, Grassian AR, Melman T, Lyssiotis CA, Mattaini KR, Bass AJ, Heffron G, Metallo CM, Muranen T, Sharfi H, Sasaki AT, Anastasiou D, Mullarky E, Vokes NI, Sasaki M, Beroukhim R, Stephanopoulos G, Ligon AH, Meyerson M, Richardson AL, Chin L, Wagner G, Asara JM, Brugge JS, Cantley LC, Vander Heiden MG (2011) Phosphoglycerate dehydrogenase diverts glycolytic flux and contributes to oncogenesis. Nat Genet 43(9):869–874. doi:10.1038/ng.890 ng.890 [pii]

Louie SM, Roberts LS, Mulvihill MM, Luo K, Nomura DK (2013) Cancer cells incorporate and remodel exogenous palmitate into structural and oncogenic signaling lipids. Biochim Biophys Acta 1831(10):1566–1572. doi:10.1016/j.bbalip.2013.07.008

Lunt SY, Vander Heiden MG (2011) Aerobic glycolysis: meeting the metabolic requirements of cell proliferation. Annu Rev Cell Dev Biol 27:441–464. doi:10.1146/annurev-cellbio-092910-154237

Maddocks OD, Berkers CR, Mason SM, Zheng L, Blyth K, Gottlieb E, Vousden KH (2013) Serine starvation induces stress and p 53-dependent metabolic remodelling in cancer cells. Nature 493(7433):542–546. doi:10.1038/nature11743

Majewski N, Nogueira V, Bhaskar P, Coy PE, Skeen JE, Gottlob K, Chandel NS, Thompson CB, Robey RB, Hay N (2004) Hexokinase-mitochondria interaction mediated by Akt is required to inhibit apoptosis in the presence or absence of Bax and Bak. Mol Cell 16(5):819–830

Manning BD, Cantley LC (2007) AKT/PKB signaling: navigating downstream. Cell 129(7): 1261–1274

Marnett LJ (1999) Lipid peroxidation-DNA damage by malondialdehyde. Mutat Res 424(1–2): 83–95

Mazurek S, Boschek CB, Hugo F, Eigenbrodt E (2005) Pyruvate kinase type M2 and its role in tumor growth and spreading. Semin Cancer Biol 15(4):300–308

Medes G, Thomas A, Weinhouse S (1953) Metabolism of neoplastic tissue. IV. a study of lipid synthesis in neoplastic tissue slices in vitro. Cancer Res 13(1):27–29

Menendez JA, Lupu R (2007) Fatty acid synthase and the lipogenic phenotype in cancer pathogenesis. Nat Rev Cancer 7(10):763–777

Metallo CM, Gameiro PA, Bell EL, Mattaini KR, Yang J, Hiller K, Jewell CM, Johnson ZR, Irvine DJ, Guarente L, Kelleher JK, Vander Heiden MG, Iliopoulos O, Stephanopoulos G (2011) Reductive glutamine metabolism by IDH1 mediates lipogenesis under hypoxia. Nature 481(7381):380–384. doi:10.1038/nature10602

Mullen AR, Wheaton WW, Jin ES, Chen PH, Sullivan LB, Cheng T, Yang Y, Linehan WM, Chandel NS, DeBerardinis RJ (2012) Reductive carboxylation supports growth in tumour cells with defective mitochondria. Nature 481(7381):385–388. doi:10.1038/nature10642

Murata T, Lin MI, Aritake K, Matsumoto S, Narumiya S, Ozaki H, Urade Y, Hori M, Sessa WC (2008) Role of prostaglandin D2 receptor DP as a suppressor of tumor hyperpermeability and angiogenesis in vivo. Proc Nat Acad Sci U.S.A. 105(50):20009–20014. doi:10.1073/pnas. 0805171105

Nakagawa Y (2004) Role of mitochondrial phospholipid hydroperoxide glutathione peroxidase (PHGPx) as an antiapoptotic factor. Biol Pharm Bull 27(7):956–960

Niki E, Yoshida Y, Saito Y, Noguchi N (2005) Lipid peroxidation: mechanisms, inhibition, and biological effects. Biochem Biophys. Res Commun 338(1):668–676. doi:10.1016/j.bbrc.2005. 08.072

Nohturfft A (2008) Regulation of lipid synthesis. Annu Rev Cell Dev Biol. doi:10.1146/annurev. cellbio.24.110707.175344

Nomura K, Imai H, Koumura T, Kobayashi T, Nakagawa Y (2000) Mitochondrial phospholipid hydroperoxide glutathione peroxidase inhibits the release of cytochrome c from mitochondria by suppressing the peroxidation of cardiolipin in hypoglycaemia-induced apoptosis. Biochem J 351(Pt 1):183–193

Nomura DK, Long JZ, Niessen S, Hoover HS, Ng SW, Cravatt BF (2010) Monoacylglycerol lipase regulates a fatty acid network that promotes cancer pathogenesis. Cell 140(1):49–61. doi:10.1016/j.cell.2009.11.027 S0092-8674(09)01439-1 [pii]

Patra KC, Wang Q, Bhaskar PT, Miller L, Wang Z, Wheaton W, Chandel N, Laakso M, Muller WJ, Allen EL, Jha AK, Smolen GA, Clasquin MF, Robey RB, Hay N (2013) Hexokinase 2 is required for tumor initiation and maintenance and its systemic deletion is therapeutic in mouse models of cancer. Cancer Cell 24(2):213–228. doi:10.1016/j.ccr.2013.06.014

Peterson TR, Sengupta SS, Harris TE, Carmack AE, Kang SA, Balderas E, Guertin DA, Madden KL, Carpenter AE, Finck BN, Sabatini DM (2011) mTOR complex 1 regulates lipin 1 localization to control the SREBP pathway. Cell 146(3):408–420. doi:10.1016/j.cell.2011.06. 034 S0092-8674(11)00709-4 [pii]

Porstmann T, Griffiths B, Chung YL, Delpuech O, Griffiths JR, Downward J, Schulze A (2005) PKB/Akt induces transcription of enzymes involved in cholesterol and fatty acid biosynthesis via activation of SREBP. Oncogene 24(43):6465–6481. doi:10.1038/sj.onc.1208802

Porstmann T, Santos CR, Griffiths B, Cully M, Wu M, Leevers S, Griffiths JR, Chung YL, Schulze A (2008) SREBP activity is regulated by mTORC1 and contributes to Akt-dependent cell growth. Cell Metab 8(3):224–236. doi:10.1016/j.cmet.2008.07.007

Possemato R, Marks KM, Shaul YD, Pacold ME, Kim D, Birsoy K, Sethumadhavan S, Woo HK, Jang HG, Jha AK, Chen WW, Barrett FG, Stransky N, Tsun ZY, Cowley GS, Barretina J, Kalaany NY, Hsu PP, Ottina K, Chan AM, Yuan B, Garraway LA, Root DE, Mino-Kenudson M, Brachtel EF, Driggers EM, Sabatini DM (2011) Functional genomics reveal that the serine synthesis pathway is essential in breast cancer. Nature 476(7360):346–350. doi:10.1038/nature10350 nature10350 [pii]

Riahi Y, Cohen G, Shamni O, Sasson S (2010) Signaling and cytotoxic functions of 4-hydroxyalkenals. Am J physiol Endocrinol Metabol 299(6):E879–E886. doi:10.1152/ajpendo.00508.2010

Rider MH, Bertrand L, Vertommen D, Michels PA, Rousseau GG, Hue L (2004) 6-phosphofructo-2-kinase/fructose-2,6-bisphosphatase: head-to-head with a bifunctional enzyme that controls glycolysis. Biochem J 381(Pt 3):561–579. doi:10.1042/BJ20040752BJ20040752 BJ20040752 [pii]

Rohr-Udilova N, Sieghart W, Eferl R, Stoiber D, Bjorkhem-Bergman L, Eriksson LC, Stolze K, Hayden H, Keppler B, Sagmeister S, Grasl-Kraupp B, Schulte-Hermann R, Peck-Radosavljevic M (2012) Antagonistic effects of selenium and lipid peroxides on growth control in early hepatocellular carcinoma. Hepatology 55(4):1112–1121. doi:10.1002/hep.24808

Ros S, Santos CR, Moco S, Baenke F, Kelly G, Howell M, Zamboni N, Schulze A (2012) Functional metabolic screen identifies 6-phosphofructo-2-kinase/fructose-2,6-bisphosphatase 4 as an important regulator of prostate cancer cell survival. Cancer Discov 2(4):328–343. doi:10.1158/2159-8290.CD-11-0234

Sales KJ, Boddy SC, Williams AR, Anderson RA, Jabbour HN (2007) F-prostanoid receptor regulation of fibroblast growth factor 2 signaling in endometrial adenocarcinoma cells. Endocrinology 148(8):3635–3644. doi:10.1210/en.2006-1517

Schneider M, Wortmann M, Mandal PK, Arpornchayanon W, Jannasch K, Alves F, Strieth S, Conrad M, Beck H (2010) Absence of glutathione peroxidase 4 affects tumor angiogenesis through increased 12/15-lipoxygenase activity. Neoplasia 12(3):254–263

Seiler A, Schneider M, Forster H, Roth S, Wirth EK, Culmsee C, Plesnila N, Kremmer E, Radmark O, Wurst W, Bornkamm GW, Schweizer U, Conrad M (2008) Glutathione peroxidase 4 senses and translates oxidative stress into 12/15-lipoxygenase dependent- and AIF-mediated cell death. Cell Metabol 8(3):237–248. doi:10.1016/j.cmet.2008.07.005

Shao W, Espenshade PJ (2012) Expanding roles for SREBP in metabolism. Cell Metab 16(4):414–419. doi:10.1016/j.cmet.2012.09.002

Sukhatme VP, Chan B (2012) Glycolytic cancer cells lacking 6-phosphogluconate dehydrogenase metabolize glucose to induce senescence. FEBS Lett 586(16):2389–2395. doi:10.1016/j.febslet.2012.05.052

Sundqvist A, Bengoechea-Alonso MT, Ye X, Lukiyanchuk V, Jin J, Harper JW, Ericsson J (2005) Control of lipid metabolism by phosphorylation-dependent degradation of the SREBP family of transcription factors by SCF(Fbw7). Cell Metab 1(6): 379–391

Telang S, Yalcin A, Clem AL, Bucala R, Lane AN, Eaton JW, Chesney J (2006) Ras transformation requires metabolic control by 6-phosphofructo-2-kinase. Oncogene 25(55):7225–7234. doi:10.1038/sj.onc.1209709 1209709 [pii]

Vanhaesebroeck B, Leevers SJ, Ahmadi K, Timms J, Katso R, Driscoll PC, Woscholski R, Parker PJ, Waterfield MD (2001) Synthesis and function of 3-phosphorylated inositol lipids. Annu Rev Biochem 70:535–602. doi:10.1146/annurev.biochem.70.1.535

Vanhaesebroeck B, Stephens L, Hawkins P (2012) PI3 K signalling: the path to discovery and understanding. Nat Rev Mol Cell Biol 13(3):195–203. doi:10.1038/nrm3290 nrm3290 [pii]

Vladykovskaya E, Sithu SD, Haberzettl P, Wickramasinghe NS, Merchant ML, Hill BG, McCracken J, Agarwal A, Dougherty S, Gordon SA, Schuschke DA, Barski OA, O'Toole T, D'Souza SE, Bhatnagar A, Srivastava S (2012) Lipid peroxidation product 4-hydroxy-trans-2-nonenal causes endothelial activation by inducing endoplasmic reticulum stress. J Biol Chem 287(14):11398–11409. doi:10.1074/jbc.M111.320416

Walker AK, Jacobs RL, Watts JL, Rottiers V, Jiang K, Finnegan DM, Shioda T, Hansen M, Yang F, Niebergall LJ, Vance DE, Tzoneva M, Hart AC, Naar AM (2011) A conserved SREBP-1/Phosphatidylcholine feedback circuit regulates lipogenesis in metazoans. Cell 147 (4):840–852. doi:10.1016/j.cell.2011.09.045 S0092-8674(11)01193-7 [pii]

Wang D, Dubois RN (2010) Eicosanoids and cancer. Nat Rev Cancer 10(3):181–193. doi:10. 1038/nrc2809

Wang D, Wang H, Shi Q, Katkuri S, Walhi W, Desvergne B, Das SK, Dey SK, DuBois RN (2004) Prostaglandin E(2) promotes colorectal adenoma growth via transactivation of the nuclear peroxisome proliferator-activated receptor delta. Cancer Cell 6(3):285–295. doi:10.1016/j.ccr. 2004.08.011

Warburg O, Posener K, Negelein E (1924) Über den Stoffwechsel der Carcinomzelle. Biochem Zeitschr 152:309–344

Welcker M, Clurman BE (2008) FBW7 ubiquitin ligase: a tumour suppressor at the crossroads of cell division, growth and differentiation. Nat Rev Cancer 8(2):83–93. doi:10.1038/nrc2290 nrc2290 [pii]

Williams KJ, Argus JP, Zhu Y, Wilks MQ, Marbois BN, York AG, Kidani Y, Pourzia AL, Akhavan D, Lisiero DN, Komisopoulou E, Henkin AH, Soto H, Chamberlain BT, Vergnes L, Jung ME, Torres JZ, Liau LM, Christofk HR, Prins RM, Mischel PS, Reue K, Graeber TG, Bensinger S (2013) An essential requirement for the SCAP/SREBP signaling axis to protect cancer cells from lipotoxicity. Cancer Res. doi:10.1158/0008-5472.CAN-13-0382-T

Wise DR, Ward PS, Shay JE, Cross JR, Gruber JJ, Sachdeva UM, Platt JM, DeMatteo RG, Simon MC, Thompson CB (2011) Hypoxia promotes isocitrate dehydrogenase-dependent carboxylation of alpha-ketoglutarate to citrate to support cell growth and viability. Proc Natl Acad Sci USA 108(49):19611–19616. doi:10.1073/pnas.1117773108

Wymann MP, Schneiter R (2008) Lipid signalling in disease. Nature Rev Mol Cell Biol 9(2):162–176. doi:10.1038/nrm2335

Yagublu V, Arthur JR, Babayeva SN, Nicol F, Post S, Keese M (2011) Expression of selenium-containing proteins in human colon carcinoma tissue. Anticancer Res 31(9): 2693–2698

Yalcin A, Telang S, Clem B, Chesney J (2009) Regulation of glucose metabolism by 6-phosphofructo-2-kinase/fructose-2,6-bisphosphatases in cancer. Exp Mol Pathol 86(3):174–179. doi:10.1016/j.yexmp.2009.01.003 S0014-4800(09)00008-2 [pii]

Yang WS, SriRamaratnam R, Welsch ME, Shimada K, Skouta R, Viswanathan VS, Cheah JH, Clemons PA, Shamji AF, Clish CB, Brown LM, Girotti AW, Cornish VW, Schreiber SL, Stockwell BR (2014) Regulation of ferroptotic cancer cell death by GPX4. Cell 156(1–2): 317–331. doi:10.1016/j.cell.2013.12.010

Yoshii Y, Furukawa T, Yoshii H, Mori T, Kiyono Y, Waki A, Kobayashi M, Tsujikawa T, Kudo T, Okazawa H, Yonekura Y, Fujibayashi Y (2009) Cytosolic acetyl-CoA synthetase affected tumor cell survival under hypoxia: the possible function in tumor acetyl-CoA/acetate metabolism. Cancer Sci 100(5):821–827

Young RM, Ackerman D, Quinn ZL, Mancuso A, Gruber M, Liu L, Giannoukos DN, Bobrovnikova-Marjon E, Diehl JA, Keith B, Simon MC (2013) Dysregulated mTORC1 renders cells critically dependent on desaturated lipids for survival under tumor-like stress. Genes Dev 27(10):1115–1131. doi:10.1101/gad.198630.112

Yue S, Li J, Lee SY, Lee HJ, Shao T, Song B, Cheng L, Masterson TA, Liu X, Ratliff TL, Cheng JX (2014) Cholesteryl ester accumulation induced by PTEN loss and PI3 K/AKT activation underlies human prostate cancer aggressiveness. Cell Metabol 19(3):393–406. doi:10.1016/j.cmet.2014.01.019

Zhang WC, Shyh-Chang N, Yang H, Rai A, Umashankar S, Ma S, Soh BS, Sun LL, Tai BC, Nga ME, Bhakoo KK, Jayapal SR, Nichane M, Yu Q, Ahmed DA, Tan C, Sing WP, Tam J, Thirugananam A, Noghabi MS, Pang YH, Ang HS, Mitchell W, Robson P, Kaldis P, Soo RA, Swarup S, Lim EH, Lim B (2012) Glycine decarboxylase activity drives non-small cell lung cancer tumor-initiating cells and tumorigenesis. Cell 148(1–2):259–272. doi:10.1016/j.cell.2011.11.050 S0092-8674(11)01444-9 [pii]

Lactate—An Integrative Mirror of Cancer Metabolism

Stefan Walenta, Nadine F. Voelxen
and Wolfgang Mueller-Klieser

Abstract

The technique of induced metabolic bioluminescence imaging (imBI) has been developed to obtain a "snapshot" of the momentary metabolic status of biological tissues. Using cryosections of snap-frozen tissue specimens, imBI combines highly specific and sensitive in situ detection of metabolites with a spatial resolution on a microscopic level and with metabolic imaging in relation to tissue histology. Here, we present the application of imBI in human colorectal cancer. Comparing the metabolic information of one biopsy with that of 2 or 3 biopsies per individual cancer, the classification into high versus low lactate tumors, reflecting different glycolytic activities, based on a single biopsy was in agreement with the result from multiple biopsies in 83 % of all cases. We further demonstrate that the metabolic status of tumor tissue can be preserved at least over 10 years by storage in liquid nitrogen, but not by storage at -80 °C. This means that tissue banking with long-term preservation of the metabolic status is possible at -180 °C, which may be relevant for studies on long-term survival of cancer patients. As with other tumor entities, tissue lactate concentration was shown to be correlated with tumor development and progression in colorectal cancer. At first-time diagnosis, lactate values were low in rectal normal tissue and adenomas, were significantly elevated to intermediate levels in non-metastatic adenocarcinomas, and were very high in carcinomas with distant metastasis. There was an inverse behavior of tissue glucose concentration under corresponding conditions. The expression level of monocarboxylate transporter-4 (MCT4) was positively correlated with the tumor lactate concentration and may thus contribute to high lactate tumors being associated with a high degree of malignancy.

S. Walenta · N.F. Voelxen · W. Mueller-Klieser (✉)
Institute of Pathophysiology, University Medical Center, Johannes Gutenberg-University of Mainz, 55128 Mainz, Germany
e-mail: mue-kli@uni-mainz.de

© Springer International Publishing Switzerland 2016
T. Cramer and C.A. Schmitt (eds.), *Metabolism in Cancer*,
Recent Results in Cancer Research 207, DOI 10.1007/978-3-319-42118-6_2

Keywords

Lactate · Induced Metabolic Bioluminescence Imaging (imBI) · Tumor Metabolism · Colorectal Adenocarcinoma · Warburg Effect · Metabolic Cryopreservation

1 Introduction

Steady-state concentrations of metabolites can mirror the metabolic status of live tissues. In most healthy organs, metabolic activity is not homogeneously distributed across the entire tissue, but exhibits a metabolic zonation, such as renal medulla versus cortex or periportal versus central venous regions in the liver. Metabolic zonation also exists in solid malignant tumors, yet in a specific way: The metabolic zones in tumors are extremely heterogeneous with regard to spatial arrangement, size, and cellular function (Aly et al. 2015; Jeng et al. 2015). There is recent evidence for this characteristic tumor heterogeneity being the major cause of therapeutic failure in medical oncology (Walther et al. 2015) (Table 1).

Tumor heterogeneity is characterized by variable intrinsic properties of cancer cells, such as variability in proliferation and invasiveness; by the coexistence of cancer cells with variable proportions of different histological cell populations, such

Table 1 From 13 patients with tumors, 29 biopsies were analyzed for lactate content (μmol/g)

Tumor	Biopsy 1	Biopsy 2	Biopsy 3	Mean	Classification
1	**5.7**	**4.9**	**3.0**	**4.5**	**L**
2	8.5	**6.1**	7.4	7.3	H
3	10.2	18.0	7.5	11.9	H
4	**4.5**	**4.1**		**4.3**	**L**
5	8.3	**4.4**		**6.3**	**L**
6	9.5	4.9		7.2	H
7	**6.1**	**5.1**		**5.6**	**L**
8	9.1	**5.1**		7.1	H
9	**2.1**	**5.7**		**3.9**	**L**
10	11.1	7.6		9.3	H
11	14.8	8.0		11.4	H
12	**3.8**	8.3		**6.0**	**L**
13	7.8	11.6		9.7	H
			Median =	7.1	

Comparing each average tumor value with the overall median (7.1 μmol/g), each malignancy was classified as "low" (L) or "high" (H) lactate tumor. Comparing the mean values of each single biopsy with the median value of the entire tumor cohort revealed an identical classification in 24 cases (=83 %) and an opposite classification in 5 cases (=17 %)

L low lactate: 6 tumors

H high lactate: 7 tumors

as fibroblasts, endothelial or immune cells; and by an extremely chaotic vascular supply with regard to microcirculatory geometry and function. Typically, the dimensions of such agglomerated heterogeneous tumor regions are in the submillimeter range. These conditions within solid tumors limit the biological significance of information that is obtained by averaging signals over macroscopic tissue regions. In other words, a selective evaluation of uniform subregions, e.g., of largely viable tumor areas, requires a spatial resolution on a microscopic level.

Imaging metabolites in cancerous tissue in a biologically significant manner requires the quantitative detection of metabolic substances within microscopic dimensions in association with the histological tissue structure. The challenges of these requirements are reflected by an experimental technique called "the laser-capture microdissection of cryosections" (Espina et al. 2009; Silasi et al. 2008; Zanni and Chan 2011). This sophisticated and elaborate method has been successfully combined with polymerase chain reaction (PCR) and matrix-assisted laser desorption/ionization (MALDI) for genomic and proteomic profiling of homogeneous tissue regions from healthy organs or malignant tumors, which has been documented in a number of original reports and recent reviews (Datta et al. 2015; Nishimura et al. 2015; Sethi et al. 2015). In contrast, approaches to a direct quantification of metabolites in microdissected tissue samples seem to be rare in the literature (Teutsch et al. 1995). It is evident from the complexity of laser-capture microdissection that this technique is not suited for prospective studies with cancer patients or for clinical routine. On the other hand, noninvasive techniques for metabolic imaging, which are suited for clinical use, such as magnetic resonance imaging (MRI), do not cope with the challenges inherent to tumor heterogeneity at present.

This situation suggests that metabolomic analysis may be performed in tumor biopsies, provided a relative simple technique for structure-associated, quantitative metabolic imaging in situ is available. We have developed such a method, which will be briefly presented below. Since tumor specimens are routinely taken in the clinic for pathohistological diagnosis, spare material is often available for scientific purposes under consideration of ethics and patient consent. There are, however, two major concerns with such tumor specimens.

Firstly, there is the question of how closely a relatively small biopsy can represent the metabolic situation of a whole tumor. So far, this old problem has not been investigated with respect to the metabolite distribution in cancers. We have addressed this issue by metabolic measurements in multiple biopsies from the same tumor and by comparing the result from one biopsy with that from all biopsies from one tumor, as will be presented below.

The second concern with clinical tumor biopsy material is related to the physiological status of metabolism. As soon as live tissue is removed from its natural site, the absence of blood flow in such a specimen is associated with a decrease in ATP and an accumulation of acids within less than 1 min. The only way to prevent such undesirable artifacts of ischemia is the rapid cryopreservation of the momentary physiological metabolic status in vivo, e.g., by immersion into liquid nitrogen. In this way, a "metabolic snapshot" of the tumor tissue can be generated. Despite all these critical issues, measurement of metabolites by enzymatic photometric assays or by mass spectrometry in extracts from snap-frozen tissue can still be considered a "gold standard" for the assessment of tissue concentrations of metabolites.

2 Metabolic Reprogramming in Malignant Tumors

Malignant transformation of normal cells is frequently accompanied by an increase in cellular glucose uptake by a factor of 5–10. This is the basis for tumor diagnosis and treatment follow-up with positron emission tomography (PET) measuring the uptake of 2-fluoro-deoxy-glucose (FDG). The increase in glucose uptake is mostly associated with an elevated cellular production and secretion of lactate, which often accumulates up to excessive concentrations (in the range of 40–50 mM) in the tumor microenvironment. This increased lactate production persists even in the presence of oxygen, which is eventually termed "aerobic glycolysis," "Warburg effect" or "Warburg metabolism" (Bayley and Devilee 2010; Bayley and Devilee 2012; Cairns et al. 2011; Herling et al. 2011; Icard et al. 2012; Israel and Schwartz 2011; Rodriguez-Enriquez et al. 2011).

During the past few years, an increasing number of signaling pathways has been identified to be closely linked to the Warburg metabolism of cancer cells. As a result, the metabolic deregulation in tumors may be recognized as a complex network of interrelated pathways that is largely variable and unpredictable in its functionality in individual tumors (Carroll et al. 2015). On the other hand, there is a common readout of cancer cell metabolism integrating over their various signaling activities, i.e., the efflux of lactate into the tumor microenvironment (Dhup et al. 2012; Hirschhaeuser et al. 2011; Luc et al. 2015).

The clinical significance of the extremely variable levels of lactate concentrations in solid tumors has been identified first by our group in a patient study in 2000 (Walenta et al. 2000) using induced metabolic bioluminescence imaging (imBI). The imBI technique for lactate quantification and assessment of the regional distribution of this metabolite within tissues of interest has been developed in our laboratory on the basis of biochemical precursor studies mainly on brain metabolism (Kim et al. 1993; Kricka 2000; Paschen et al. 1981; Paschen 1985). The current status of imBI, its advantages, and limitations has been reviewed recently (Walenta et al. 2014), and we refer the reader to this article for further methodological details.

Up to now, tissue concentrations of glycolytic key metabolites, such as glucose, pyruvate, or lactate, cannot be measured noninvasively, e.g., by proton (magnetic resonance spectroscopy) MRS or PET, with a simultaneous combination of appropriate *sensitivity* and *spatial resolution* in the clinical setting. Some optimism may be justified considering magnetic resonance spectroscopy imaging (MRSI) with hyperpolarized C-13, but this technique is still in an early phase of clinical testing (Bohndiek et al. 2012; Witney et al. 2011; Gallagher et al. 2009; Day et al. 2007; Reed et al. 2012; Hu et al. 2011; Kurhanewicz et al. 2011). There is, however, a compelling evidence for the relevance of Warburg metabolism in malignant disease to be evaluated systematically, among other approaches, by quantitative measurements of glycolysis-associated metabolites.

3 The imBI Technique

In the imBI technique, as developed in our laboratory, mixtures of definite exogenous enzymes and cofactors are used as self-made kits for the detection of metabolites with high specificity and sensitivity in cryosections from snap-frozen tissue. These enzymes may be considered "molecular metabolic sensors" for the assessment of tissue concentrations of specific metabolites. For example, lactate dehydrogenase is used as a "biosensor" of lactate. The enzymatic activity specifically induced by the metabolite of interest as a substrate is biochemically coupled to luciferases. This leads to the emission of light, i.e., to bioluminescence via the NAD $(P)H + H^+$ redox system. The biochemical reactions can be adjusted in a way that the bioluminescence intensity is proportional to the respective metabolite concentration. Details on the enzyme cocktails have been summarized recently in a comprehensive manner (Walenta et al. 2014).

Figure 1 illustrates the imBI processing routine for the tissue cryobiopsy. For the assessment of the spatial distribution of the metabolites, serial cryosections are generated from the snap-frozen tumor specimens. A specially designed sandwich technique provides a homogeneous contact between enzyme cocktail and tissue section. The sandwich is put on the thermostated stage of a microscope which is equipped with an ultrasensitive back-illuminated EM-CCD camera (iXonEM+ DU-888; Andor Technology PLC, Belfast, UK), the signal of which is transferred to a computer for image analysis. By integrating the light emission intensities over a selected time interval, a two-dimensional density profile is generated, which represents the two-dimensional metabolite distribution across the tissue section.

Using appropriate standards, which are handled in exactly the same way as the tissue of interest, bioluminescence intensities can be transformed into absolute tissue concentrations of the respective metabolite, e.g., in micromole per gram of tissue (µmol/g), which corresponds approximately to mM in solution. The spatial distribution of tissue concentrations is routinely displayed in a color-coded manner, as indicated in Fig. 1.

High sensitivity and high spatial resolution are to some extent contradicting demands, but the advantage of imBI is its versatility in adjusting the measuring conditions toward either resolution or sensitivity depending on the requirements of an actual experiment. For most measurements in tumor biopsies, experimental settings are chosen in a way that the minimum detectable substrate concentration is around 100 nmol/g with a spatial linear resolution of around 100 µm. Keeping the former sensitivity of detection, the spatial resolution can be adjusted to 20 µm.

Using serial sections, a section can be stained for the histological structure, e.g., with hematoxylin and eosin (H&E) followed by sections stained for the various metabolites (see Fig. 2). Before sectioning, two parallel channels are driven through

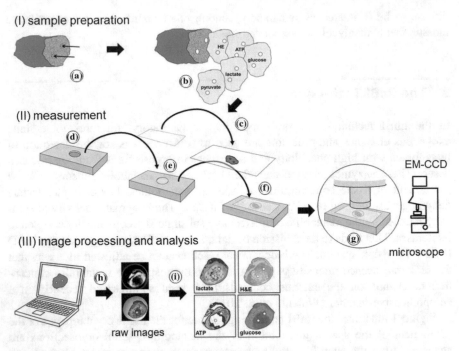

Fig. 1 Flowchart of the specimen processing for the imBI technique [according to (Walenta et al. 2014)]. **a** Driving 2 parallel pin channels into the cryobiopsy, **b** serial sectioning of the biopsy perpendicular to the pin channels generating pinholes for exact overlay, **c** putting each section on a cover glass, **d** glass block with circular casting mold, **e** pipetting the bioluminescence enzyme solution into the casting mold, **f** putting cover glass upside down on glass block (=sandwich) leading to contact between section and enzyme solution, **g** putting sandwich on thermostated microscope stage starting bioluminescence reaction and low light detection via a cooled CCD camera, **h** computerized signal and image processing, **i** generation of calibrated 2D metabolite distributions including structure-associated evaluation

the frozen specimen with a specially designed "microfork." Adjusting the sectioning plane perpendicular to the two channels generates two holes in each section which allows for an exact computer-assisted overlay between the sections. Such an overlay makes it possible to detect the bioluminescence signals in a structure-associated way, i.e., within selected histological areas, such as viable tumor regions, stromal areas, or necrosis. For methodological details, the reader is referred to Walenta et al. (Walenta et al. 2014).

In the clinical setting, the imBI technique has been applied to patients with cervix carcinomas (Walenta et al. 2000), squamous cell carcinomas of the head and neck (Brizel et al. 2001), rectal adenocarcinomas (Walenta et al. 2003), and glioblastomas (Walenta et al. 2014). Here, we summarize our previous findings from rectal carcinomas, supplemented by new data and new aspects of data evaluation and interpretation.

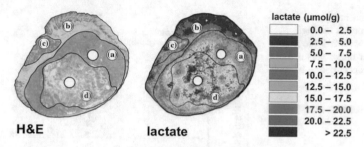

Fig. 2 Illustration of structure-associated evaluation of the color-coded lactate concentration distribution from imBI based on an adjacent H- and E-stained cryosection. Templates from circumscribed histological areas can be digitally overlaid on the corresponding imBI image to obtain regional metabolite concentrations [according to (Walenta et al. 2014)]. **a** Viable tumor region, **b** stroma/skin, **c** skeletal muscle, and **d** tumor necrosis

4 Representativeness of One Versus Multiple Biopsies

Considering the aforementioned tumor heterogeneity, it is not surprising that the representativeness of data obtained from one biopsy with regard to the whole tumor is a persisting critical issue in experimental cancer research as well as in clinical oncology. Colorectal cancer is one of the tumor entities which may give access to two or multiple biopsies in the clinical routine. In a separate study on rectal cancer, we were able to collect and metabolically characterize such multiple biopsies with imBI. We were then using appropriate strategies for data acquisition and evaluation to address this respective issue.

All tumor samples, presented in this report, were obtained from the Department of Surgery at the University of Regensburg, Germany, after approval by the local ethical committee and following the receipt of informed consent by the patients. Biopsies were taken within diagnostic routines for first-time diagnosis of primary rectal cancer before conventional surgical resection therapy. Tumor specimens were immersed into liquid nitrogen within 10 s after removal from the patient. After final diagnosis in the Department of Pathology in parts of the biopsies, spare material was used for imBI measurements. Patient data, such as metastatic status or survival, were obtained in an anonymous manner from the Regensburg Tumor Center to be correlated with metabolic data.

In the previous studies on clinical cancer, a classification of tumors into two classes regarding the tissue lactate concentration proved to be useful and clinically relevant. For practical purposes, the median lactate concentration of a given cohort of tumors was chosen for separation into low and high lactate tumors. Interestingly, the median for lactate consistently was in the range of 7–9 µmol/g in a number of independent studies including different tumor entities and applying different modifications of the imBI technique. As common findings in these various prospective studies, the incidence of metastasis was significantly higher and patient survival

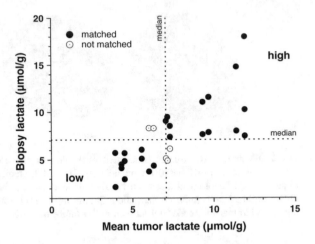

Fig. 3 Lactate concentrations in single biopsies with 2–3 biopsies taken per tumor versus mean lactate concentration of each tumor

was significantly reduced in high compared to low lactate cancer (summarized in Walenta and Mueller-Klieser 2004).

This way of classification was applied in the small cohort of 13 patients with rectum adenocarcinomas. From each of the patient with tumors, 2 or 3 biopsies were removed resulting in a total number of 29 tumor biopsies. The mean lactate concentration of each biopsy was than compared with the median lactate concentration of the entire tumor cohort. The results show that the classification into high and low lactate tumors based on only 1 biopsy differed from that based on the average from 2 or 3 biopsies per tumor in 5 of 29 cases, i.e., in 17 % of all biopsies considered.

This result does not completely discourage biopsy-based investigations on tumors, and on the other hand, it is not entirely inspiring either. Considering, however, the mean lactate values of the five cases of discrepancy in relation to their deviation from the cohort mean, it is obvious that all five means are close to the median. This is illustrated by plotting the biopsy mean values over the global tumor means (Fig. 3). The 2 dotted lines represent the cohort median delimiting the upper right quadrant as the high lactate tumor field and the lower left quadrant as the field of low lactate tumors. The five circles represent the average lactate values of the biopsies of discrepancy visualizing their proximity to the median.

The finding suggests that the border line lactate concentration should not be considered a cutting edge, but rather a transitional zone with an inherent uncertainty with regard to the respective classification. It has to be mentioned in this connection that in all imBI studies on clinical tumors, the intratumoral variation of lactate was much less than that of the interindividual differences between tumors.

5 Metabolic Asservation in Liquid Nitrogen

It seems obvious that the physiological situation of metabolites in solution within a live tissue cannot be preserved in a paraffinized asservate. On the other hand, there is a demand for the long-term preservation of the physiological state in natural tissue, and so far, this can only be done by rapid freezing. Immersion of the biopsy into liquid nitrogen immediately after removal from the tumor turned out to be appropriate for most of our metabolic measurements. Since tumor metabolism is a booming field in both experimental and clinical oncology, numerous research institutions and clinics start to build up tumor banks with tissues in liquid nitrogen. This raises, however, the question of the time span for which the biopsy can be asservated at −180 °C without disturbances in the metabolic state. Also, it seems important to know whether preservation at −80 °C would be as appropriate as storage in liquid nitrogen.

We have systematically registered concentrations of metabolites in freshly frozen tissue versus storage in the −80 °C freezer for 3–6 months or for more than 2 years. Similar measurements were performed in samples stored in liquid nitrogen for around 10 years. Representative data are graphically displayed in Fig. 4. The results clearly show that there are no major changes in metabolite concentrations within up to 6 months of storage under either condition. In contrast, storage in −80 °C for several years leads to significant decreases of the concentration values far below the initial levels. Such a decrease does not occur in liquid nitrogen. In conclusion, a long-term metabolic asservation of freeze-clamped tissue over a period of several years up to at least one decade is possible, but requires storage at −180 °C.

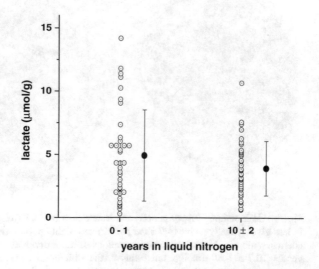

Fig. 4 Mean tumor lactate concentration in biopsies which were fresh or stored for no longer than 1 year (0–1 year) compared to lactate concentrations in the same cohort following storage in liquid nitrogen for 8–12 years (10 ± 2 years)

6 Metabolic Imaging with ImBI in Rectum Cancer

Tissue specimens from patients were obtained from the University of Regensburg, as outlined above. In a cohort of 24 patients with rectal adenocarcinoma, imBI measurements were performed on the basis of 33 cryobiopsies. Of these specimens, 7 contained only normal tissue, 3 contained adenoma tissue, and 33 contained tumor material. The latter group included 5 patients with synchronous distant liver metastasis; i.e., they already had distant metastasis at first-time diagnosis.

Figure 5 shows a representative documentation of data from routine imBI measurements in a glycolytically less active (Fig. 5, left-hand panels) and in a highly glycolytic (Fig. 5, right-hand panels) human colorectal adenocarcinomas. The upper

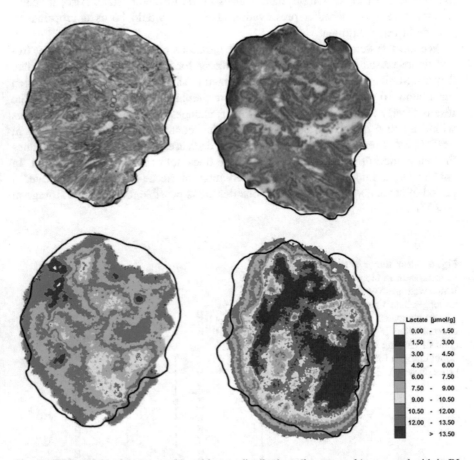

Fig. 5 H&E staining (*upper panels*) and lactate distributions (*lower panels*) measured with imBI in less glycolytically active (*left-hand panels*) and highly glycolytic (*right-hand panels*) colorectal adenocarcinomas. Mean lactate concentrations of the individual cancers were 4.9 ± 1.11 μmol/g versus 10.7 ± 0.80 μmol/g. Interestingly, the difference between the highest regional lactate concentration measured in these tumors was even drastically larger than that between the means, i.e., 10.1 μmol/g versus 31.0 μmol/g

panels represent staining with H&E, and the lower panels, the color-coded, regional distribution of lactate. Mean lactate concentrations averaged over whole individual cancers were quite different, as shown by the respective values for the distribution in Fig. 5, i.e., 4.9 ± 1.11 µmol/g versus 10.7 ± 0.80 µmol/g. Interestingly, the

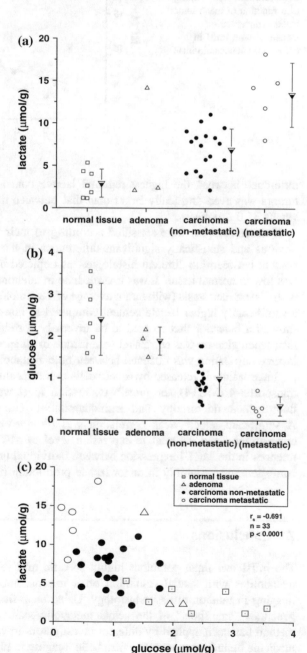

Fig. 6 Results of imBI measurements in colorectal adenocarcinomas, adenomas, and tumor-adjacent normal tissue [modified according to (Walenta et al. 2003)]. **a** Lactate, **b** glucose, and **c** lactate versus glucose

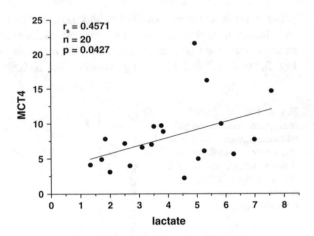

Fig. 7 Relative expression level of monocarboxylate transporters-4 (MCT4), as obtained by immunofluorescent staining, as a function of mean tumor lactate concentration measured with imBI in colorectal adenocarcinomas

difference between the highest regional lactate concentration measured in these tumors was even drastically larger than that between the means, i.e., 10.1 μmol/g versus 31.0 μmol/g.

If tissue samples were classified according to their histology, there were quite obvious and statistically significant differences in the mean lactate and glucose content between the different histologies, as depicted in Figs. 6a, b. While lactate was low in normal tissue, it was intermediate in adenomas and very high in tumors with distant metastasis (with tumor average of 18 μmol/g). Metastatic primaries had a significantly higher lactate content compared to non-metastatic tumors. Glucose showed a behavior that seemed to be inversely correlated with that of lactate. In fact, when glucose was correlated with lactate in all specimens, a highly significant inverse correlation was obtained between both metabolites (see Fig. 6c).

Since lactate is released by cancer cells predominantly by the monocarboxylate transporter-4 (MCT4), the protein expression level was determined by immunofluorescence microscopy and semiquantitative computerized image analysis. A significant, albeit weak correlation was obtained between the tumor lactate concentration and the relative expression level of MCT4. This indicates that differences in the MCT4 expression between individual tumors may contribute to the corresponding variability in tumor lactate production (Fig. 7).

7 Conclusions

The imBI technique combines highly specific and sensitive in situ detection of metabolites with spatial resolution on a microscopic level and with metabolic imaging in relation to tissue histology. Using snap-frozen tissue specimens, imBI generates "snapshots" of the actual metabolic status of biological tissues. The method has been applied by different investigators in various fields of biology and medicine including quantitative metabolic imaging in plant seedlings, 2D or 3D cell cultures, blood vessel walls, tissue wounds, various animal organs, experimental tumors, and cancers in patients.

The imBI technique has been used in several independent clinical studies on cancer of the cervix uteri and the head and neck region, including different clinics and clinical settings, various experimenters, and different modifications of the imBI method, such as several updates of the equipment for very low light-level detection. Although the technique is restricted to prospective studies with inherently low number of patients, these various studies currently enable an overview on data from several hundreds of patients. In summary, these studies uniformly identify the lactate content of primary tumors as an indicator of tumor progression and malignant traits (Hirschhaeuser et al. 2011). This has been illustrated here by imBI data obtained in colorectal cancer, showing tissue lactate content to be low in normal tissue and adenomas, intermediate in non-metastatic disease, and extremely high in primary cancer with distant metastasis. A separate study on the expression of MCT4 in colorectal cancer is indicative of this membrane transporter to contribute to increased cellular lactate secretion and tissue accumulation. This finding is in accordance with a general function of MCT4 as a transporter which preferentially exports lactate out of the cell, in contrast to MCT1 transporting into or out of the cells (Pinheiro et al. 2012).

It has often been questioned whether a biopsy can be representative of cancerous tissue considering the pronounced heterogeneity of malignant tumors. In a subset of colorectal cancers, we got access to tumors from which 2 or 3 biopsies were taken. In the previous metabolic studies on malignant tumors, a classification into tumors with a low or high lactate content with the median lactate concentration being the discriminating criterion appeared to be useful, since low and high lactate tumors potentially represent tumors with a less or a highly active Warburg metabolism, respectively. The cohort of rectal cancers with multiple biopsies made it possible to compare the classification based on one biopsy only with that based on 2 or 3 biopsies. The agreement between both approaches in 83 % of all cases and the fact that in all cases of discrepancies, the lactate concentration was close to the border line between the two classes may justify the respective classification from one biopsy.

A number of larger clinics have started to build up tumor banks storing tissue in liquid nitrogen. Since we have documented a change in metabolites of tissue being stored at -80 °C for a few years, we have measured metabolites in tumor specimens before and after storage in liquid nitrogen for around 10 years. This may be relevant, for example, during long-term follow-ups of cancer patients where a situation may emerge that may require the determination of a tumor metabolite, which has not been measured initially. Our data suggest that the metabolic status can be cryopreserved at least for one decade.

In summary, biologically and clinically relevant and significant results were obtained with imBI most likely due to the "natural way" of detecting metabolites, i.e., using them as substrates of specific enzymes in selected histological tissue regions. Based on the latter potential, imBI may cope with challenges related to tumor heterogeneity, which is nowadays considered one of the major causes of tumor treatment failure (Aly et al. 2015; Jeng et al. 2015).

Acknowledgments This work was supported by the Deutsche Forschungsgemeinschaft: Mu 576/15-1, 15-2, and by the German Federal Ministry of Education and Research ("ISIMEP"; 02NUK016A).

References

Aly A, Mullins CD, Hussain A (2015) Understanding heterogeneity of treatment effect in prostate cancer. Curr Opin Oncol 27(3):209–216

Bayley JP, Devilee P (2010) Warburg tumours and the mechanisms of mitochondrial tumour suppressor genes. Barking up the right tree? Curr Opin Genet Dev 20(3):324–329

Bayley JP, Devilee P (2012) The Warburg effect in 2012. Curr Opin Oncol 24(1):62–67

Bohndiek SE, Kettunen MI, Hu DE, Brindle KM (2012) Hyperpolarized (13)C spectroscopy detects early changes in tumor vasculature and metabolism after VEGF neutralization. Cancer Res 72(4):854–864

Brizel DM, Schroeder T, Scher RL, Walenta S, Clough RW, Dewhirst MW, Mueller-Klieser W (2001) Elevated tumor lactate concentrations predict for an increased risk of metastases in head-and-neck cancer. Int J Radiat Oncol Biol Phys 51(2):349–353

Cairns RA, Harris IS, Mak TW (2011) Regulation of cancer cell metabolism. Nat Rev Cancer 11 (2):85–95

Carroll PA, Diolaiti D, McFerrin L, Gu H, Djukovic D, Du J, Cheng PF, Anderson S, Ulrich M, Hurley JB, Raftery D, Ayer DE, Eisenman RN (2015) Deregulated Myc requires MondoA/Mlx for metabolic reprogramming and tumorigenesis. Cancer Cell 27(2):271–285

Datta S, Malhotra L, Dickerson R, Chaffee S, Sen CK, Roy S (2015) Laser capture microdissection: Big data from small samples. Histol Histopathol 30(11):1255–1269

Day SE, Kettunen MI, Gallagher FA, Hu DE, Lerche M, Wolber J, Golman K, Ardenkjaer-Larsen JH, Brindle KM (2007) Detecting tumor response to treatment using hyperpolarized 13C magnetic resonance imaging and spectroscopy. Nat Med 13 (11):1382–1387

Dhup S, Dadhich RK, Porporato PE, Sonveaux P (2012) Multiple biological activities of lactic acid in cancer: influences on tumor growth, angiogenesis and metastasis. Curr Pharm Des 18 (10):1319–1330

Espina V, Wulfkuhle J, Liotta LA (2009) Application of laser microdissection and reverse-phase protein microarrays to the molecular profiling of cancer signal pathway networks in the tissue microenvironment. Clin Lab Med 29(1):1–13

Gallagher FA, Kettunen MI, Hu DE, Jensen PR, Zandt RI, Karlsson M, Gisselsson A, Nelson SK, Witney TH, Bohndiek SE, Hansson G, Peitersen T, Lerche MH, Brindle KM (2009) Production of hyperpolarized [1,4-13C2]malate from [1,4-13C2]fumarate is a marker of cell necrosis and treatment response in tumors. Proc Natl Acad Sci USA 106(47):19801–19806

Herling A, Konig M, Bulik S, Holzhutter HG (2011) Enzymatic features of the glucose metabolism in tumor cells. FEBS J 278(14):2436–2459

Hirschhaeuser F, Sattler UG, Mueller-Klieser W (2011) Lactate: a metabolic key player in cancer. Cancer Res 71(22):6921–6925

Hu S, Zhu M, Yoshihara HA, Wilson DM, Keshari KR, Shin P, Reed G, von MC, Bok R, Larson PE, Kurhanewicz J, Vigneron DB (2011) In vivo measurement of normal rat intracellular pyruvate and lactate levels after injection of hyperpolarized [1-(13)C]alanine. Magn Reson Imaging 29(8):1035–1040

Icard P, Poulain L, Lincet H (2012) Understanding the central role of citrate in the metabolism of cancer cells. Biochim Biophys Acta 1825(1):111–116

Israel M, Schwartz L (2011) The metabolic advantage of tumor cells. Mol Cancer 10:70

Jeng KS, Chang CF, Jeng WJ, Sheen IS, Jeng CJ (2015) Heterogeneity of hepatocellular carcinoma contributes to cancer progression. Crit Rev Oncol Hematol 94(3):337–347

Kim JK, Haselgrove JC, Shapiro IM (1993) Measurement of metabolic events in the avian epiphyseal growth cartilage using a bioluminescence technique. J Histochem Cytochem 41 (5):693–702

Kricka LJ (2000) Application of bioluminescence and chemiluminescence in biomedical sciences. Methods Enzymol 305:333–345

Kurhanewicz J, Vigneron DB, Brindle K, Chekmenev EY, Comment A, Cunningham CH, Deberardinis RJ, Green GG, Leach MO, Rajan SS, Rizi RR, Ross BD, Warren WS, Malloy CR (2011) Analysis of cancer metabolism by imaging hyperpolarized nuclei: prospects for translation to clinical research. Neoplasia 13(2):81–97

Luc R, Tortorella SM, Ververis K, Karagiannis TC (2015) Lactate as an insidious metabolite due to the Warburg effect. Mol Biol Rep 42(4):835–840

Nishimura Y, Tomita Y, Yuno A, Yoshitake Y, Shinohara M (2015) Cancer immunotherapy using novel tumor-associated antigenic peptides identified by genome-wide cDNA microarray analyses. Cancer Sci 106(5):505–511

Paschen W (1985) Regional quantitative determination of lactate in brain sections. A bioluminescent approach. J Cereb Blood Flow Metab 5(4):609–612

Paschen W, Niebuhr I, Hossmann KA (1981) A bioluminescence method for the demonstration of regional glucose distribution in brain slices. J Neurochem 36(2):513–517

Pinheiro C, Longatto-Filho A, zevedo-Silva J, Casal M, Schmitt FC, Baltazar F (2012) Role of monocarboxylate transporters in human cancers: state of the art. J Bioenerg Biomembr 44(1): 127–139

Reed GD, Larson PE, Morze C, Bok R, Lustig M, Kerr AB, Pauly JM, Kurhanewicz J, Vigneron DB (2012) A method for simultaneous echo planar imaging of hyperpolarized (13)C pyruvate and (13)C lactate. J Magn Reson 217:41–47

Rodriguez-Enriquez S, Gallardo-Perez JC, Marin-Hernandez A, guilar-Ponce JL, Mandujano-Tinoco EA, Meneses A, Moreno-Sanchez R (2011) Oxidative phosphorylation as a target to arrest malignant neoplasias. Curr Med Chem 18(21):3156–3167

Sethi S, Chourasia D, Parhar IS (2015) Approaches for targeted proteomics and its potential applications in neuroscience. J Biosci 40(3):607–627

Silasi DA, Alvero AB, Mor J, Chen R, Fu HH, Montagna MK, Mor G (2008) Detection of cancer-related proteins in fresh-frozen ovarian cancer samples using laser capture microdissection. Methods Mol Biol 414:35–45

Teutsch HF, Goellner A, Mueller-Klieser W (1995) Glucose levels and succinate and lactate dehydrogenase activity in EMT6/Ro tumor spheroids. Eur J Cell Biol 66(3):302–307

Walenta S, Mueller-Klieser WF (2004) Lactate: mirror and motor of tumor malignancy. Semin Radiat Oncol 14(3):267–274

Walenta S, Wetterling M, Lehrke M, Schwickert G, Sundfor K, Rofstad EK, Mueller-Klieser W (2000) High lactate levels predict likelihood of metastases, tumor recurrence, and restricted patient survival in human cervical cancers. Cancer Res 60(4):916–921

Walenta S, Chau TV, Schroeder T, Lehr HA, Kunz-Schughart LA, Fuerst A, Mueller-Klieser W (2003) Metabolic classification of human rectal adenocarcinomas: a novel guideline for clinical oncologists? J Cancer Res Clin Oncol 129(6):321–326

Walenta S, Voelxen NF, Sattler UGA, Mueller-Klieser W (2014) Localizing and quantifying metabolites in situ with luminometry: induced metabolic bioluminescence imaging (imBI). In: Hirrlinger J, Waagepetersen HS (eds) Brain energy metabolism. Humana Press (Springer), New York, pp 195–216

Walther V, Hiley CT, Shibata D, Swanton C, Turner PE, Maley CC (2015) Can oncology recapitulate paleontology? Lessons from species extinctions. Nat Rev Clin Oncol 12(5):273–285

Witney TH, Kettunen MI, Brindle KM (2011) Kinetic modeling of hyperpolarized 13C label exchange between pyruvate and lactate in tumor cells. J Biol Chem 286(28):24572–24580

Zanni KL, Chan GK (2011) Laser capture microdissection: understanding the techniques and implications for molecular biology in nursing research through analysis of breast cancer tumor samples. Biol Res Nurs 13(3):297–305

Metabolic Reprogramming by the PI3K-Akt-mTOR Pathway in Cancer

Evan C. Lien, Costas A. Lyssiotis and Lewis C. Cantley

Abstract

In the past decade, there has been a resurgence of interest in elucidating how metabolism is altered in cancer cells and how such dependencies can be targeted for therapeutic gain. At the core of this research is the concept that metabolic pathways are reprogrammed in cancer cells to divert nutrients toward anabolic processes to facilitate enhanced growth and proliferation. Importantly, physiological cellular signaling mechanisms normally tightly regulate the ability of cells to gain access to and utilize nutrients, posing a fundamental barrier to transformation. This barrier is often overcome by aberrations in cellular signaling that drive tumor pathogenesis by enabling cancer cells to make critical cellular decisions in a cell-autonomous manner. One of the most frequently altered pathways in human cancer is the PI3K-Akt-mTOR signaling pathway. Here, we describe mechanisms by which this

E.C. Lien
Department of Pathology, Beth Israel Deaconess Medical Center,
Harvard Medical School, 330 Brookline Avenue, EC/CLS-628C,
Boston, MA 02215, USA
e-mail: elien@fas.harvard.edu

C.A. Lyssiotis
Department of Molecular and Integrative Physiology, University of Michigan,
1150 E. Medical Center Drive, Room 6308, Ann Arbor, MI 48109, USA
e-mail: clyssiot@med.umich.edu

C.A. Lyssiotis
Department of Internal Medicine, Division of Gastroenterology, University of Michigan,
1150 E. Medical Center Drive, Room 6308, Ann Arbor, MI 48109, USA

L.C. Cantley (✉)
Department of Medicine, the Cancer Center, Weill Cornell Medical College,
The Belfer Research Building, 413 East 69th Street, Floor 13 Room BB-1362,
New York, NY 10021, USA
e-mail: lcantley@med.cornell.edu

© Springer International Publishing Switzerland 2016
T. Cramer and C.A. Schmitt (eds.), *Metabolism in Cancer*,
Recent Results in Cancer Research 207, DOI 10.1007/978-3-319-42118-6_3

signaling network is responsible for controlling cellular metabolism. Through both the post-translational regulation and the induction of transcriptional programs, the PI3K-Akt-mTOR pathway coordinates the uptake and utilization of multiple nutrients, including glucose, glutamine, nucleotides, and lipids, in a manner best suited for supporting the enhanced growth and proliferation of cancer cells. These regulatory mechanisms illustrate how metabolic changes in cancer are closely intertwined with oncogenic signaling pathways that drive tumor initiation and progression.

Keywords

PI3K · Akt · mTOR · Cancer · Signaling · Metabolism

Abbreviations

PI3K	Phosphoinositide 3-kinase
mTOR	Mammalian target of rapamycin
RTKs	Receptor tyrosine kinases
GPCRs	G-protein-coupled receptors
PIP$_3$	Phosphatidylinositol-3,4,5-trisphosphate
PIP$_2$	Phosphatidylinositol-4,5-bisphosphate
PTEN	Phosphatase and tensin homolog
PH	Pleckstrin homology
PDK1	Phosphoinositide-dependent protein kinase 1
mTORC1/2	Mammalian target of rapamycin complex 1/2
HK	Hexokinase
PFK1/2	Phosphofructokinase 1/2
TSC1/2	Tuberous sclerosis 1/2
GAP	GTPase-activating protein
S6K1/2	S6 kinase 1/2
4E-BP1/2	eIF4E (eukaryotic initiation factor 4E)-binding protein 1/2
FoxO	Forkhead box O
GSK-3β	Glycogen synthase kinase-3β
HIF	Hypoxia-inducible factor
SREBP	Sterol regulatory element-binding protein
SRE	Sterol regulatory elements
TXNIP	Thioredoxin-interacting protein
AMPK	AMP-dependent protein kinase
^{18}FDG	18-fluoro-deoxyglucose
PET	Positron emission tomography
G6P	Glucose 6-phosphate
VDAC	Voltage-dependent anion channel
OMM	Outer mitochondrial membrane
PDK1	Pyruvate dehydrogenase kinase 1
PDH	Pyruvate dehydrogenase

TCA	Tricarboxylic acid
NADH	Nicotinamide adenine dinucleotide
LDHA	Lactate dehydrogenase A
GPI	Glucose phosphate isomerase
GAPDH	Glyceraldehyde-3-phosphate dehydrogenase
TPI	Triosephosphate isomerase
PGK1	Phosphoglycerate kinase 1
ENO1	Enolase
GCK	Glucokinase
ALDOB	Aldolase B
PK1	Pyruvate kinase 1
PPP	Pentose phosphate pathway
NADPH	Nicotinamide adenine dinucleotide phosphate
Ru5P	Ribulose 5-phosphate
R5P	Ribose 5-phosphate
F6P	Fructose 6-phosphate
Ga3P	Glyceraldehyde 3-phosphate
G6PD	Glucose 6-phosphate dehydrogenase
TKT	Transketolase
PGD	6-phosphogluconate dehydrogenase
RPE	Ribulose 5-phosphate epimerase
RPIA	Ribulose 5-phosphate isomerase
TALDO1	Transaldolase 1
CAD	Carbamoyl-phosphate synthetase 2, aspartate transcarbamylase, and dihydroorotase
PRPP	Phosphoribosyl pyrophosphate
ACL	ATP citrate lyase
αKG	α-ketoglutarate
GLS	Glutaminase
GLUL	Glutamine synthetase
GDH	Glutamate dehydrogenase
ASNS	Asparagine synthetase
GFPT1	Glutamine–fructose 6-phosphate transaminase 1

1 Introduction

A fundamental property of cells in multicellular organisms is the ability to process and integrate a variety of extracellular signals, such as growth factors, cytokines, and nutrients, in order to make decisions about cell fate, including proliferation, growth, survival, differentiation, motility, and metabolism. To accomplish this, cells have developed complex signaling networks that allow them to transduce extra-cellular signals into cellular decisions. Over the past few decades, much research has been dedicated toward elucidating how cellular signaling pathways are wired, and it has become clear that aberrations in cellular signaling underlie a wide variety of human diseases. In particular, our increasing understanding of the genomic landscape of human cancers over the past few years has revealed that cancer cells are frequently selected for mutations that occur in key nodes of various signaling pathways. These genetic alterations enable cancer cells to escape regulation by extrinsic signals and instead make critical cellular decisions in a cell-autonomous manner to promote uncontrollable cell survival, growth, and proliferation.

In the past decade, there has been a resurgence of interest in elucidating how metabolism is altered in cancer cells, based on observations that components of signal transduction pathways frequently regulate nutrient metabolism. At the core of this research is the concept that metabolic pathways are reprogrammed in cancer cells to divert nutrients toward anabolic processes to facilitate enhanced growth and proliferation. Importantly, physiological cellular signaling mechanisms normally tightly regulate the ability of cells to gain access to and utilize nutrients, posing a fundamental barrier to transformation. Therefore, it is becoming increasingly clear that cancer cells frequently select for mutations that enhance signal transduction through pathways that converge upon a common set of metabolic processes to promote nutrient uptake and utilization in cell building processes. It is the hope that an integrated understanding of oncogenic signaling and its modulation of cancer metabolism will reveal metabolic dependencies that are unique to cancer cells, thereby opening a therapeutic window that can be exploited to selectively target tumor cells.

The phosphoinositide 3-kinase (PI3K)-Akt-mammalian target of rapamycin (mTOR) signaling pathway is one of the most frequently altered pathways in human cancer. The importance of the pathway in cancer first became apparent in the 1980s, when it was discovered that a PI kinase activity was associated with transformation mediated by viral oncogenes such as the SRC tyrosine kinase and polyomavirus middle T antigen (Whitman et al. 1985) and that this enzyme carried out a previously unknown reaction, the phosphorylation of PI at the 3 position of the inositol ring (Whitman et al. 1988). Subsequent work placed PI3K at the head of a signaling network in which PI3K transduces upstream signals from receptor tyrosine kinases (RTKs) to generate critical lipid second messengers that serve to further activate downstream signaling effectors such as Akt and mTOR (Toker and Cantley 1997; Toker 2012). Multiple components of the pathway have since been demonstrated to be *bona fide* oncogenes and tumor suppressors, and activation of this pathway has a critical role in controlling most of the hallmarks of cancer, including proliferation, cell survival, motility, genomic instability, angiogenesis, and metabolism (Hanahan and Weinberg 2011). Studies in cancer genomics over the past decade have reinforced the importance of the PI3K-Akt-mTOR signaling pathway in human cancer, as genetic lesions that drive tumorigenesis are frequently found within components of the signaling network (see Table 1). As a result, numerous drugs have been developed in the past decade that target various nodes of

Table 1 Common genetic alterations of the PI3K-Akt-mTOR signaling pathway in human cancers

Gene	Alteration	Common tumor types	Selected references
PIK3CA	GOF mutation	Breast, ovarian, endometrial, cervical, colorectal, lung, glioblastoma	COSMIC, Samuels et al. (2004)
	Amplification	Breast, cervical, gastric, lung, ovarian, prostate	Campbell et al. (2004), Sun et al. (2009), Rácz et al. (1999), Ma et al. (2000)
PTEN	LOF mutation	Breast, ovarian, endometrial, colorectal, cervical, brain, bladder	COSMIC, Li et al. (1997)
	Deletion	Leukemia, breast, prostate, ovarian, liver, lung, melanoma	COSMIC, Li et al. (1997)
	Epigenetic silencing	Breast, colorectal	Goel et al. (2004), García et al. (2004)
AKT1	GOF mutation (E17K)	Breast, colorectal, ovarian, lung, bladder, endometrial, urothelial, prostate	Carpten et al. (2007), Shoji et al. (2009), Malanga et al. (2008), Boormans et al. (2010)
	Amplification	Gastric	Staal (1987)
AKT2	GOF mutation (E17K, others)	Breast, colorectal, lung	Sasaki et al. (2008); Stephens et al. (2012), Parsons et al. (2005)
	Amplification	Breast, colorectal, ovarian, pancreatic	Bellacosa et al. (1995), Cheng et al. (1996)
AKT3	GOF mutation (E17K, MAGI3-AKT3 fusion)	Breast, melanoma	Davies et al. (2008), Banerji et al. (2012)
	Amplification	Breast	Chin et al. (2014)
PDPK1	GOF mutation	Colorectal	Parsons et al. (2005)
TSC1/2	LOF mutation	Tuberous sclerosis	Jones et al. (1999)
MTOR	GOF mutation	Lung, renal, endometrial, melanoma	Grabiner et al. (2014)

Abbreviations: *COSMIC* Catalog of somatic mutations in cancer (http://cancer.sanger.ac.uk), *GOF* gain of function, *LOF* loss of function

the pathway, many of which are currently in clinical trials (Engelman 2009; Fruman and Rommel 2014; Fruman and Cantley 2014).

Given the critical role of PI3K-Akt-mTOR signaling in driving tumorigenesis, the importance of the pathway in modulating metabolism in cancer cells has also become apparent in recent years. Here, we will focus on mechanisms by which the PI3K-Akt-mTOR signaling pathway is responsible for coordinating nutrient uptake and utilization, and how cancer cells exploit these metabolic processes to support enhanced growth and proliferation. In doing so, we hope to illustrate how metabolic changes in cancer are closely intertwined with oncogenic signaling pathways that drive tumor initiation and progression.

2 Signaling Through the PI3K-Akt-mTOR Pathway

The PI3K-Akt-mTOR pathway coordinates insulin signaling during organismal growth and is a critical mediator of a variety of physiological processes, including glucose homeostasis, protein synthesis, cell proliferation, cell growth, survival, and angiogenesis. In fact, its central role in cellular physiology is underscored by the conservation of the pathway back to *Drosophila melanogaster* and *Caenorhabditis elegans*; furthermore, mTOR, but not PI3K/Akt, is conserved as far back as yeast (Vivanco and Sawyers 2002). Given the importance of PI3K-Akt-mTOR signaling in controlling organismal growth, it is not unexpected that this pathway is one of the most frequently dysregulated signaling pathways in human cancers (Engelman 2009).

2.1 PI3K-Akt Pathway Activation

Activation of RTKs and G-protein-coupled receptors (GPCRs) by extracellular stimuli, including growth factors, leads to PI3K activation (Fig. 1). As indicated above, PI3K is a lipid kinase that phosphorylates the 3'-hydroxyl group of the inositol ring of phosphatidylinositol to generate 3'-phosphoinositides (Whitman et al. 1988; Auger et al. 1989; Cantley 2002). Accumulation of these lipid second messengers subsequently activates pathways that control fundamental cellular processes. One key 3'-phosphoinositide is phosphatidylinositol-3,4,5-trisphosphate (PI(3,4,5)P_3; PIP$_3$), which is produced from phosphatidylinositol-4,5-bisphosphate (PI(4,5)P_2; PIP$_2$) by Class I PI3Ks. Negative regulation of the pathway is conferred by the lipid phosphatase, phosphatase and tensin homolog (PTEN), which converts PIP$_3$ back into PIP$_2$ (Li et al. 1997; Maehama and Dixon 1998; Cantley and Neel 1999). PIP$_3$ recruits inactive cytosolic molecules containing pleckstrin homology (PH) domains to the plasma membrane to promote their activation (Vanhaesebroeck and Waterfield 1999). While these PIP$_3$-regulated effectors include multiple proteins such as PREX1, PREX2, and GRP1 that may all have important roles in cancer metabolism, here we focus on the serine/threonine AGC-family protein kinase Akt. PIP$_3$ binds to the PH domain of Akt, recruiting it to the plasma membrane where it is phosphorylated at Thr308 by another PIP$_3$-recruited molecule, phosphoinositide-dependent protein kinase 1 (PDPK1) (Franke et al. 1997; Alessi et al. 1997; Stephens et al. 1998; Storz and Toker 2002). For complete activation, Akt is also phosphorylated on Ser473 by the mammalian target of rapamycin complex 2 (mTORC2) (Bellacosa et al. 2005; Saci et al. 2011). In recent years, further mechanisms that account for full Akt activation have also been described, including Akt ubiquitination, sumoylation, and regulation by the cell cycle (Yang et al. 2009b, 2010, 2013; Chan et al. 2012; Fan et al. 2013a; Li

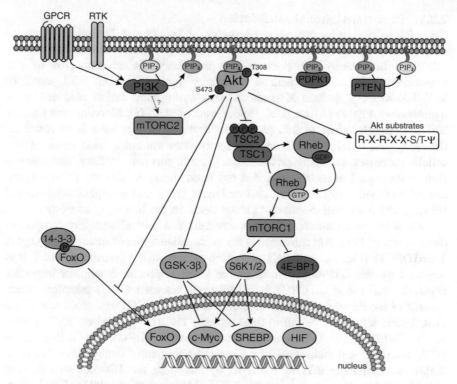

Fig. 1 PI3K-Akt-mTOR signaling pathway

et al. 2013; Risso et al. 2013; de la Cruz-Herrera et al. 2015; Liu et al. 2014). As more details behind these additional layers of regulation are clarified, it is becoming clear that the regulation of signaling downstream of PI3K is highly complex and multifaceted.

2.2 Downstream Effectors

PI3K-Akt pathway activation controls a wide variety of cellular processes. Broadly, the majority of the downstream effectors of Akt are regulated by two general mechanisms. First, Akt can acutely regulate various cellular processes in a post-translational manner by phosphorylating and controlling the activity, localization, or stability of its protein substrates. Second, Akt can establish longer-term changes in cellular behavior by phosphorylating and thereby influencing the activity of various transcription factors. It will become evident as we progress through our discussion of metabolism that in many cases, changes in metabolic processes are initiated by post-translational responses and then further reinforced by the induction of transcriptional programs.

2.2.1 Post-translational Modulation

Upon full activation by dual phosphorylation at Ser473 and Thr308, Akt translocates to distinct subcellular compartments and phosphorylates various downstream substrates. Importantly, Akt preferentially phosphorylates substrate proteins in a sequence-specific context, with a minimally required amino acid motif of R-X-R-X-X-S/T-Ψ (where X tends to be a hydrophilic amino acid and Ψ is hydrophobic; Fig. 1) (Alessi et al. 1996; Obata et al. 2000; Manning and Cantley 2007). Currently, close to 200 proposed substrates of Akt have been identified (Manning and Cantley 2007), and these substrates control a wide range of key cellular processes, including proliferation, growth, survival, motility, and metabolism. With regard to metabolism, Akt has been shown to directly phosphorylate several metabolic enzymes such as hexokinase (HK) and phosphofructokinase 2 (PFK2), which we will describe in greater detail in the following sections.

One of the major effectors that has emerged as a critical signaling component downstream of PI3K-Akt activation is the mammalian target of rapamycin complex 1 (mTORC1) (Fig. 1). mTORC1 activation occurs at the lysosome, where it is activated by the GTP-bound form of the small G-protein Rheb. An important negative regulator of mTORC1 is the tuberous sclerosis (TSC) complex, which consists of the tumor suppressors tuberous sclerosis 1 (TSC1), tuberous sclerosis 2 (TSC2), and a recently described third protein, TBC1D7 (Tee et al. 2002; Dibble et al. 2012). The TSC complex suppresses mTORC1 by acting as a GTPase-activating protein (GAP) for Rheb (Manning and Cantley 2003; Tee et al. 2003). Akt stimulates mTORC1 activity by inhibiting the TSC complex through phosphorylation of multiple sites on TSC2 (Manning et al. 2002). TSC2 phosphorylation leads to dissociation of the TSC complex from the lysosome, thus relieving its inhibition of Rheb and resulting in mTORC1 activation (Menon et al. 2014). Akt can also directly phosphorylate mTOR and an mTOR-associated protein, PRAS40 (AKT1S1), and these events may also promote activation of mTORC1 (Nave et al. 1999; Sekulić et al. 2000; Vander Haar et al. 2007). Importantly, mTORC1 activation requires both active PI3K-Akt signaling and sufficient intracellular pools of amino acids. Hence, mTORC1 is considered to be a key integrator of systemic signals (such as growth factors) and local signals (nutrient availability), with its activation culminating in anabolic processes that promote growth such as protein synthesis and metabolism (Dibble and Manning 2013; Bar-Peled and Sabatini 2014). Indeed, the most well-studied substrates of mTORC1 are the ribosomal S6 kinases, S6K1 and S6K2, and the eIF4E (eukaryotic initiation factor 4E)-binding proteins, 4E-BP1 and 4E-BP2, which all have critical roles in regulating mRNA translation to enhance cellular growth and proliferation (Ma and Blenis 2009). In addition, recent emerging evidence has also demonstrated that signaling downstream of mTORC1 regulates anabolic metabolic processes, through translational and even direct post-translational modulation of metabolic enzymes.

2.2.2 Transcriptional Output

In addition to acutely regulating various cellular processes through post-translational modifications, PI3K-Akt-mTOR signaling can also establish longer-term changes in cellular behavior by influencing the activity of various transcription factors (Fig. 1). Here, we will briefly describe how PI3K-Akt-mTOR signaling regulates a few of these transcription factors, with particular emphasis on those that have a major role in modulating cellular metabolism.

FoxO

The forkhead box O (FoxO) transcription factor family, which consists of FoxO1, FoxO3, FoxO4, and FoxO6, functions primarily downstream of the insulin and insulin-like growth factor receptors to regulate diverse cellular processes, including proliferation, apoptosis, antioxidant responses, longevity, and metabolism (Tzivion et al. 2011). As a major effector of insulin signaling, Akt was identified in early genetic and biochemical studies as a key regulator of FoxO function (Paradis and Ruvkun 1998). FoxO1/3/4 each contains three Akt phosphorylation sites, while FoxO6 contains two. These phosphorylation events primarily serve as docking points for interactions with 14-3-3 proteins (Tzivion et al. 2011). Upon binding, 14-3-3 proteins regulate the cellular localization of FoxO, both by increasing nuclear export of FoxO and by sequestering FoxO in the cytoplasm (Brunet et al. 2002). 14-3-3 also masks the DNA-binding domain of FoxO, thus preventing it from binding to the promoters of its target genes (Obsil et al. 2003; Silhan et al. 2009). Together, the functional consequence of phosphorylation by Akt and subsequent 14-3-3 binding is the inhibition of FoxO-mediated transcriptional activity.

c-Myc

c-Myc is a key transcription factor encoded by the *MYC* oncogene that controls a number of physiological cellular processes, including cell growth, proliferation, apoptosis, and energy metabolism (Eilers and Eisenman 2008). c-Myc can also cooperate with PI3K to promote cell proliferation and transformation (Zhao et al. 2003). Given the potential for this adverse consequence, multiple PI3K-mediated regulatory mechanisms exist to modulate and fine-tune c-Myc activity. For example, activation of Akt relieves FoxO-mediated inhibition of c-Myc target genes and enables the transformative capacity of c-Myc (Bouchard et al. 2004). PI3K-Akt signaling also controls c-Myc protein stability through glycogen synthase kinase-3β (GSK-3β), which is directly phosphorylated and inhibited by Akt. Inhibition of GSK-3β prevents c-Myc phosphorylation at Thr58 and its subsequent degradation, since phosphorylation at this site promotes its ubiquitination by the SCF[Fbw7] ubiquitin ligase followed by proteasomal degradation (Gregory et al. 2003; Welcker et al. 2004). Finally, S6K phosphorylates Ser145 on Mad1, which is an antagonist of c-Myc. Mad1 phosphorylation promotes its ubiquitination and degradation, leading to the upregulation of Myc-mediated transcription (Zhu et al. 2008).

HIFs

The hypoxia-inducible factor (HIF) family of transcription factors, which consists of three members, HIF1, HIF2, and HIF3, are activated in response to hypoxia in order to induce genes that facilitate adaptation to the hypoxic environment. These gene programs enable adaptation by directing glucose away from oxidative phosphorylation toward glycolysis and lactate production and by the stimulation of angiogenesis to restore oxygen homeostasis. Each HIF consists of an oxygen-dependent α subunit and a constitutively expressed β subunit, and the active heterodimer is stabilized under hypoxic conditions (reviewed in Brahimi-Horn et al. 2007). PI3K-Akt-mTOR signaling can also regulate HIF1 independently of oxygen concentration to stimulate its increased accumulation and activity in tumor cells (Zundel et al. 2000; Zhong et al. 2000; Blancher et al. 2001; Chan et al. 2002; Bardos and Ashcroft 2004). This regulation occurs primarily at the level of both transcription and translation by mTORC1. Transcriptionally, both HIF1α and HIF2α message levels are elevated in an mTORC1-dependent manner in TSC2-null cells (Düvel et al. 2010). mTORC1 also selectively increases the translation of HIF1α through the 5′-untranslated region of the HIF1α mRNA by inhibiting 4E-BP1 to activate cap-dependent translation (Laughner et al. 2001; Thomas et al. 2006; Düvel et al. 2010).

SREBPs

The sterol regulatory element-binding proteins (SREBPs), which consist of three isoforms SREBP-1a, SREBP-1c, and SREBP2, are transcription factors that function as central regulators of lipid homeostasis. SREBPs reside in the endoplasmic reticulum membrane in their inactive, full-length forms. In response to sterol depletion, the SREBPs are trafficked to the Golgi, where they are proteolytically activated. The cleaved N-terminal domain of SREBPs subsequently translocates to the nucleus and binds to sterol regulatory elements (SRE) to upregulate genes involved in the uptake and biosynthesis of cholesterol, fatty acids, triglycerides, and phospholipids (reviewed in Espenshade and Hughes 2007). PI3K-Akt-mTOR signaling has been implicated in SREBP activation, since signaling through S6K1 promotes the accumulation of the active form of SREBP1 (Porstmann et al. 2008; Düvel et al. 2010). Furthermore, inhibition of GSK-3β by Akt also enhances the stability of SREBP, which can be phosphorylated by GSK-3β and directed for ubiquitination by the SCF^{Fbw7} ubiquitin ligase and subsequent proteasomal degradation (Kim et al. 2004b; Sundqvist et al. 2005).

3 The PI3K-Akt-mTOR Pathway Regulates Glucose Metabolism and the Warburg Effect

The altered metabolism of cancer cells was first noted in 1924, when Otto Warburg observed that tumors in rats take up more than ten times as much glucose as compared to corresponding normal tissue, and that this glucose is primarily converted to lactate (Warburg 1956). This led to the hypothesis that through a process known as aerobic glycolysis, or the "Warburg effect," cancer cells promote the glycolytic conversion of glucose into lactate even in the presence of sufficient oxygen to support mitochondrial oxidative phosphorylation. Recently, it has become evident that the switch to aerobic glycolysis provides tumor cells with a proliferative advantage (Vander Heiden et al. 2009). Hence, there has been intense investigation into the mechanisms by which this process is activated and regulated in an effort to exploit this pathway for therapeutic gain.

One major mechanism by which this occurs is through the oncogenic activation of PI3K-Akt-mTOR signaling. For example, immortalized hematopoietic cells transformed by a constitutively active mutant of Akt display aerobic glycolysis— that is, higher rates of glycolysis without effects on the rate of oxidative phosphorylation (Elstrom et al. 2004). Similar results were found in human glioblastoma cells possessing constitutive Akt activity. Importantly, these cells are dependent on aerobic glycolysis for growth and survival, since they are more susceptible to cell death after glucose withdrawal. Together, these findings indicate that PI3K-Akt-mTOR signaling is sufficient to stimulate the switch to aerobic glycolysis. Since the studies described above, it has become clear that this switch is achieved through a number of mechanisms that converge on the regulation of glycolysis, which we describe below.

3.1 Glucose Uptake

Enhanced glucose uptake is a requisite event for aerobic glycolysis. Under normal physiological conditions, the PI3K-Akt-mTOR signaling pathway is a major regulator of glucose uptake, through both post-translational and transcriptional effects. At the post-translational level, glucose uptake is stimulated primarily through the regulation of glucose transporter trafficking (Fig. 2). For example, acute insulin treatment in adipocytes and muscle cells leads to the translocation of the glucose transporter GLUT4 to the plasma membrane (Huang and Czech 2007), in a manner that is dependent on PI3K and Akt (Okada et al. 1994; Kotani et al. 1995; Kohn et al. 1996; Cong et al. 1997). Interestingly, insulin-stimulated GLUT4 translocation is specifically dependent on the Akt2 isoform, since translocation is impaired in adipocytes upon Akt2 knockout or knockdown and cannot be rescued by Akt1 expression (Katome et al. 2003; Bae et al. 2003). Following insulin stimulation of adipocytes, Akt2 is preferentially localized to the plasma membrane, where it phosphorylates and inhibits the RabGAP AS160 (Gonzalez and McGraw 2009). Inhibition of AS160 is required for GLUT4 vesicle docking and fusion with the plasma membrane (Sano et al. 2003; Larance et al. 2005). As a consequence, insulin stimulation leads to GLUT4 localization at the plasma membrane, which enables glucose uptake. Another contributing mechanism is the activating phosphorylation of the phosphatidylinositol 3-phosphate 5-kinase PIKfyve by Akt (Berwick et al. 2004). PIKfyve participates in endosomal trafficking through the generation of $PI(3,5)P_2$ from $PI(3)P$, and these lipid species appear to positively regulate GLUT4 trafficking to the plasma membrane through a poorly understood mechanism (Ikonomov et al. 2002; Sbrissa et al. 2004; Berwick et al. 2004; Ikonomov et al. 2007; Shisheva 2008; Ikonomov et al. 2009).

Although regulation of GLUT4 by Akt has been relatively well studied, it is important to note that GLUT4 is a muscle- and fat-cell-specific glucose transporter. In other words, most cancer cells do not express GLUT4, and rather express the

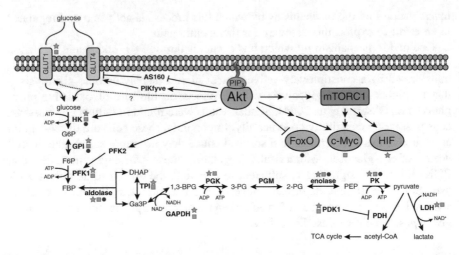

Fig. 2 PI3K-Akt-mTOR signaling regulates glucose uptake and glycolysis. *Filled green square* target genes for c-Myc; *filled green star* target genes for HIF; *filled red circle* genes that are repressed by FoxO activation. *HK* hexokinase; *G6P* glucose 6-phosphate; *GPI* glucose 6-phosphate isomerase; *F6P* fructose 6-phosphate; *PFK1/2* phosphofructokinase 1/2; *FBP* fructose 1,6-bisphosphate; *DHAP* dihydroxyacetone phosphate; *Ga3P* glyceraldehyde 3-phosphate; *TPI* triose phosphate isomerase; *GAPDH* glyceraldehyde 3-phosphate dehydrogenase; *1,3-BPG* 1,3-bisphosphoglycerate; *PGK* phosphoglycerate kinase; *3-PG* 3-phosphoglycerate; *PGM* phosphoglycerate mutase; *2-PG* 2-phosphoglycerate; *PEP* phosphoenolpyruvate; *PK* pyruvate kinase; *LDH* lactate dehydrogenase; *PDH* pyruvate dehydrogenase; *PDK1* pyruvate dehydrogenase kinase 1; *ATP* adenosine triphosphate; *ADP* adenosine diphosphate; *NADH* nicotinamide adenine dinucleotide

embryonic glucose transporter isoform, GLUT1. Notably, GLUT1 has a high affinity for glucose and may be selected for by cancer cells to increase glucose transport efficiency (Ganapathy et al. 2009). PI3K-Akt signaling has also been proposed to regulate GLUT1 trafficking to the plasma membrane (Rathmell et al. 2003; Bentley et al. 2003). However, the specific mechanisms by which such regulation occurs remains an important open question. Of interest, thioredoxin-interacting protein (TXNIP) was recently shown to negatively regulate glucose uptake by binding to GLUT1 and inducing its internalization through clathrin-coated pits. Upon energy stress, TXNIP is phosphorylated by AMP-dependent protein kinase (AMPK), which leads to its rapid degradation and subsequent induction of GLUT1 activity (Parikh et al. 2007; Stoltzman et al. 2008; Wu et al. 2013). Given this regulation of GLUT1 trafficking by AMPK signaling, it would be interesting to evaluate whether TXNIP plays a role in GLUT1 regulation downstream of PI3K-Akt-mTOR signaling.

The glucose transporters are also regulated by PI3K-Akt-mTOR signaling at the transcriptional level (Fig. 2). Constitutive Akt activation results in elevated GLUT1 mRNA and protein levels, but not GLUT4 (Kohn et al. 1996; Barthel et al. 1999). Increased expression of GLUT1 is dependent on mTORC1 and its subsequent upregulation of HIF1α levels and activity (Wieman et al. 2007; Zhou et al. 2008;

Düvel et al. 2010). GLUT1 can also be transcriptionally upregulated by c-Myc (Osthus et al. 2000; Ying et al. 2012), which lies downstream of PI3K.

Taken together, oncogenic activation of the PI3K-Akt-mTOR signaling network promotes post-translational regulation of glucose transporter trafficking to the plasma membrane, which is reinforced by transcriptional upregulation of the glucose transporter genes. These mechanisms combine to robustly stimulate glucose uptake in cancer cells, thus providing them with sufficient amounts of the substrate required for aerobic glycolysis.

3.1.1 Glucose Uptake in the Clinic

Well before the mechanistic details concerning growth factor- and/or oncogene-mediated glucose uptake were understood, an appreciation for the glucose avidity of cancer cells was exploited clinically. Indeed, for more than 30 years now, uptake of the radioactive glucose analog 18-fluoro-deoxyglucose (^{18}FDG) has been used to diagnose and stage tumors and to monitor response to treatment. This is done using a technique that allows tumors to be imaged by positron emission tomography (PET), where cellular uptake of ^{18}FDG is directly proportional to PET response (Ben-Haim and Ell 2009). Importantly, given the relationship between PI3K-Akt-mTOR signaling and glucose uptake, ^{18}FDG-PET has been used as a method to monitor tumor responses to drugs that target the PI3K-Akt-mTOR pathway. For example, in a mouse model of lung cancer driven by the oncogenic *PIK3CA*(H1047R) mutation, ^{18}FDG-PET imaging of lung tumors following administration of a dual PI3K-mTOR inhibitor for 4 days revealed that the ^{18}FDG signal was significantly reduced, consistent with a measurable decrease in tumor size (Engelman et al. 2008). ^{18}FDG-PET was similarly used in a triple-negative breast cancer mouse model to demonstrate that tumors show a decrease in ^{18}FDG uptake in response to PI3K inhibition (Juvekar et al. 2012). Significantly, these results suggest that inhibition of ^{18}FDG uptake by tumors may be an early and predictive pharmacodynamic marker for the efficacy of PI3K pathway inhibitors in the clinic. Indeed, a recent phase I study showed that the subset of patients who obtained a clinical benefit (partial tumor shrinkage) from PI3K inhibitor therapy exhibited a decline in ^{18}FDG-PET signal within two weeks of initiating drug treatment, while those whose tumors did not show this early decline in ^{18}FDG-PET had no tumor shrinkage (Mayer et al. 2014). Thus, early changes in ^{18}FDG-PET might be more effective than mutational analysis in predicting which patients will respond to PI3K inhibitors (see Rodon et al. 2013 for review).

3.2 Regulation of Glycolytic Enzymes

PI3K-Akt-mTOR signaling also promotes aerobic glycolysis by directly regulating multiple nodes in glycolysis. For example, an early study demonstrated that both Akt and S6K can phosphorylate the bifunctional enzyme PFK-2/FBPase-2 (PFK2) in vitro, resulting in increased PFK-2 activity (Deprez et al. 1997). PFK2 regulates glycolysis by generating fructose 2,6-bisphosphate, which is the most potent known allosteric activator of PFK1, a rate-limiting enzyme for glycolysis (Fig. 2). Hence, PI3K-Akt signaling can indirectly activate PFK1 and stimulate glycolysis through PFK2 phosphorylation. It should be noted that PFK2 phosphorylation and glycolytic flux studies for PFK2 activity in the context of PI3K-Akt-mTOR signaling have not been carefully done in vivo. Furthermore, although one study has shown in prostate cancer cells that androgen treatment stimulates glycolysis by increasing PFK2 activity in a PI3K-dependent manner (Moon et al. 2011), a more comprehensive understanding of the broader applicability of Akt- or S6K-mediated phosphorylation of PFK2 in the reprogrammed metabolism of cancer cells remains to be determined.

Akt activation has also been implicated in increasing the activity of hexokinase (HK), which is overexpressed in multiple cancers and has been shown to be a key mediator of aerobic glycolysis, increased cell proliferation, and therapeutic resistance (Pastorino et al. 2005; Mathupala et al. 2006; Ahn et al. 2009; Wolf et al. 2011). HK catalyzes the first, and first rate-limiting, step of glycolysis by phosphorylating glucose to form glucose 6-phosphate (G6P) (Rathmell et al. 2003) (Fig. 2). Akt promotes the association of hexokinase I (HK1) and hexokinase II (HK2) with mitochondria through interactions with voltage-dependent anion channels (VDAC) and the outer mitochondrial membrane (OMM) (Majewski et al. 2004a; Robey and Hay 2009). The specific mechanisms by which this occurs are not yet entirely clear, but, at least for HK2, mitochondrial association is promoted in part by Akt-mediated phosphorylation of HK2 at Thr473 (Roberts et al. 2013). Akt also indirectly promotes mitochondrial HK2 association by inhibiting GSK-3β, which when active phosphorylates VDAC and inhibits its ability to bind HK2 (Pastorino et al. 2005).

Mitochondrial HK association serves several functions. First, it has been suggested to increase HK activity and to direct G6P toward glycolysis, as opposed to glycogen synthesis (Gottlob et al. 2001; John et al. 2011; Yeo et al. 2013). Mitochondrial HKs also directly couple mitochondrial ATP synthesis to glycolysis by preferentially using intramitochondrial ATP to phosphorylate glucose, thus efficiently regenerating ADP that can be directed back to support oxidative phosphorylation (Gottlob et al. 2001; Robey and Hay 2009). Finally, mitochondrial HKs are required for Akt-mediated inhibition of apoptosis in response to a variety of apoptotic stimuli, since they maintain mitochondrial integrity and prevent cytochrome c release by competing with the binding of Bax and Bak on the OMM (Pastorino et al. 2002, Majewski et al. 2004a, b; Miyamoto et al. 2008). In this way, the anti-apoptotic role of Akt is closely linked with its regulation of glucose metabolism, which provides a mechanism by which metabolism and cell survival are coupled.

In addition to post-translational regulation of glycolytic enzymes, PI3K-Akt-mTOR signaling also controls the expression of glycolytic genes (Fig. 2). One of the primary transcription factors through which this is regulated is HIF1. mTORC1-mediated upregulation of HIF1α increases the expression of GLUT1 and GLUT3, as well as almost all genes involved in glycolysis (Semenza et al. 1994, 1996; Denko 2008; Düvel et al. 2010). Interestingly, in addition to upregulating glycolysis, HIF1 also inhibits mitochondrial respiration, primarily by modulating the fate of pyruvate. HIF1 induces the expression of pyruvate dehydrogenase kinase 1 (PDK1), which phosphorylates and inhibits pyruvate dehydrogenase (PDH) (Kim et al. 2006; Papandreou et al. 2006). PDH is responsible for the oxidation of pyruvate by converting it into acetyl-CoA, which enters the tricarboxylic acid (TCA) cycle and condenses with oxaloacetate to form citrate (Fig. 2). Therefore, PDH inhibition by PDK1 shunts pyruvate away from mitochondrial metabolism and oxidative phosphorylation. Another consequence of this, however, is that the reduced nicotinamide adenine dinucleotide (NADH) produced from glycolysis is no longer recycled by oxidative phosphorylation to regenerate NAD^+, which is a critical cofactor for glycolytic activity. This is resolved by the upregulation of lactate dehydrogenase A (LDHA) by HIF1 (Semenza et al. 1996). LDHA regenerates NAD^+ from NADH in the process of converting pyruvate into lactate, thus explaining the increased lactate production that is a hallmark of aerobic glycolysis.

Other transcription factors in the PI3K-Akt-mTOR signaling network, such as c-Myc and FoxO, also contribute to glycolytic gene expression. Like HIF1, c-Myc can directly regulate almost all the genes involved in glycolysis, including GLUT1, glucose phosphate isomerase (GPI), PFK1, glyceraldehyde 3-phosphate dehydrogenase (GAPDH), triosephosphate isomerase (TPI), phosphoglycerate kinase 1 (PGK1), and enolase (ENO1) (Osthus et al. 2000; Kim et al. 2004a). It is important to note that such regulation can be highly fine-tuned and context dependent, as c-Myc activation does not necessarily always upregulate all glycolytic genes in a given tumor. c-Myc can also increase the expression of LDHA and PDK1, which shifts glycolysis toward lactate production as opposed to mitochondrial metabolism, as described above (Shim et al. 1997; Dang 2007). Finally, activation of the FoxO transcription factors results in the decreased expression of several glycolytic genes, including glucokinase (GCK), GPI, aldolase B (ALDOB), ENO1, and pyruvate kinase 1 (PK1) (Zhang et al. 2006). Hence, inhibition of FoxO by Akt relieves the suppression of glycolytic gene expression. Interestingly, an Akt-independent mechanism of FoxO regulation that impinges on glycolysis has also been demonstrated in glioblastoma (Masui et al. 2013). Specifically, mTORC2 promotes the inactivating phosphorylation of several histone deacetylases, which results in the acetylation and inactivation of FoxO. The regulation of glycolysis by FoxO occurs, at least in part, through transcriptional upregulation of TSC1 by FoxO, which attenuates mTORC1 activation and its effects on glycolytic gene expression (Khatri et al. 2010). FoxO also antagonizes c-Myc, in part through the increased expression of miR-145 and miR-34c, which limits c-Myc mRNA stability and translation (Gan et al. 2010; Kress et al. 2011; Peck et al. 2013).

Taken together, activation of the PI3K-Akt-mTOR network initiates a variety of mechanisms at both the post-translational and the transcriptional levels that converge to strongly stimulate the metabolism of glucose in glycolysis and to divert glucose carbon from being oxidized in the mitochondria. Several of the alternative biosynthetic fates of this carbon are described below.

4 Beyond Aerobic Glycolysis: Coordination with Anabolic Metabolism

Despite clear evidence that cancer cells reprogram glucose metabolism to favor aerobic glycolysis, the precise molecular reasons for this shift remain to be determined. Initial explanations proposed that aerobic glycolysis constituted a shift to glycolytic ATP production and energy dependence, as opposed to generation of ATP through mitochondrial oxidative phosphorylation. In fact, Warburg initially hypothesized that most cancer cells likely have mitochondrial defects that impair oxidative phosphorylation, thus forcing them to rely on glycolysis for energy production (Warburg 1956). However, several studies have shown that mitochondrial function is intact in most cancer cells (Weinhouse 1976; Fantin et al. 2006; Moreno-Sanchez et al. 2007).

The predominant logic in the field now posits that aerobic glycolysis is activated to provide cells with unrestricted access to basic cellular building blocks, in the form of glucose carbon, from which to draw for biosynthetic purposes (Vander Heiden et al. 2009). In particular, cancer cells must promote an anabolic state to produce the nucleotides, amino acids, and lipids needed to grow and divide. In other words, the enhanced glycolytic flux in cancer cells functions primarily to catabolize glucose into biosynthetic intermediates that are used for anabolic processes to meet the increased demands of enhanced growth and proliferation.

This concept is particularly evident when considering the integration of oncogenic PI3K-Akt-mTOR signaling with metabolic reprogramming in cancer. Activation of the signaling pathway does not merely stimulate aerobic glycolysis, but also tightly coordinates it with downstream anabolic processes to synthesize metabolites required for malignant cell growth and proliferation. The specific mechanism by which the PI3K-Akt-mTOR network regulates the metabolism of cellular building blocks such as nucleotides, lipids, and amino acids is an active area of investigation, and much remains to be delineated. Here, we describe a few mechanisms by which PI3K-Akt-mTOR signaling controls these processes.

4.1 Pentose Phosphate Pathway

The pentose phosphate pathway (PPP) branches off from glycolysis and uses glycolytic intermediates to generate reducing power in the form of nicotinamide adenine dinucleotide phosphate (NADPH) and 5-carbon sugars. There are two distinct but related phases of the pathway: the oxidative branch and the nonoxidative branch. In the oxidative PPP, glucose is oxidized to facilitate the production of NADPH, a reducing equivalent that contributes both to maintaining redox homeostasis and to providing sufficient reducing power for various anabolic processes. The reduction of $NADP^+$ in the oxidative arm of the PPP occurs in two steps, ultimately converting glucose 6-phosphate into ribulose 5-phosphate (Ru5P), which is then directed into ribose 5-phosphate (R5P) (Fig. 3). This R5P can be used as a precursor in RNA and DNA biosyntheses, or it can be recycled back into

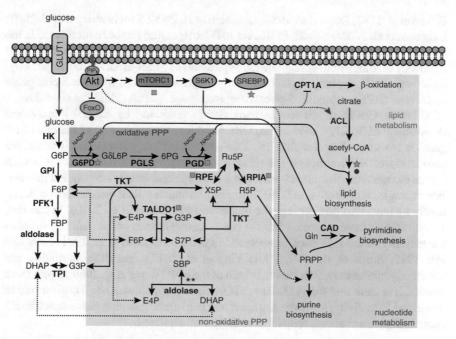

Fig. 3 PI3K-Akt-mTOR signaling coordinates the pentose phosphate pathway with lipid and nucleotide metabolism. *Filled Green square* genes that are upregulated by mTORC1 activation; *filled green star* target genes for SREBP1; *filled red circle* genes that are repressed by FoxO activation. For the purposes of representing these pathways schematically, information about the subcellular location of enzymes or signaling proteins is not conveyed. *PPP* pentose phosphate pathway; *G6PD* glucose 6-phosphate dehydrogenase; *GδL6P* 6-phosphogluconolactone; *PGLS* 6-phosphogluconolactonase; *6PG* 6-phosphogluconate; *PGD* 6-phosphogluconate dehydrogenase; *NADPH* nicotinamide adenine dinucleotide phosphate; *Ru5P* ribulose 5-phosphate; *RPE* ribulose 5-phosphate epimerase; *RPIA* ribulose 5-phosphate isomerase; *X5P* xylulose 5-phosphate; *R5P* ribose 5-phosphate; *TKT* transketolase; *S7P* sedoheptulose 7-phosphate; *E4P* erythrose 4-phosphate; *TALDO1* transaldolase 1; *SBP* sedoheptulose-1,7-bisphosphate; *CPT1A* carnitine palmitoyltransferase 1A; *ACL* ATP citrate lyase; *Gln* glutamine; *CAD* carbamoyl-phosphate synthetase 2, aspartate transcarbamylase, and dihydroorotase; *PRPP* phosphoribosyl pyrophosphate. *Double asterisk* this reaction (SBP synthesis) has not been demonstrated in mammalian cells and has thus far only been observed in yeast

glycolytic intermediates through the nonoxidative arm of the PPP. In certain contexts, the nonoxidative arm can be used to generate R5P from the glycolytic intermediates fructose 6-phosphate (F6P) and glyceraldehyde 3-phosphate (Ga3P), thereby bypassing the oxidative arm. This pathway has been described to be operative when the oxidative arm is inhibited and/or when alternative sources of NADPH generation dominate (Ying et al. 2012; Stanton 2012; Son et al. 2013).

Both branches of the PPP have been shown to be important in human cancers. For example, activity of the rate-limiting enzyme glucose 6-phosphate dehydrogenase (G6PD) in the oxidative arm and expression of transketolase (TKT) in the nonoxidative arm are frequently elevated in a variety of cancers (Zampella et al. 1982;

Bokun et al. 1987; Dessì et al. 1988; Langbein et al. 2006; Krockenberger et al. 2007; Langbein et al. 2008). Indeed, PI3K-Akt-mTOR signaling has been implicated in the regulation of the pentose phosphate pathway (Fig. 3). For example, activation of mTORC1 is sufficient to upregulate the expression of genes involved in the pathway, including G6PD, 6-phosphogluconate dehydrogenase (PGD), ribulose 5-phosphate epimerase (RPE), ribulose 5-phosphate isomerase (RPIA), and transaldolase 1 (TALDO1). G6PD expression in particular is mediated by SREBP1 activation downstream of S6K1, while how mTORC1 induces the expression of the other PPP genes is unknown (Düvel et al. 2010). PI3K-Akt-mTOR signaling also stimulates the production of R5P, although the relative contribution from the oxidative branch versus the nonoxidative branch varies in a cell-type and context-dependent manner. For example, constitutive mTORC1 activity stimulates R5P generation predominantly by increased flux through the oxidative PPP (Düvel et al. 2010). In contrast, several studies have suggested that the nonoxidative branch of the pathway supplies the majority of the R5P used in nucleotide synthesis (Raivio et al. 1981; Boss and Pilz 1985; Boros et al. 1997, 2001; Ying et al. 2012), and PI3K inhibition can selectively inhibit the nonoxidative branch of the PPP (Wang et al. 2009). From these studies, it is clear that the PI3K-Akt-mTOR signaling pathway has a critical role in dictating PPP activity; however, the exact mechanisms explaining tissue-specific and context-dependent effects remain to be fully characterized.

4.2 Nucleotide Biosynthesis

By stimulating the production of R5P through the PPP, the PI3K-Akt-mTOR network provides the building block for de novo nucleotide biosynthesis (Fig. 3). As such, it would seem logical that there would be coordination between PI3K-Akt-mTOR signaling-mediated increases in R5P levels and enhanced nucleotide synthesis. Indeed, mTORC1 acutely stimulates flux through the de novo pyrimidine biosynthetic pathway through S6K1. S6K1 directly phosphorylates S1859 on CAD (carbamoyl-phosphate synthetase 2, aspartate transcarbamylase, and dihydroorotase), the rate-limiting multifunctional enzyme that mediates the first three steps of de novo pyrimidine synthesis. This phosphorylation event drives the activity of CAD to increase pyrimidine biosynthetic flux (Ben-Sahra et al. 2013; Robitaille et al. 2013). Whether CAD phosphorylation by S6K1 is necessary to sustain the enhanced growth and proliferation of cancer cells in the context of oncogenic signaling is an important question that will need to be addressed.

There is also evidence that PI3K-Akt signaling may regulate purine metabolism, although the mechanisms are not as clear. In C2C12 cells, inhibition of PI3K or Akt, but not mTORC1, decreased flux through both the de novo purine biosynthetic pathway and the purine salvage pathway, in part through reducing the availability of phosphoribosyl pyrophosphate (PRPP), which is required for the first step of both purine biosynthesis and purine salvage. Reduced PRPP levels are likely due to inhibition of R5P production through the PPP, since R5P is incorporated into PRPP (Fig. 3). Interestingly, it was also found in cell extracts that the activity of ATIC, which catalyzes the last two steps of de novo purine synthesis, is reduced by PI3K inhibition (Wang et al. 2009). ATIC does not have a consensus Akt substrate motif, however, suggesting that the potential regulation of ATIC enzymatic activity by Akt is indirect. Furthermore, the expression of the genes involved in purine synthesis and salvage was not assessed upon PI3K inhibition. It is clear that much remains to be understood regarding how the PI3K-Akt-mTOR network regulates purine metabolism, and especially in the context of oncogenic signaling in cancer cells.

4.3　Lipid Synthesis

Most normal cells have low rates of lipid synthesis, and only a few specialized cell types, such as hepatocytes and adipocytes, are designed to produce lipids. On the other hand, nearly all tumor cells have an increased rate of de novo lipid synthesis (Menendez and Lupu 2007). This metabolic feature allows cancer cells to autonomously generate lipids to form new membranes and to support other features associated with cell division and proliferation.

PI3K-Akt-mTOR signaling also has an important role in initiating both post-translational and transcriptional programs to stimulate de novo lipid synthesis. At the post-translational level, Akt has been proposed to phosphorylate Ser454 on ATP citrate lyase (ACL), which converts citrate to cytosolic acetyl-CoA that is required for lipid biosynthesis (Fig. 3) (Berwick et al. 2002; Sale et al. 2006; Migita et al. 2008). ACL phosphorylation on Ser454 increases its enzymatic activity in vitro (Potapova et al. 2000), although careful in vivo studies evaluating the contribution of phosphorylation at this site still need to be performed. It is also important to note that the proposed Akt phosphorylation site lacks the critical Arg at the −5 position, suggesting that there may be an intermediate kinase between Akt and ACL. Nevertheless, there is clear genetic evidence linking ACL phosphorylation to PI3K-Akt signaling. For instance, in nonsmall cell lung cancer, ACL phosphorylation is associated with PI3K-Akt activation, since inhibition of the pathway blocks ACL phosphorylation and activity. Direct knockdown of ACL induces cell-cycle arrest in vitro and inhibits tumor growth in vivo (Migita et al. 2008). ACL knockdown also impairs Akt-mediated leukemogenesis (Bauer et al. 2005). Taken together, ACL is a critical downstream effector of PI3K-Akt signaling that is post-translationally regulated to enhance de novo lipid synthesis.

The PI3K-Akt-mTOR network also establishes multiple transcriptional programs that converge to stimulate fatty acid and lipid synthesis. First, as previously described, mTORC1-mediated upregulation of the oxidative PPP generates NADPH. NADPH is required for lipid synthesis, as it provides the reducing power necessary for extending the carbon units on fatty acid chains (Fig. 3). mTORC1 also upregulates genes involved in the cholesterol and fatty acid biosynthetic pathways, primarily through activation of SREBP (Porstmann et al. 2005, 2008; Düvel et al. 2010). Indeed, one key target of SREBP, fatty acid synthase (FASN), is highly expressed in many types of human cancer (Van de Sande et al. 2002, 2005; Menendez and Lupu 2007). Moreover, at least in the liver, FoxO transcription factors inhibit lipogenesis by suppressing the expression of lipogenic genes, in part through their suppression of SREBP-1c (Zhang et al. 2006; Deng et al. 2012). Therefore, Akt-mediated inhibition of FoxO relieves this suppression of lipid synthesis. Finally, in addition to stimulating lipid synthesis, PI3K-Akt signaling also suppresses β-oxidation, the process by which lipids are catabolized, by decreasing the expression of the β-oxidation enzyme carnitine palmitoyltransferase 1A (CPT1A) (Deberardinis et al. 2006). The PI3K-regulated transcription factor responsible for this effect, however, has not yet been identified. Taken together, these transcriptional programs combine with the post-translational levels of regulation described above to robustly enhance lipogenesis in cancer cells driven by oncogenic PI3K-Akt-mTOR signaling.

4.4 Glutamine Metabolism

Due to a metabolic shift to aerobic glycolysis in cancer cells, many of the central glucose-derived intermediates are directed away from the mitochondria, where the TCA cycle and oxidative phosphorylation occur. Biosynthetic activity in the TCA cycle, however, is essential to proliferating cells as it provides intermediates for amino acid, nucleotide, and lipid biosynthesis. As a result, cancer cells must coordinate a shift to aerobic glycolysis with mechanisms to compensate for the depletion of TCA cycle intermediates, in a process known as anaplerosis. One major pathway by which anaplerosis occurs is through the utilization of glutamine, which can be converted into α-ketoglutarate (αKG) and fed into the TCA cycle (Fig. 4) (Hensley et al. 2013). In fact, some cancer cells generate more than half of their ATP from glutamine-derived αKG (Reitzer et al. 1979; Fan et al. 2013b).

Since PI3K-Akt-mTOR signaling stimulates aerobic glycolysis, it must, in some way, also initiate mechanisms to facilitate anaplerosis. Early studies indicated that insulin can acutely stimulate glutamine transport, at least in primary cultures of rat skeletal muscle cells and hepatocytes (Gebhardt and Kleemann 1987; Tadros et al. 1993; Low et al. 1996). IGF-1 has also been shown to stimulate glutamine uptake in a human neuroblastoma cell line when extracellular glutamine concentrations are limited (Wasa et al. 2001). In addition, mTORC1 was recently shown to increase glutamine uptake in TSC2-null MEFs and in various human epithelial tumor cell lines (Csibi et al. 2013). Although one study in *Xenopus* oocytes suggested that the stability of the glutamine transporter SN1 may be promoted by Akt and the related serum- and glucocorticoid-dependent kinases SGK1 and SGK3 (Boehmer et al. 2003), the detailed mechanisms behind the acute stimulation of glutamine transport by insulin are not known. At the transcriptional level, glutamine uptake may be stimulated in part by upregulation of the glutamine transporters ASCT2 and SN2 by c-Myc (Wise et al. 2008; Li and Simon 2013), which can be activated by the PI3K-Akt-mTOR network.

The first and rate-limiting step for anaplerotic utilization of glutamine carbon within the mitochondria is its deamination to glutamate (Fig. 4). This reaction is mediated by the enzyme glutaminase (GLS). The reverse reaction is catalyzed by

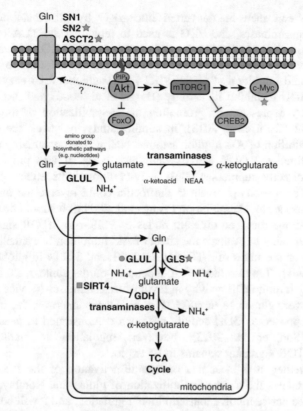

Fig. 4 PI3K-Akt-mTOR signaling regulates glutamine uptake and utilization. *Filled green square* target genes for CREB2; *filled green star* target genes for c-Myc; *filled red circle* target genes for FoxO. *Gln* glutamine; *GLUL* glutamine synthetase; *GLS* glutaminase; *GDH* glutamate dehydrogenase; *NEAA* nonessential amino acid

glutamine synthetase (GLUL), which condenses glutamate and ammonia in an ATP-dependent reaction to form glutamine. Both of these enzymes can be regulated transcriptionally by PI3K-Akt-mTOR signaling. GLS is a c-Myc target gene (Wise et al. 2008; Gao et al. 2009; Li and Simon 2013), while GLUL expression is increased by FoxO activation. Hence, upon PI3K-Akt-mTOR pathway activation, c-Myc activation and FoxO inhibition result in the upregulation of GLS concurrent with the downregulation of GLUL (van der Vos et al. 2012). Together, this promotes increased glutaminolysis and subsequent utilization of glutamate.

It is worth noting that glutamine can also be converted to glutamate by a number of other mechanisms, including transaminase-mediated nitrogen transfer. Such transaminases include asparagine synthetase (ASNS) and glutamine-fructose-6-phosphate transaminase 1 (GFPT1), which utilize the amide nitrogen of glutamine for anabolic processes such as nucleotide base synthesis and protein glycosylation, respectively (Ying et al. 2012; Balasubramanian et al. 2013). Whether the PI3K-Akt-mTOR pathway regulates these glutamine transaminases is not known.

Glutamate can then be converted into αKG by glutamate dehydrogenase (GDH) or transaminases, and αKG is used to replenish the TCA cycle (Fig. 4). Consistent with increased utilization of glutamine, mTORC1 activation elevates GDH activity in MEFs and human epithelial tumor cell lines by suppressing SIRT4-mediated inhibition of GDH. SIRT4 is a mitochondrial enzyme that ADP ribosylates GDH to inhibit its activity (Haigis et al. 2006). mTORC1 suppresses SIRT4 mRNA expression by promoting the destabilization of the transcription factor CREB2 (Csibi et al. 2013). In another study, however, the opposite was observed: Inhibition of Akt in glioblastoma cells, which presumably also inhibited mTORC1, stimulated GDH activity, and vice versa. It was proposed that Akt activation indirectly suppresses GDH activity through its effects on glycolysis, since adding exogenous pyruvate to the media could reverse the increased GDH activity in response to Akt inhibition (Yang et al. 2009a). It should be noted that the two studies above focus on different nodes of PI3K-Akt-mTOR signaling, and it would be interesting to evaluate the effect of Akt inhibition in epithelial cancer cell lines, as well as the status of mTORC1, CREB2, and SIRT4 in glioblastoma cells, on GDH activity. It is also likely that depending on the context, PI3K-Akt-mTOR signaling may inhibit GDH activity in favor of transaminases (or vice versa), which can also convert glutamate to αKG (Fig. 4). Dependence on the transamination pathway as opposed to GDH activity has been demonstrated in pancreatic adeno-carcinoma (Son et al. 2013); however, regulation of transaminases by PI3K-Akt-mTOR signaling remains to be explored.

Taken together, it is clear that oncogenic activation of the PI3K-Akt-mTOR pathway modulates the uptake and utilization of glutamine. Notably, the observations described above involve transcriptional regulation, and it will be interesting to assess whether PI3K-Akt-mTOR signaling exerts any post-translational effects on metabolic enzymes that mediate the uptake and utilization of glutamine. The precise molecular mechanisms underlying this regulation remain an important avenue of investigation, especially in light of the emerging perspective that glutamine metabolism may be an appealing target for novel clinical strategies to detect, monitor, and treat cancer (Hensley et al. 2013).

5 Concluding Remarks

The reprogramming of metabolic processes in cancer cells by the PI3K-Akt-mTOR signaling network clearly illustrates the close relationship between oncogenic cellular signaling and cancer metabolism. As the mechanisms underlying the regulation of metabolism by PI3K-Akt-mTOR signaling are explored more deeply, an emerging theme is the concept that the metabolic differences observed in cancer cells do not occur in isolation. Rather, there exists a layer of regulation established by oncogenic signaling processes that coordinate uptake and utilization of multiple nutrients in a manner best suited for supporting survival, growth, and proliferation. Furthermore, the mode of regulation is often multifaceted, with robust changes in metabolism being initiated by post-translational responses and then further reinforced by the induction of transcriptional programs.

Many open questions remain in our understanding of how the PI3K-Akt-mTOR signaling pathway governs metabolic reprogramming in cancer. In particular, it is important to note that the complexity of cellular metabolism extends beyond glucose, glutamine, nucleotide, and lipid metabolism. In fact, the Human Metabolome Database (www.hmdb.ca) lists more than 29,000 endogenous human metabolites involved in more than 6000 reactions, many of which have not been explored in cancer. Future studies will no doubt reveal more metabolic pathways that are regulated by PI3K-Akt-mTOR signaling.

Another key challenge is that several of the regulatory mechanisms that have been described in this chapter were studied in the context of physiological metabolism in nontumorigenic cells. It will be valuable to assess whether these same mechanisms are functional and relevant in the context of cancer cells. Even more importantly, it will be critical to gain an understanding of how the mechanisms of metabolic regulation by oncogenic PI3K-Akt-mTOR signaling confer cancer cells with metabolic dependencies and vulnerabilities that can be exploited as a therapeutic intervention. Particularly exciting are the prospects of identifying metabolic pathways that share a synthetic lethal relationship with oncogenic PI3K-Akt-mTOR pathway activation, as well as the identification of metabolic enzymes that may potentially be pharmacologically targeted in combination with PI3K-Akt-mTOR pathway inhibitors currently in development. As we increase our understanding of oncogenic signaling and its modulation of cancer metabolism, it is the hope that cancer-specific metabolic dependences will open novel therapeutic strategies that can be exploited to selectively and effectively target tumor cells.

References

Ahn KJ, Hwang HS, Park JH, Bang SH, Kang WJ, Yun M, Lee JD (2009) Evaluation of the role of hexokinase type II in cellular proliferation and apoptosis using human hepatocellular carcinoma cell lines. J Nucl Med 50:1525–1532

Alessi DR, Caudwell FB, Andjelkovic M, Hemmings BA, Cohen P (1996) Molecular basis for the substrate specificity of protein kinase B; comparison with MAPKAP kinase-1 and p70 S6 kinase. FEBS Lett 399:333–338

Alessi DR, James SR, Downes CP, Holmes AB, Gaffney PR, Reese CB, Cohen P (1997) Characterization of a 3-phosphoinositide-dependent protein kinase which phosphorylates and activates protein kinase Balpha. Curr Biol 7:261–269

Auger KR, Serunian LA, Soltoff SP, Libby P, Cantley LC (1989) PDGF-dependent tyrosine phosphorylation stimulates production of novel polyphosphoinositides in intact cells. Cell 57:167–175

Bae SS, Cho H, Mu J, Birnbaum MJ (2003) Isoform-specific regulation of insulin-dependent glucose uptake by Akt/protein kinase B. J Biol Chem 278:49530–49536

Balasubramanian MN, Butterworth EA, Kilberg MS (2013) Asparagine synthetase: regulation by cell stress and involvement in tumor biology. Am J Physiol Endocrinol Metab 304:E789–E799

Banerji S, Cibulskis K, Rangel-Escareno C, Brown KK, Carter SL, Frederick AM, Lawrence MS, Sivachenko AY, Sougnez C, Zou L, Cortes ML, Fernandez-Lopez JC, Peng S, Ardlie KG, Auclair D, Bautista-Pina V, Duke F, Francis J, Jung J, Maffuz-Aziz A, Onofrio RC, Parkin M,

Pho NH, Quintanar-Jurado V, Ramos AH, Rebollar-Vega R, Rodriguez-Cuevas S, Romero-Cordoba SL, Schumacher SE, Stransky N, Thompson KM, Uribe-Figueroa L, Baselga J, Beroukhim R, Polyak K, Sgroi DC, Richardson AL, Jimenez-Sanchez G, Lander ES, Gabriel SB, Garraway LA, Golub TR, Melendez-Zajgla J, Toker A, Getz G, Hidalgo-Miranda A, Meyerson M (2012) Sequence analysis of mutations and translocations across breast cancer subtypes. Nature 486:405–409

Bar-Peled L, Sabatini DM (2014) Regulation of mTORC1 by amino acids. Trends Cell Biol 24:400–406

Bardos JI, Ashcroft M (2004) Hypoxia-inducible factor-1 and oncogenic signalling. BioEssays 26:262–269

Barthel A, Okino ST, Liao J, Nakatani K, Li J, Whitlock JPJ, Roth RA (1999) Regulation of GLUT1 gene transcription by the serine/threonine kinase Akt1. J Biol Chem 274:20281–20286

Bauer DE, Hatzivassiliou G, Zhao F, Andreadis C, Thompson CB (2005) ATP citrate lyase is an important component of cell growth and transformation. Oncogene 24:6314–6322

Bellacosa A, de Feo D, Godwin AK, Bell DW, Cheng JQ, Altomare DA, Wan M, Dubeau L, Scambia G, Masciullo V, Ferrandina G, Benedetti Panici P, Mancuso S, Neri G, Testa JR (1995) Molecular alterations of the AKT2 oncogene in ovarian and breast carcinomas. Int J Cancer 64:280–285

Bellacosa A, Kumar CC, Di Cristofano A, Testa JR (2005) Activation of AKT kinases in cancer: implications for therapeutic targeting. Adv Cancer Res 94:29–86

Ben-Haim S, Ell P (2009) 18F-FDG PET and PET/CT in the evaluation of cancer treatment response. J Nucl Med 50:88–99

Ben-Sahra I, Howell JJ, Asara JM, Manning BD (2013) Stimulation of de novo pyrimidine synthesis by growth signaling through mTOR and S6K1. Science 339:1323–1328

Bentley J, Itchayanan D, Barnes K, McIntosh E, Tang X, Downes CP, Holman GD, Whetton AD, Owen-Lynch PJ, Baldwin SA (2003) Interleukin-3-mediated cell survival signals include phosphatidylinositol 3-kinase-dependent translocation of the glucose transporter GLUT1 to the cell surface. J Biol Chem 278:39337–39348

Berwick DC, Dell GC, Welsh GI, Heesom KJ, Hers I, Fletcher LM, Cooke FT, Tavare JM (2004) Protein kinase B phosphorylation of PIKfyve regulates the trafficking of GLUT4 vesicles. J Cell Sci 117:5985–5993

Berwick DC, Hers I, Heesom KJ, Moule SK, Tavare JM (2002) The identification of ATP-citrate lyase as a protein kinase B (Akt) substrate in primary adipocytes. J Biol Chem 277:33895–33900

Blancher C, Moore JW, Robertson N, Harris AL (2001) Effects of ras and von Hippel-Lindau (VHL) gene mutations on hypoxia-inducible factor (HIF)-1alpha, HIF-2alpha, and vascular endothelial growth factor expression and their regulation by the phosphatidylinositol 3′-kinase/Akt signaling pathway. Cancer Res 61:7349–7355

Boehmer C, Okur F, Setiawan I, Broer S, Lang F (2003) Properties and regulation of glutamine transporter SN1 by protein kinases SGK and PKB. Biochem Biophys Res Commun 306:156–162

Bokun R, Bakotin J, Milasinović D (1987) Semiquantitative cytochemical estimation of glucose-6-phosphate dehydrogenase activity in benign diseases and carcinoma of the breast. Acta Cytol 31:249–252

Boormans JL, Korsten H, Ziel-van der Made AC, van Leenders GJ, Verhagen PC, Trapman J (2010) E17K substitution in AKT1 in prostate cancer. Br J Cancer 102:1491–1494

Boros LG, Bassilian S, Lim S, Lee WN (2001) Genistein inhibits nonoxidative ribose synthesis in MIA pancreatic adenocarcinoma cells: a new mechanism of controlling tumor growth. Pancreas 22:1–7

Boros LG, Puigjaner J, Cascante M, Lee WN, Brandes JL, Bassilian S, Yusuf FI, Williams RD, Muscarella P, Melvin WS, Schirmer WJ (1997) Oxythiamine and dehydroepiandrosterone inhibit the nonoxidative synthesis of ribose and tumor cell proliferation. Cancer Res 57:4242–4248

Boss GR, Pilz RB (1985) Phosphoribosylpyrophosphate synthesis from glucose decreases during amino acid starvation of human lymphoblasts. J Biol Chem 260:6054–6059

Bouchard C, Marquardt J, Bras A, Medema RH, Eilers M (2004) Myc-induced proliferation and transformation require Akt-mediated phosphorylation of FoxO proteins. EMBO J 23:2830–2840

Brahimi-Horn MC, Chiche J, Pouyssegur J (2007) Hypoxia signalling controls metabolic demand. Curr Opin Cell Biol 19:223–229

Brunet A, Kanai F, Stehn J, Xu J, Sarbassova D, Frangioni JV, Dalal SN, DeCaprio JA, Greenberg ME, Yaffe MB (2002) 14-3-3 transits to the nucleus and participates in dynamic nucleocytoplasmic transport. J Cell Biol 156:817–828

Campbell IG, Russell SE, Choong DY, Montgomery KG, Ciavarella ML, Hooi CS, Cristiano BE, Pearson RB, Phillips WA (2004) Mutation of the PIK3CA gene in ovarian and breast cancer. Cancer Res 64:7678–7681

Cantley LC (2002) The phosphoinositide 3-kinase pathway. Science 296:1655–1657

Cantley LC, Neel BG (1999) New insights into tumor suppression: PTEN suppresses tumor formation by restraining the phosphoinositide 3-kinase/AKT pathway. Proc Natl Acad Sci USA 96:4240–4245

Carpten JD, Faber AL, Horn C, Donoho GP, Briggs SL, Robbins CM, Hostetter G, Boguslawski S, Moses TY, Savage S, Uhlik M, Lin A, Du J, Qian YW, Zeckner DJ, Tucker-Kellogg G, Touchman J, Patel K, Mousses S, Bittner M, Schevitz R, Lai MH, Blanchard KL, Thomas JE (2007) A transforming mutation in the pleckstrin homology domain of AKT1 in cancer. Nature 448:439–444

Chan CH, Li CF, Yang WL, Gao Y, Lee SW, Feng Z, Huang HY, Tsai KK, Flores LG, Shao Y, Hazle JD, Yu D, Wei W, Sarbassov D, Hung MC, Nakayama KI, Lin HK (2012) The Skp2-SCF E3 ligase regulates Akt ubiquitination, glycolysis, herceptin sensitivity, and tumorigenesis. Cell 149:1098–1111

Chan DA, Sutphin PD, Denko NC, Giaccia AJ (2002) Role of prolyl hydroxylation in oncogenically stabilized hypoxia-inducible factor-1alpha. J Biol Chem 277:40112–40117

Cheng JQ, Ruggeri B, Klein WM, Sonoda G, Altomare DA, Watson DK, Testa JR (1996) Amplification of AKT2 in human pancreatic cells and inhibition of AKT2 expression and tumorigenicity by antisense RNA. Proc Natl Acad Sci USA 93:3636–3641

Chin YR, Yoshida T, Marusyk A, Beck AH, Polyak K, Toker A (2014) Targeting Akt3 signaling in triple-negative breast cancer. Cancer Res 74:964–973

Cong LN, Chen H, Li Y, Zhou L, McGibbon MA, Taylor SI, Quon MJ (1997) Physiological role of Akt in insulin-stimulated translocation of GLUT4 in transfected rat adipose cells. Mol Endocrinol 11:1881–1890

Csibi A, Fendt SM, Li C, Poulogiannis G, Choo AY, Chapski DJ, Jeong SM, Dempsey JM, Parkhitko A, Morrison T, Henske EP, Haigis MC, Cantley LC, Stephanopoulos G, Yu J, Blenis J (2013) The mTORC1 pathway stimulates glutamine metabolism and cell proliferation by repressing SIRT4. Cell 153:840–854

Dang CV (2007) The interplay between MYC and HIF in the Warburg effect. In: Ernst Schering Foundation Symposium Proceedings, pp 35–53

Davies MA, Stemke-Hale K, Tellez C, Calderone TL, Deng W, Prieto VG, Lazar AJ, Gershenwald JE, Mills GB (2008) A novel AKT3 mutation in melanoma tumours and cell lines. Br J Cancer 99:1265–1268

de la Cruz-Herrera CF, Campagna M, Lang V, del Carmen González-Santamaría J, Marcos-Villar L, Rodríguez MS, Vidal A, Collado M, Rivas C (2015) SUMOylation regulates AKT1 activity. Oncogene 34:1442–1450

Deberardinis RJ, Lum JJ, Thompson CB (2006) Phosphatidylinositol 3-kinase-dependent modulation of carnitine palmitoyltransferase 1A expression regulates lipid metabolism during hematopoietic cell growth. J Biol Chem 281:37372–37380

Deng X, Zhang W, O-Sullivan I, Williams JB, Dong Q, Park EA, Raghow R, Unterman TG, Elam MB (2012) FoxO1 inhibits sterol regulatory element-binding protein-1c (SREBP-1c) gene expression via transcription factors Sp1 and SREBP-1c. J Biol Chem 287:20132–20143

Denko NC (2008) Hypoxia, HIF1 and glucose metabolism in the solid tumour. Nat Rev Cancer 8:705–713

Deprez J, Vertommen D, Alessi DR, Hue L, Rider MH (1997) Phosphorylation and activation of heart 6-phosphofructo-2-kinase by protein kinase B and other protein kinases of the insulin signaling cascades. J Biol Chem 272:17269–17275

Dessì S, Batetta B, Cherchi R, Onnis R, Pisano M, Pani P (1988) Hexose monophosphate shunt enzymes in lung tumors from normal and glucose-6-phosphate-dehydrogenase-deficient subjects. Oncology 45:287–291

Dibble CC, Elis W, Menon S, Qin W, Klekota J, Asara JM, Finan PM, Kwiatkowski DJ, Murphy LO, Manning BD (2012) TBC1D7 is a third subunit of the TSC1-TSC2 complex upstream of mTORC1. Mol Cell 47:535–546

Dibble CC, Manning BD (2013) Signal integration by mTORC1 coordinates nutrient input with biosynthetic output. Nat Cell Biol 15:555–564

Düvel K, Yecies JL, Menon S, Raman P, Lipovsky AI, Souza AL, Triantafellow E, Ma Q, Gorski R, Cleaver S, Vander Heiden MG, MacKeigan JP, Finan PM, Clish CB, Murphy LO, Manning BD (2010) Activation of a metabolic gene regulatory network downstream of mTOR complex 1. Mol Cell 39:171–183

Eilers M, Eisenman RN (2008) Myc's broad reach. Genes Dev 22:2755–2766

Elstrom RL, Bauer DE, Buzzai M, Karnauskas R, Harris MH, Plas DR, Zhuang H, Cinalli RM, Alavi A, Rudin CM, Thompson CB (2004) Akt stimulates aerobic glycolysis in cancer cells. Cancer Res 64:3892–3899

Engelman JA (2009) Targeting PI3K signalling in cancer: opportunities, challenges and limitations. Nat Rev Cancer 9:550–562

Engelman JA, Chen L, Tan X, Crosby K, Guimaraes AR, Upadhyay R, Maira M, McNamara K, Perera SA, Song Y, Chirieac LR, Kaur R, Lightbown A, Simendinger J, Li T, Padera RF, Garcia-Echeverria C, Weissleder R, Mahmood U, Cantley LC, Wong KK (2008) Effective use of PI3K and MEK inhibitors to treat mutant Kras G12D and PIK3CA H1047R murine lung cancers. Nat Med 14:1351–1356

Espenshade PJ, Hughes AL (2007) Regulation of sterol synthesis in eukaryotes. Annu Rev Genet 41:401–427

Fan CD, Lum MA, Xu C, Black JD, Wang X (2013a) Ubiquitin-dependent regulation of phospho-AKT dynamics by the ubiquitin E3 ligase, NEDD4-1, in the insulin-like growth factor-1 response. J Biol Chem 288:1674–1684

Fan J, Kamphorst JJ, Mathew R, Chung MK, White E, Shlomi T, Rabinowitz JD (2013b) Glutamine-driven oxidative phosphorylation is a major ATP source in transformed mammalian cells in both normoxia and hypoxia. Mol Syst Biol 9:712

Fantin VR, St-Pierre J, Leder P (2006) Attenuation of LDH-A expression uncovers a link between glycolysis, mitochondrial physiology, and tumor maintenance. Cancer Cell 9:425–434

Franke TF, Kaplan DR, Cantley LC, Toker A (1997) Direct regulation of the Akt proto-oncogene product by phosphatidylinositol-3,4-bisphosphate. Science 275:665–668

Fruman DA, Cantley LC (2014) Idelalisib–a PI3Kδ inhibitor for B-cell cancers. N Engl J Med 370:1061–1062

Fruman DA, Rommel C (2014) PI3K and cancer: lessons, challenges and opportunities. Nat Rev Drug Discov 13:140–156

Gan B, Lim C, Chu G, Hua S, Ding Z, Collins M, Hu J, Jiang S, Fletcher-Sananikone E, Zhuang L, Chang M, Zheng H, Wang YA, Kwiatkowski DJ, Kaelin WGJ, Signoretti S, DePinho RA (2010) FoxOs enforce a progression checkpoint to constrain mTORC1-activated renal tumorigenesis. Cancer Cell 18:472–484

Ganapathy V, Thangaraju M, Prasad PD (2009) Nutrient transporters in cancer: relevance to Warburg hypothesis and beyond. Pharmacol Ther 121:29–40

Gao P, Tchernyshyov I, Chang TC, Lee YS, Kita K, Ochi T, Zeller KI, De Marzo AM, Van Eyk JE, Mendell JT, Dang CV (2009) c-Myc suppression of miR-23a/b enhances mitochondrial glutaminase expression and glutamine metabolism. Nature 458:762–765

García JM, Silva J, Peña C, Garcia V, Rodríguez R, Cruz MA, Cantos B, Provencio M, España P, Bonilla F (2004) Promoter methylation of the PTEN gene is a common molecular change in breast cancer. Genes Chromosom Cancer 41:117–124

Gebhardt R, Kleemann E (1987) Hormonal regulation of amino acid transport system N in primary cultures of rat hepatocytes. Eur J Biochem 166:339–344

Goel A, Arnold CN, Niedzwiecki D, Carethers JM, Dowell JM, Wasserman L, Compton C, Mayer RJ, Bertagnolli MM, Boland CR (2004) Frequent inactivation of PTEN by promoter hypermethylation in microsatellite instability-high sporadic colorectal cancers. Cancer Res 64:3014–3021

Gonzalez E, McGraw TE (2009) Insulin-modulated Akt subcellular localization determines Akt isoform-specific signaling. Proc Natl Acad Sci USA 106:7004–7009

Gottlob K, Majewski N, Kennedy S, Kandel E, Robey RB, Hay N (2001) Inhibition of early apoptotic events by Akt/PKB is dependent on the first committed step of glycolysis and mitochondrial hexokinase. Genes Dev 15:1406–1418

Grabiner BC, Nardi V, Birsoy K, Possemato R, Shen K, Sinha S, Jordan A, Beck AH, Sabatini DM (2014) A diverse array of cancer-associated MTOR mutations are hyperactivating and can predict rapamycin sensitivity. Cancer Discov 4:554–563

Gregory MA, Qi Y, Hann SR (2003) Phosphorylation by glycogen synthase kinase-3 controls c-myc proteolysis and subnuclear localization. J Biol Chem 278:51606–51612

Haigis MC, Mostoslavsky R, Haigis KM, Fahie K, Christodoulou DC, Murphy AJ, Valenzuela DM, Yancopoulos GD, Karow M, Blander G, Wolberger C, Prolla TA, Weindruch R, Alt FW, Guarente L (2006) SIRT4 inhibits glutamate dehydrogenase and opposes the effects of calorie restriction in pancreatic beta cells. Cell 126:941–954

Hanahan D, Weinberg RA (2011) Hallmarks of cancer: the next generation. Cell 144:646–674

Hensley CT, Wasti AT, DeBerardinis RJ (2013) Glutamine and cancer: cell biology, physiology, and clinical opportunities. J Clin Invest 123:3678–3684

Huang S, Czech MP (2007) The GLUT4 glucose transporter. Cell Metab 5:237–252

Ikonomov OC, Sbrissa D, Dondapati R, Shisheva A (2007) ArPIKfyve-PIKfyve interaction and role in insulin-regulated GLUT4 translocation and glucose transport in 3T3-L1 adipocytes. Exp Cell Res 313:2404–2416

Ikonomov OC, Sbrissa D, Mlak K, Shisheva A (2002) Requirement for PIKfyve enzymatic activity in acute and long-term insulin cellular effects. Endocrinology 143:4742–4754

Ikonomov OC, Sbrissa D, Shisheva A (2009) YM201636, an inhibitor of retroviral budding and PIKfyve-catalyzed PtdIns(3,5)P2 synthesis, halts glucose entry by insulin in adipocytes. Biochem Biophys Res Commun 382:566–570

John S, Weiss JN, Ribalet B (2011) Subcellular localization of hexokinases I and II directs the metabolic fate of glucose. PLoS ONE 6:e17674

Jones AC, Shyamsundar MM, Thomas MW, Maynard J, Idziaszczyk S, Tomkins S, Sampson JR, Cheadle JP (1999) Comprehensive mutation analysis of TSC1 and TSC2-and phenotypic correlations in 150 families with tuberous sclerosis. Am J Hum Genet 64:1305–1315

Juvekar A, Burga LN, Hu H, Lunsford EP, Ibrahim YH, Balmana J, Rajendran A, Papa A, Spencer K, Lyssiotis CA, Nardella C, Pandolfi PP, Baselga J, Scully R, Asara JM, Cantley LC, Wulf GM (2012) Combining a PI3K inhibitor with a PARP inhibitor provides an effective therapy for BRCA1-related breast cancer. Cancer Discov 2:1048–1063

Katome T, Obata T, Matsushima R, Masuyama N, Cantley LC, Gotoh Y, Kishi K, Shiota H, Ebina Y (2003) Use of RNA interference-mediated gene silencing and adenoviral overexpression to elucidate the roles of AKT/protein kinase B isoforms in insulin actions. J Biol Chem 278:28312–28323

Khatri S, Yepiskoposyan H, Gallo CA, Tandon P, Plas DR (2010) FOXO3a regulates glycolysis via transcriptional control of tumor suppressor TSC1. J Biol Chem 285:15960–15965

Kim JW, Tchernyshyov I, Semenza GL, Dang CV (2006) HIF-1-mediated expression of pyruvate dehydrogenase kinase: a metabolic switch required for cellular adaptation to hypoxia. Cell Metab 3:177–185

Kim JW, Zeller KI, Wang Y, Jegga AG, Aronow BJ, O'Donnell KA, Dang CV (2004a) Evaluation of myc E-box phylogenetic footprints in glycolytic genes by chromatin immunoprecipitation assays. Mol Cell Biol 24:5923–5936

Kim KH, Song MJ, Yoo EJ, Choe SS, Park SD, Kim JB (2004b) Regulatory role of glycogen synthase kinase 3 for transcriptional activity of ADD1/SREBP1c. J Biol Chem 279:51999–52006

Kohn AD, Summers SA, Birnbaum MJ, Roth RA (1996) Expression of a constitutively active Akt Ser/Thr kinase in 3T3-L1 adipocytes stimulates glucose uptake and glucose transporter 4 translocation. J Biol Chem 271:31372–31378

Kotani K, Carozzi AJ, Sakaue H, Hara K, Robinson LJ, Clark SF, Yonezawa K, James DE, Kasuga M (1995) Requirement for phosphoinositide 3-kinase in insulin-stimulated GLUT4 translocation in 3T3-L1 adipocytes. Biochem Biophys Res Commun 209:343–348

Kress TR, Cannell IG, Brenkman AB, Samans B, Gaestel M, Roepman P, Burgering BM, Bushell M, Rosenwald A, Eilers M (2011) The MK5/PRAK kinase and Myc form a negative feedback loop that is disrupted during colorectal tumorigenesis. Mol Cell 41:445–457

Krockenberger M, Honig A, Rieger L, Coy JF, Sutterlin M, Kapp M, Horn E, Dietl J, Kammerer U (2007) Transketolase-like 1 expression correlates with subtypes of ovarian cancer and the presence of distant metastases. Int J Gynecol Cancer 17:101–106

Langbein S, Frederiks WM, zur Hausen A, Popa J, Lehmann J, Weiss C, Alken P, Coy JF (2008) Metastasis is promoted by a bioenergetic switch: new targets for progressive renal cell cancer. Int J Cancer 122:2422–2428

Langbein S, Zerilli M, Zur Hausen A, Staiger W, Rensch-Boschert K, Lukan N, Popa J, Ternullo MP, Steidler A, Weiss C, Grobholz R, Willeke F, Alken P, Stassi G, Schubert P, Coy JF (2006) Expression of transketolase TKTL1 predicts colon and urothelial cancer patient survival: Warburg effect reinterpreted. Br J Cancer 94:578–585

Larance M, Ramm G, Stockli J, van Dam EM, Winata S, Wasinger V, Simpson F, Graham M, Junutula JR, Guilhaus M, James DE (2005) Characterization of the role of the Rab GTPase-activating protein AS160 in insulin-regulated GLUT4 trafficking. J Biol Chem 280:37803–37813

Laughner E, Taghavi P, Chiles K, Mahon PC, Semenza GL (2001) HER2 (neu) signaling increases the rate of hypoxia-inducible factor 1alpha (HIF-1alpha) synthesis: novel mechanism for HIF-1-mediated vascular endothelial growth factor expression. Mol Cell Biol 21:3995–4004

Li B, Simon MC (2013) Molecular pathways: targeting MYC-induced metabolic reprogramming and oncogenic stress in cancer. Clin Cancer Res 19:5835–5841

Li J, Yen C, Liaw D, Podsypanina K, Bose S, Wang SI, Puc J, Miliaresis C, Rodgers L, McCombie R, Bigner SH, Giovanella BC, Ittmann M, Tycko B, Hibshoosh H, Wigler MH, Parsons R (1997) PTEN, a putative protein tyrosine phosphatase gene mutated in human brain, breast, and prostate cancer. Science 275:1943–1947

Li R, Wei J, Jiang C, Liu D, Deng L, Zhang K, Wang P (2013) Akt SUMOylation regulates cell proliferation and tumorigenesis. Cancer Res 73:5742–5753

Liu P, Begley M, Michowski W, Inuzuka H, Ginzberg M, Gao D, Tsou P, Gan W, Papa A, Kim BM, Wan L, Singh A, Zhai B, Yuan M, Wang Z, Gygi SP, Lee TH, Lu KP, Toker A, Pandolfi PP, Asara JM, Kirschner MW, Sicinski P, Cantley L, Wei W (2014) Cell-cycle-regulated activation of Akt kinase by phosphorylation at its carboxyl terminus. Nature 508:541–545

Low SY, Taylor PM, Rennie MJ (1996) Responses of glutamine transport in cultured rat skeletal muscle to osmotically induced changes in cell volume. J Physiol 492:877–885

Ma XM, Blenis J (2009) Molecular mechanisms of mTOR-mediated translational control. Nat Rev Mol Cell Biol 10:307–318

Ma YY, Wei SJ, Lin YC, Lung JC, Chang TC, Whang-Peng J, Liu JM, Yang DM, Yang WK, Shen CY (2000) PIK3CA as an oncogene in cervical cancer. Oncogene 19:2739–2744

Maehama T, Dixon JE (1998) The tumor suppressor, PTEN/MMAC1, dephosphorylates the lipid second messenger, phosphatidylinositol 3,4,5-trisphosphate. J Biol Chem 273:13375–13378

Majewski N, Nogueira V, Bhaskar P, Coy PE, Skeen JE, Gottlob K, Chandel NS, Thompson CB, Robey RB, Hay N (2004a) Hexokinase-mitochondria interaction mediated by Akt is required to inhibit apoptosis in the presence or absence of Bax and Bak. Mol Cell 16:819–830

Majewski N, Nogueira V, Robey RB, Hay N (2004b) Akt inhibits apoptosis downstream of BID cleavage via a glucose-dependent mechanism involving mitochondrial hexokinases. Mol Cell Biol 24:730–740

Malanga D, Scrima M, De Marco C, Fabiani F, De Rosa N, De Gisi S, Malara N, Savino R, Rocco G, Chiappetta G, Franco R, Tirino V, Pirozzi G, Viglietto G (2008) Activating E17 K mutation in the gene encoding the protein kinase AKT1 in a subset of squamous cell carcinoma of the lung. Cell Cycle 7:665–669

Manning BD, Cantley LC (2003) United at last: the tuberous sclerosis complex gene products connect the phosphoinositide 3-kinase/Akt pathway to mammalian target of rapamycin (mTOR) signalling. Biochem Soc Trans 31:573–578

Manning BD, Cantley LC (2007) AKT/PKB signaling: navigating downstream. Cell 129:1261–1274

Manning BD, Tee AR, Logsdon MN, Blenis J, Cantley LC (2002) Identification of the tuberous sclerosis complex-2 tumor suppressor gene product tuberin as a target of the phosphoinositide 3-kinase/akt pathway. Mol Cell 10:151–162

Masui K, Tanaka K, Akhavan D, Babic I, Gini B, Matsutani T, Iwanami A, Liu F, Villa GR, Gu Y, Campos C, Zhu S, Yang H, Yong WH, Cloughesy TF, Mellinghoff IK, Cavenee WK, Shaw RJ, Mischel PS (2013) mTOR complex 2 controls glycolytic metabolism in glioblastoma through FoxO acetylation and upregulation of c-Myc. Cell Metab 18:726–739

Mathupala SP, Ko YH, Pedersen PL (2006) Hexokinase II: cancer's double-edged sword acting as both facilitator and gatekeeper of malignancy when bound to mitochondria. Oncogene 25:4777–4786

Mayer IA, Abramson VG, Isakoff SJ, Forero A, Balko JM, Kuba MG, Sanders ME, Yap JT, Van den Abbeele AD, Li Y, Cantley LC, Winer E, Arteaga CL (2014) Stand up to cancer phase Ib study of pan-phosphoinositide-3-kinase inhibitor buparlisib with letrozole in estrogen receptor-positive/human epidermal growth factor receptor 2-negative metastatic breast cancer. J Clin Oncol 32:1202–1209

Menendez JA, Lupu R (2007) Fatty acid synthase and the lipogenic phenotype in cancer pathogenesis. Nat Rev Cancer 7:763–777

Menon S, Dibble CC, Talbott G, Hoxhaj G, Valvezan AJ, Takahashi H, Cantley LC, Manning BD (2014) Spatial control of the TSC complex integrates insulin and nutrient regulation of mTORC1 at the lysosome. Cell 156:771–785

Migita T, Narita T, Nomura K, Miyagi E, Inazuka F, Matsuura M, Ushijima M, Mashima T, Seimiya H, Satoh Y, Okumura S, Nakagawa K, Ishikawa Y (2008) ATP citrate lyase: activation and therapeutic implications in non-small cell lung cancer. Cancer Res 68:8547–8554

Miyamoto S, Murphy AN, Brown JH (2008) Akt mediates mitochondrial protection in cardiomyocytes through phosphorylation of mitochondrial hexokinase-II. Cell Death Differ 15:521–529

Moon JS, Jin WJ, Kwak JH, Kim HJ, Yun MJ, Kim JW, Park SW, Kim KS (2011) Androgen stimulates glycolysis for de novo lipid synthesis by increasing the activities of hexokinase 2 and 6-phosphofructo-2-kinase/fructose-2,6-bisphosphatase 2 in prostate cancer cells. Biochem J 433:225–233

Moreno-Sanchez R, Rodriguez-Enriquez S, Marin-Hernandez A, Saavedra E (2007) Energy metabolism in tumor cells. FEBS J 274:1393–1418

Nave BT, Ouwens M, Withers DJ, Alessi DR, Shepherd PR (1999) Mammalian target of rapamycin is a direct target for protein kinase B: identification of a convergence point for opposing effects of insulin and amino-acid deficiency on protein translation. Biochem J 344(Pt 2):427–431

Obata T, Yaffe MB, Leparc GG, Piro ET, Maegawa H, Kashiwagi A, Kikkawa R, Cantley LC (2000) Peptide and protein library screening defines optimal substrate motifs for AKT/PKB. J Biol Chem 275:36108–36115

Obsil T, Ghirlando R, Anderson DE, Hickman AB, Dyda F (2003) Two 14-3-3 binding motifs are required for stable association of Forkhead transcription factor FOXO4 with 14-3-3 proteins and inhibition of DNA binding. Biochemistry 42:15264–15272

Okada T, Kawano Y, Sakakibara T, Hazeki O, Ui M (1994) Essential role of phosphatidylinositol 3-kinase in insulin-induced glucose transport and antilipolysis in rat adipocytes. Studies with a selective inhibitor wortmannin. J Biol Chem 269:3568–3573

Osthus RC, Shim H, Kim S, Li Q, Reddy R, Mukherjee M, Xu Y, Wonsey D, Lee LA, Dang CV (2000) Deregulation of glucose transporter 1 and glycolytic gene expression by c-Myc. J Biol Chem 275:21797–21800

Papandreou I, Cairns RA, Fontana L, Lim AL, Denko NC (2006) HIF-1 mediates adaptation to hypoxia by actively downregulating mitochondrial oxygen consumption. Cell Metab 3:187–197

Paradis S, Ruvkun G (1998) Caenorhabditis elegans Akt/PKB transduces insulin receptor-like signals from AGE-1 PI3 kinase to the DAF-16 transcription factor. Genes Dev 12:2488–2498

Parikh H, Carlsson E, Chutkow WA, Johansson LE, Storgaard H, Poulsen P, Saxena R, Ladd C, Schulze PC, Mazzini MJ, Jensen CB, Krook A, Bjornholm M, Tornqvist H, Zierath JR, Ridderstrale M, Altshuler D, Lee RT, Vaag A, Groop LC, Mootha VK (2007) TXNIP regulates peripheral glucose metabolism in humans. PLoS Med 4:e158

Parsons DW, Wang TL, Samuels Y, Bardelli A, Cummins JM, DeLong L, Silliman N, Ptak J, Szabo S, Willson JK, Markowitz S, Kinzler KW, Vogelstein B, Lengauer C, Velculescu VE (2005) Colorectal cancer: mutations in a signalling pathway. Nature 436:792

Pastorino JG, Hoek JB, Shulga N (2005) Activation of glycogen synthase kinase 3beta disrupts the binding of hexokinase II to mitochondria by phosphorylating voltage-dependent anion channel and potentiates chemotherapy-induced cytotoxicity. Cancer Res 65:10545–10554

Pastorino JG, Shulga N, Hoek JB (2002) Mitochondrial binding of hexokinase II inhibits Bax-induced cytochrome c release and apoptosis. J Biol Chem 277:7610–7618

Peck B, Ferber EC, Schulze A (2013) Antagonism between FOXO and MYC Regulates Cellular Powerhouse. Front Oncol 3:96

Porstmann T, Griffiths B, Chung YL, Delpuech O, Griffiths JR, Downward J, Schulze A (2005) PKB/Akt induces transcription of enzymes involved in cholesterol and fatty acid biosynthesis via activation of SREBP. Oncogene 24:6465–6481

Porstmann T, Santos CR, Griffiths B, Cully M, Wu M, Leevers S, Griffiths JR, Chung YL, Schulze A (2008) SREBP activity is regulated by mTORC1 and contributes to Akt-dependent cell growth. Cell Metab 8:224–236

Potapova IA, El-Maghrabi MR, Doronin SV, Benjamin WB (2000) Phosphorylation of recombinant human ATP:citrate lyase by cAMP-dependent protein kinase abolishes homotropic allosteric regulation of the enzyme by citrate and increases the enzyme activity. Allosteric activation of ATP:citrate lyase by phosphorylated sugars. Biochemistry 39:1169–1179

Rácz A, Brass N, Heckel D, Pahl S, Remberger K, Meese E (1999) Expression analysis of genes at 3q26-q27 involved in frequent amplification in squamous cell lung carcinoma. Eur J Cancer 35:641–646

Raivio KO, Lazar CS, Krumholz HR, Becker MA (1981) The phosphogluconate pathway and synthesis of 5-phosphoribosyl-1-pyrophosphate in human fibroblasts. Biochim Biophys Acta 678:51–57

Rathmell JC, Fox CJ, Plas DR, Hammerman PS, Cinalli RM, Thompson CB (2003) Akt-directed glucose metabolism can prevent Bax conformation change and promote growth factor-independent survival. Mol Cell Biol 23:7315–7328

Reitzer LJ, Wice BM, Kennell D (1979) Evidence that glutamine, not sugar, is the major energy source for cultured HeLa cells. J Biol Chem 254:2669–2676

Risso G, Pelisch F, Pozzi B, Mammi P, Blaustein M, Colman-Lerner A, Srebrow A (2013) Modification of Akt by SUMO conjugation regulates alternative splicing and cell cycle. Cell Cycle 12:3165–3174

Roberts DJ, Tan-Sah VP, Smith JM, Miyamoto S (2013) Akt phosphorylates HK-II at Thr-473 and increases mitochondrial HK-II association to protect cardiomyocytes. J Biol Chem 288:23798–23806

Robey RB, Hay N (2009) Is Akt the "Warburg kinase"?-Akt-energy metabolism interactions and oncogenesis. Semin Cancer Biol 19:25–31

Robitaille AM, Christen S, Shimobayashi M, Cornu M, Fava LL, Moes S, Prescianotto-Baschong C, Sauer U, Jenoe P, Hall MN (2013) Quantitative phosphoproteomics reveal mTORC1 activates de novo pyrimidine synthesis. Science 339:1320–1323

Rodon J, Dienstmann R, Serra V, Tabernero J (2013) Development of PI3K inhibitors: lessons learned from early clinical trials. Nat Rev Clin Oncol 10:143–153

Saci A, Cantley LC, Carpenter CL (2011) Rac1 regulates the activity of mTORC1 and mTORC2 and controls cellular size. Mol Cell 42:50–61

Sale EM, Hodgkinson CP, Jones NP, Sale GJ (2006) A new strategy for studying protein kinase B and its three isoforms. Role of protein kinase B in phosphorylating glycogen synthase kinase-3, tuberin, WNK1, and ATP citrate lyase. Biochemistry 45:213–223

Samuels Y, Wang Z, Bardelli A, Silliman N, Ptak J, Szabo S, Yan H, Gazdar A, Powell SM, Riggins GJ, Willson JK, Markowitz S, Kinzler KW, Vogelstein B, Velculescu VE (2004) High frequency of mutations of the PIK3CA gene in human cancers. Science 304:554

Sano H, Kane S, Sano E, Miinea CP, Asara JM, Lane WS, Garner CW, Lienhard GE (2003) Insulin-stimulated phosphorylation of a Rab GTPase-activating protein regulates GLUT4 translocation. J Biol Chem 278:14599–14602

Sasaki H, Okuda K, Kawano O, Yukiue H, Yano M, Fujii Y (2008) AKT1 and AKT2 mutations in lung cancer in a Japanese population. Mol Med Rep 1:663–666

Sbrissa D, Ikonomov OC, Strakova J, Shisheva A (2004) Role for a novel signaling intermediate, phosphatidylinositol 5-phosphate, in insulin-regulated F-actin stress fiber breakdown and GLUT4 translocation. Endocrinology 145:4853–4865

Sekulić A, Hudson CC, Homme JL, Yin P, Otterness DM, Karnitz LM, Abraham RT (2000) A direct linkage between the phosphoinositide 3-kinase-AKT signaling pathway and the mammalian target of rapamycin in mitogen-stimulated and transformed cells. Cancer Res 60:3504–3513

Semenza GL, Jiang BH, Leung SW, Passantino R, Concordet JP, Maire P, Giallongo A (1996) Hypoxia response elements in the aldolase A, enolase 1, and lactate dehydrogenase A gene promoters contain essential binding sites for hypoxia-inducible factor 1. J Biol Chem 271:32529–32537

Semenza GL, Roth PH, Fang HM, Wang GL (1994) Transcriptional regulation of genes encoding glycolytic enzymes by hypoxia-inducible factor 1. J Biol Chem 269:23757–23763

Shim H, Dolde C, Lewis BC, Wu CS, Dang G, Jungmann RA, Dalla-Favera R, Dang CV (1997) c-Myc transactivation of LDH-A: implications for tumor metabolism and growth. Proc Natl Acad Sci U S A 94:6658–6663

Shisheva A (2008) PIKfyve: partners, significance, debates and paradoxes. Cell Biol Int 32:591–604

Shoji K, Oda K, Nakagawa S, Hosokawa S, Nagae G, Uehara Y, Sone K, Miyamoto Y, Hiraike H, Hiraike-Wada O, Nei T, Kawana K, Kuramoto H, Aburatani H, Yano T, Taketani Y (2009) The oncogenic mutation in the pleckstrin homology domain of AKT1 in endometrial carcinomas. Br J Cancer 101:145–148

Silhan J, Vacha P, Strnadova P, Vecer J, Herman P, Sulc M, Teisinger J, Obsilova V, Obsil T (2009) 14-3-3 protein masks the DNA binding interface of forkhead transcription factor FOXO4. J Biol Chem 284:19349–19360

Son J, Lyssiotis CA, Ying H, Wang X, Hua S, Ligorio M, Perera RM, Ferrone CR, Mullarky E, Shyh-Chang N, Kang Y, Fleming JB, Bardeesy N, Asara JM, Haigis MC, DePinho RA, Cantley LC, Kimmelman AC (2013) Glutamine supports pancreatic cancer growth through a KRAS-regulated metabolic pathway. Nature 496:101–105

Staal SP (1987) Molecular cloning of the akt oncogene and its human homologues AKT1 and AKT2: amplification of AKT1 in a primary human gastric adenocarcinoma. Proc Natl Acad Sci USA 84:5034–5037

Stanton RC (2012) Glucose-6-phosphate dehydrogenase, NADPH, and cell survival. IUBMB Life 64:362–369

Stephens L, Anderson K, Stokoe D, Erdjument-Bromage H, Painter GF, Holmes AB, Gaffney PR, Reese CB, McCormick F, Tempst P, Coadwell J, Hawkins PT (1998) Protein kinase B kinases that mediate phosphatidylinositol 3,4,5-trisphosphate-dependent activation of protein kinase B. Science 279:710–714

Stephens PJ, Tarpey PS, Davies H, Van Loo P, Greenman C, Wedge DC, Nik-Zainal S, Martin S, Varela I, Bignell GR, Yates LR, Papaemmanuil E, Beare D, Butler A, Cheverton A, Gamble J, Hinton J, Jia M, Jayakumar A, Jones D, Latimer C, Lau KW, McLaren S, McBride DJ, Menzies A, Mudie L, Raine K, Rad R, Chapman MS, Teague J, Easton D, Langerod A, Lee MT, Shen CY, Tee BT, Huimin BW, Broeks A, Vargas AC, Turashvili G, Martens J, Fatima A, Miron P, Chin SF, Thomas G, Boyault S, Mariani O, Lakhani SR, van de Vijver M, van 't Veer L, Foekens J, Desmedt C, Sotiriou C, Tutt A, Caldas C, Reis-Filho JS, Aparicio SA, Salomon AV, Borresen-Dale AL, Richardson AL, Campbell PJ, Futreal PA, Stratton MR (2012) The landscape of cancer genes and mutational processes in breast cancer. Nature 486, 400–404

Stoltzman CA, Peterson CW, Breen KT, Muoio DM, Billin AN, Ayer DE (2008) Glucose sensing by MondoA: Mlx complexes: a role for hexokinases and direct regulation of thioredoxin-interacting protein expression. Proc Natl Acad Sci USA 105:6912–6917

Storz P, Toker A (2002) 3'-phosphoinositide-dependent kinase-1 (PDK-1) in PI 3-kinase signaling. Front Biosci 7:d886–d902

Sun X, Huang J, Homma T, Kita D, Klocker H, Schafer G, Boyle P, Ohgaki H (2009) Genetic alterations in the PI3 K pathway in prostate cancer. Anticancer Res 29:1739–1743

Sundqvist A, Bengoechea-Alonso MT, Ye X, Lukiyanchuk V, Jin J, Harper JW, Ericsson J (2005) Control of lipid metabolism by phosphorylation-dependent degradation of the SREBP family of transcription factors by SCF(Fbw7). Cell Metab 1:379–391

Tadros LB, Taylor PM, Rennie MJ (1993) Characteristics of glutamine transport in primary tissue culture of rat skeletal muscle. Am J Physiol 265:E135–E144

Tee AR, Fingar DC, Manning BD, Kwiatkowski DJ, Cantley LC, Blenis J (2002) Tuberous sclerosis complex-1 and -2 gene products function together to inhibit mammalian target of rapamycin (mTOR)-mediated downstream signaling. Proc Natl Acad Sci USA 99:13571–13576

Tee AR, Manning BD, Roux PP, Cantley LC, Blenis J (2003) Tuberous sclerosis complex gene products, Tuberin and Hamartin, control mTOR signaling by acting as a GTPase-activating protein complex toward Rheb. Curr Biol 13:1259–1268

Thomas GV, Tran C, Mellinghoff IK, Welsbie DS, Chan E, Fueger B, Czernin J, Sawyers CL (2006) Hypoxia-inducible factor determines sensitivity to inhibitors of mTOR in kidney cancer. Nat Med 12:122–127

Toker A (2012) Phosphoinositide 3-kinases-a historical perspective. Subcell Biochem 58:95–110

Toker A, Cantley LC (1997) Signalling through the lipid products of phosphoinositide-3-OH kinase. Nature 387:673–676

Tzivion G, Dobson M, Ramakrishnan G (2011) FoxO transcription factors; Regulation by AKT and 14-3-3 proteins. Biochim Biophys Acta 1813:1938–1945

Van de Sande T, De Schrijver E, Heyns W, Verhoeven G, Swinnen JV (2002) Role of the phosphatidylinositol 3'-kinase/PTEN/Akt kinase pathway in the overexpression of fatty acid synthase in LNCaP prostate cancer cells. Cancer Res 62:642–646

Van de Sande T, Roskams T, Lerut E, Joniau S, Van Poppel H, Verhoeven G, Swinnen JV (2005) High-level expression of fatty acid synthase in human prostate cancer tissues is linked to activation and nuclear localization of Akt/PKB. J Pathol 206:214–219

van der Vos KE, Eliasson P, Proikas-Cezanne T, Vervoort SJ, van Boxtel R, Putker M, van Zutphen IJ, Mauthe M, Zellmer S, Pals C, Verhagen LP, Groot Koerkamp MJ, Braat AK, Dansen TB, Holstege FC, Gebhardt R, Burgering BM, Coffer PJ (2012) Modulation of glutamine metabolism by the PI(3)K-PKB-FOXO network regulates autophagy. Nat Cell Biol 14:829–837

Vander Haar E, Lee SI, Bandhakavi S, Griffin TJ, Kim DH (2007) Insulin signalling to mTOR mediated by the Akt/PKB substrate PRAS40. Nat Cell Biol 9:316–323

Vander Heiden MG, Cantley LC, Thompson CB (2009) Understanding the Warburg effect: the metabolic requirements of cell proliferation. Science 324:1029–1033

Vanhaesebroeck B, Waterfield MD (1999) Signaling by distinct classes of phosphoinositide 3-kinases. Exp Cell Res 253:239–254

Vivanco I, Sawyers CL (2002) The phosphatidylinositol 3-Kinase AKT pathway in human cancer. Nat Rev Cancer 2:489–501

Wang W, Fridman A, Blackledge W, Connelly S, Wilson IA, Pilz RB, Boss GR (2009) The phosphatidylinositol 3-kinase/akt cassette regulates purine nucleotide synthesis. J Biol Chem 284:3521–3528

Warburg O (1956) On the origin of cancer cells. Science 123:309–314

Wasa M, Wang HS, Tazuke Y, Okada A (2001) Insulin-like growth factor-I stimulates amino acid transport in a glutamine-deprived human neuroblastoma cell line. Biochim Biophys Acta 1525:118–124

Weinhouse S (1976) The Warburg hypothesis fifty years later. Z Krebsforsch Klin Onkol Cancer Res Clin Oncol 87:115–126

Welcker M, Orian A, Jin J, Grim JE, Harper JW, Eisenman RN, Clurman BE (2004) The Fbw7 tumor suppressor regulates glycogen synthase kinase 3 phosphorylation-dependent c-Myc protein degradation. Proc Natl Acad Sci USA 101:9085–9090

Whitman M, Downes CP, Keeler M, Keller T, Cantley L (1988) Type I phosphatidylinositol kinase makes a novel inositol phospholipid, phosphatidylinositol-3-phosphate. Nature 332:644–646

Whitman M, Kaplan DR, Schaffhausen B, Cantley L, Roberts TM (1985) Association of phosphatidylinositol kinase activity with polyoma middle-T competent for transformation. Nature 315:239–242

Wieman HL, Wofford JA, Rathmell JC (2007) Cytokine stimulation promotes glucose uptake via phosphatidylinositol-3 kinase/Akt regulation of Glut1 activity and trafficking. Mol Biol Cell 18:1437–1446

Wise DR, DeBerardinis RJ, Mancuso A, Sayed N, Zhang XY, Pfeiffer HK, Nissim I, Daikhin E, Yudkoff M, McMahon SB, Thompson CB (2008) Myc regulates a transcriptional program that stimulates mitochondrial glutaminolysis and leads to glutamine addiction. Proc Natl Acad Sci USA 105:18782–18787

Wolf A, Agnihotri S, Micallef J, Mukherjee J, Sabha N, Cairns R, Hawkins C, Guha A (2011) Hexokinase 2 is a key mediator of aerobic glycolysis and promotes tumor growth in human glioblastoma multiforme. J Exp Med 208:313–326

Wu N, Zheng B, Shaywitz A, Dagon Y, Tower C, Bellinger G, Shen CH, Wen J, Asara J, McGraw TE, Kahn BB, Cantley LC (2013) AMPK-dependent degradation of TXNIP upon energy stress leads to enhanced glucose uptake via GLUT1. Mol Cell 49:1167–1175

Yang C, Sudderth J, Dang T, Bachoo RM, McDonald JG, DeBerardinis RJ (2009a) Glioblastoma cells require glutamate dehydrogenase to survive impairments of glucose metabolism or Akt signaling. Cancer Res 69:7986–7993

Yang WL, Jin G, Li CF, Jeong YS, Moten A, Xu D, Feng Z, Chen W, Cai Z, Darnay B, Gu W, Lin HK (2013) Cycles of ubiquitination and deubiquitination critically regulate growth factor-mediated activation of Akt signaling. Sci Signal 6:ra3

Yang WL, Wang J, Chan CH, Lee SW, Campos AD, Lamothe B, Hur L, Grabiner BC, Lin X, Darnay BG, Lin HK (2009b) The E3 ligase TRAF6 regulates Akt ubiquitination and activation. Science 325:1134–1138

Yang WL, Wu CY, Wu J, Lin HK (2010) Regulation of Akt signaling activation by ubiquitination. Cell Cycle 9:487–497

Yeo H, Lyssiotis CA, Zhang Y, Ying H, Asara JM, Cantley LC, Paik JH (2013) FoxO3 coordinates metabolic pathways to maintain redox balance in neural stem cells. EMBO J 32:2589–2602

Ying H, Kimmelman AC, Lyssiotis CA, Hua S, Chu GC, Fletcher-Sananikone E, Locasale JW, Son J, Zhang H, Coloff JL, Yan H, Wang W, Chen S, Viale A, Zheng H, Paik JH, Lim C, Guimaraes AR, Martin ES, Chang J, Hezel AF, Perry SR, Hu J, Gan B, Xiao Y, Asara JM, Weissleder R, Wang YA, Chin L, Cantley LC, DePinho RA (2012) Oncogenic Kras maintains pancreatic tumors through regulation of anabolic glucose metabolism. Cell 149:656–670

Zampella EJ, Bradley EL, Pretlow TG (1982) Glucose-6-phosphate dehydrogenase: a possible clinical indicator for prostatic carcinoma. Cancer 49:384–387

Zhang W, Patil S, Chauhan B, Guo S, Powell DR, Le J, Klotsas A, Matika R, Xiao X, Franks R, Heidenreich KA, Sajan MP, Farese RV, Stolz DB, Tso P, Koo SH, Montminy M, Unterman TG (2006) FoxO1 regulates multiple metabolic pathways in the liver: effects on gluconeogenic, glycolytic, and lipogenic gene expression. J Biol Chem 281:10105–10117

Zhao JJ, Gjoerup OV, Subramanian RR, Cheng Y, Chen W, Roberts TM, Hahn WC (2003) Human mammary epithelial cell transformation through the activation of phosphatidylinositol 3-kinase. Cancer Cell 3:483–495

Zhong H, Chiles K, Feldser D, Laughner E, Hanrahan C, Georgescu MM, Simons JW, Semenza GL (2000) Modulation of hypoxia-inducible factor 1alpha expression by the epidermal growth factor/phosphatidylinositol 3-kinase/PTEN/AKT/FRAP pathway in human prostate cancer cells: implications for tumor angiogenesis and therapeutics. Cancer Res 60:1541–1545

Zhou QL, Jiang ZY, Holik J, Chawla A, Hagan GN, Leszyk J, Czech MP (2008) Akt substrate TBC1D1 regulates GLUT1 expression through the mTOR pathway in 3T3-L1 adipocytes. Biochem J 411:647–655

Zhu J, Blenis J, Yuan J (2008) Activation of PI3 K/Akt and MAPK pathways regulates Myc-mediated transcription by phosphorylating and promoting the degradation of Mad1. Proc Natl Acad Sci USA 105:6584–6589

Zundel W, Schindler C, Haas-Kogan D, Koong A, Kaper F, Chen E, Gottschalk AR, Ryan HE, Johnson RS, Jefferson AB, Stokoe D, Giaccia AJ (2000) Loss of PTEN facilitates HIF-1-mediated gene expression. Genes Dev 14:391–396

MYC, Metabolic Synthetic Lethality, and Cancer

Annie L. Hsieh and Chi V. Dang

Abstract

The MYC oncogene plays a pivotal role in the development and progression of human cancers. It encodes a transcription factor that has broad reaching effects on many cellular functions, most importantly in driving cell growth through regulation of genes involved in ribosome biogenesis, metabolism, and cell cycle. Upon binding DNA with its partner MAX, MYC recruits factors that release paused RNA polymerases to drive transcription and amplify gene expression. At physiologic levels of MYC, occupancy of high-affinity DNA-binding sites drives 'house-keeping' metabolic genes and those involved in ribosome and mitochondrial biogenesis for biomass accumulation. At high oncogenic levels of MYC, invasion of low-affinity sites and enhancer sequences alter the transcriptome and cause metabolic imbalances, which activates stress response and checkpoints such as p53. Loss of checkpoints unleashes MYC's full oncogenic potential to couple metabolism with neoplastic cell growth and division. Cells

A.L. Hsieh · C.V. Dang (✉)
Abramson Family Cancer Research Institute, Perelman School of Medicine, University of Pennsylvania, Philadelphia, PA, USA
e-mail: dangvchi@exchange.upenn.edu

A.L. Hsieh · C.V. Dang
Abramson Cancer Center, Perelman School of Medicine, University of Pennsylvania, Philadelphia, PA, USA

A.L. Hsieh
Department of Pathology, Johns Hopkins University School of Medicine, Baltimore, MD, USA

C.V. Dang
Division of Hematology-Oncology, Department of Medicine, Perelman School of Medicine, University of Pennsylvania, Philadelphia, PA, USA

© Springer International Publishing Switzerland 2016
T. Cramer and C.A. Schmitt (eds.), *Metabolism in Cancer*,
Recent Results in Cancer Research 207, DOI 10.1007/978-3-319-42118-6_4

that overexpress MYC, however, are vulnerable to metabolic perturbations that provide potential new avenues for cancer therapy.

Keywords

MYC oncogene · Metabolism · Glucose · Glutamine · Ribosome biogenesis · Cancer therapy

1 Background

The classical metabolic pathway maps appear intimidating due to their complexity and the number of pathways that crisscross over one another, connecting substrates to products. However, the metabolic map is in fact quite elegant, illustrating the central bioenergetic metabolic pathways involving glycolysis and the Krebs or tricarboxylic acid (TCA) cycle. Flanking these pathways are those involved in carbohydrate, amino acid, lipid, and nucleotide metabolism, which are all connected to glycolysis and the TCA cycle. In addition, other organelles carry specific functions, such as peroxisomes, that have their unique compartmentalized pathways and chloroplasts that carry out reactions involved in photosynthesis in plants. A global view of the beauty of these metabolic pathways reveals that they have evolved to capture energy from sunlight by plants and store this energy as organized life substances, most importantly glucose, to feed other organisms on earth. Organisms use the other pathways for homeostasis in nonproliferative cells by generating energy in the form of ATP to maintain membrane potentials and support maintenance biosynthesis to replace damaged or aged molecules resulting partly from oxidative stress. Part-and-parcel of cell metabolism is the generation of by-products, chief among which are reactive oxygen species (ROS), which must be controlled and managed for proper cell homeostasis and cell proliferation or otherwise ROS could wreak havoc on cells.

Resting cells use metabolic pathways for maintenance biosynthesis that contrasts with metabolic rewiring that supports cell growth and proliferation (Dang 2013). Cell growth is defined herein as the ability of cells to increase cell mass and volume, whereas cell proliferation refers to cell division (Jorgensen and Tyers 2004). Normal cells use these metabolic pathways as do cancer cells' however, Otto Warburg observed over 90 years ago that cancer cells or tissues tend to convert high amounts of glucose to lactate, termed aerobic glycolysis, which occurs in the presence of oxygen (Warburg 1956a, b). Normal tissues tend to use glucose oxidation for maintenance and resort to anaerobic glycolysis or the conversion of glucose to lactate under hypoxic conditions, whereas many cancer tissues undergo high aerobic glycolysis or the Warburg effect (Warburg 1956a, b). The Warburg effect had been a matter of obsession for many investigators over many decades, but confusion blurred the field of cancer metabolism when Warburg, in his later years, dogmatically claimed that the cause of cancer is due to dysfunctional mitochondria that results in aerobic glycolysis (Warburg 1956b). Now, with a significant expansion of interest and resurgence of research in cancer metabolism, we know that the Warburg effect does not cause cancer per se, but it is characteristic of many growing, proliferating cancer cells. The oversimplification of the role of the Warburg effect in cancer is now replaced by new insights that connect oncogenes and tumor suppressors with metabolic pathways (Kroemer and Pouyssegur 2008). In fact, there are specific instances where alterations in metabolism can and do initiate tumorigenesis. The intimate relationships between metabolism, biosynthesis, signaling, and epigenetic control of gene expression are being uncovered in the past decade, yielding a much deeper appreciation of the regulation of metabolism and its role in cancer.

To understand cancer metabolism, we need to return to the basic principles of the regulation of cell growth and proliferation and nutrient sensing. The unicellular eukaryotic yeast, *Saccharomyces cerevisiae*, has been deeply studied and it has sensing mechanisms responding to exposure to glucose or glutamine that triggers the RAS and TOR pathways, which inactivates repressors of ribosome biogenesis genes (Lippman and Broach 2009). Activation of ribosome synthesis accounts for the bulk of nutrient influx into the growing yeast cells. Nutrients are taken up and converted through glycolysis and the TCA cycle to generate ATP and building blocks, such as amino acids, fatty acids, cholesterol, and nucleotides, for the growing cell. An increase in ribosomes allows growing cells to translate mRNAs to produce components of the cells, as instructed by the DNA sequence through transcriptional activation, which amplifies the signal for growth upon nutrient exposure. Ribosomes are not only factories of cellular macromolecules, but they sit at the crossroad of protein synthesis and nucleic acid synthesis, made from many protein subunits and several structural and catalytic RNAs (van Riggelen et al. 2010).

Mammalian cells in adult animals are largely nonproliferative. But even in the proliferative compartments, which comprise less than 1 % of cells, mammalian cells do not proliferate in response to nutrients, but rather also require extracellular cues in the form of growth factors or contact with the extracellular matrix to initiate a cell growth program. Indeed, it appears that both growth factors and adequate nutrients are required for mammalian cell proliferation (Dang 2012a). In this regard, once stimulated with growth factors, mammalian cells behave like yeast cells exposed to nutrients. Nutrients are sensed by mTOR, the mammalian ortholog of the yeast TOR (Howell and Manning 2011; Efeyan et al. 2012). mTOR, in essence, serves as an acute sensor of amino acids that triggers a program of growth through protein kinase cascades which stimulate protein synthesis, lipid, and nucleic acid synthesis. mTOR signaling is also instrumental in amplifying the effects of growth factors by catalyzing the synthesis of transcription factors that are activated by growth factors at the transcriptional level. In particular, the MYC proto-oncogene responds to growth factors by producing more transcripts, whose translation is regulated and stimulated by mTOR. Thus, upon exposure to growth factors and nutrients, mTOR is activated acutely by nutrient import and the growth program is propagated by the transcriptional activation of key transcription factors, such as MYC, to further drive cell growth in the G1 phase of the cell cycle until a critical cell mass is reached for cells to enter into S-phase of the cell cycle and initiate DNA replication.

2 Regulation of MYC and Its Function

MYC is probably the most frequently amplified human oncogene as gleaned from
The Cancer Genome Atlas (TCGA) (Ciriello et al. 2013). The MYC
proto-oncogene is downstream of growth factor signaling pathways, whereby ERK
is likely to play a role in its transcriptional activation (McCubrey et al. 2007).
Concurrent mTOR activation by nutrients increases MYC mRNA translation and
the protein is activated by GSK3-mediated phosphorylation at T58 that is followed
by a subsequent phosphorylation by ERK at S62, which is required for MYC's
transcriptional activity (Gregory et al. 2003).

MYC dimerizes through its helix-loop-helix-leucine zipper (HLH-Zip) domain
with the basic-HLH-Zip partner, MAX, to bind DNA sequences enriched in promoter regions of genes (Dang 2012b). The high-affinity sites are termed E-boxes
with the consensus sequence 5'-CACGTG-3' (Fernandez et al. 2003; Blackwell
et al. 1993). MYC-MAX could also bind noncanonical binding sites as well as
lower-affinity nonconsensus sites. Upon DNA binding, the unstructured
DNA-binding basic domains take on helical shapes to anchor onto the DNA major
grooves. After DNA binding, MYC recruits pTEFb to the promoter region and
activates RNA polymerase and releases it from a pausing state to increase the rate of
gene transcription (Rahl et al. 2010).

Several studies suggested that MYC amplifies the expression of all genes
expressed in a cell whether it is resting or proliferating (Lin et al. 2012; Nie et al.
2012). However, it stands to reason that there must be selective, varied quantitative
increases in mRNA production that ensures stoichiometric increases in the synthesis
of components of the cell for growth. Further, the expression of cell growth arrest
genes in resting cells would not be amplified as cells exit from a resting state and
enter a proliferative state. Recent and past studies support the notion that while
MYC could amplify gene expression, it does so in collaboration with other transcription factors and subject to chromatin structure and accessibility for MYC
binding (Lin et al. 2012; Nie et al. 2012; Walz et al. 2014; Sabo et al. 2014; Chen
et al. 2008). For example, computational analysis of MYC target genes revealed
enriched transcription factor-binding motifs associated with MYC binding (Elkon
et al. 2004). Further, the transcriptional modulator WDR5 can guide MYC target
gene recognition and contribute to tumorigenesis (Thomas et al. 2015). The notion
of selective amplification of gene expression by MYC is further underscored by
studying how viruses usurp cellular transcription to control metabolism. Specifically, adenovirus infection was found to enhance glycolysis of infected MCF10A
breast epithelial cells (Thai et al. 2014). Through mutational analysis of the adenoviral genome, E4ORF1 was documented to be sufficient to drive glycolysis.
Intriguingly, E4ORF1 directly interacts with MYC and enhances its binding to
selective targets. Selective activation of MYC target genes, such as hexokinase 2,
drives increased glycolysis and promotes nucleotide metabolism.

Collectively, it is intriguing to note that MYC binds to many more sites and genes than it can induce changes in mRNA levels and its occupancy is heavily determined by relative levels of MYC (Fig. 1) (Wolf et al. 2014). As cells are stimulated to proliferate, the induction of moderate normal levels of MYC would hypothetically result in the occupancy of high-affinity E-box containing genomic sites. However, not all MYC-occupied binding sites result in transcriptional activation, which appears to be influenced by other transcription factors nearby or enhancer sequences far away. As MYC protein levels exceed levels found in normal cells, it could then bind lower-affinity binding sites and invade enhancer sequences (Wolf et al. 2014). This promiscuous activity results in illegitimate nonlinear increases in gene expression (Fig. 1) that hypothetically result in disruption of the circadian molecular clock, imbalanced metabolism, and non-stoichiometric production of cellular components, leading to cellular stress such as the unfolded protein response (UPR) and increased ROS (Hart et al. 2012; Altman et al. 2015). Thus, overexpression of MYC can trigger cell growth arrest or apoptosis through the activation of checkpoints such as ARF and p53 (Eischen et al. 1999). But in multistep tumorigenesis, loss of tumor suppressors such as p53 or PTEN is common (Kim et al. 2012). Loss of p53 can unleash MYC's full oncogenic transcriptional activity by diminishing cell death. Loss of PTEN allows for PI3 K activation to collaborate with MYC. Indeed, the classical MYC-driven Burkitt's lymphoma is associated with loss of p53 in up to 40 % of cases and the PI3 K pathways are frequently activated (Farrell et al. 1991).

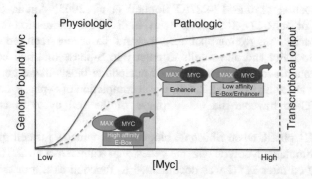

Fig. 1 Hypothetical illustration of the effect of Myc level on genome bound Myc and the transcriptional output of its target genes. *Solid line* represents the number of genes in the genome that are bound by Myc and the *dashed line* represents the level of transcriptional output from Myc-bound genes. In the physiologic condition, when the concentration of Myc is low, MYC-MAX heterodimer preferentially binds to the high-affinity E-box containing genes and transactivates the expression of some, but not all, genes bound by Myc. Conversely, in the pathological setting, such as cancer, where Myc level is elevated, MYC-MAX heterodimer binds to the enhancers ('enhancer invasion') and the low-affinity E-box containing genes in addition to the high-affinity target genes to induce a broader and greater transcriptional output for cell growth and proliferation

3   MYC Target Genes and Metabolism

The model that MYC amplifies expression of genes that are expressed is consistent with the ability of MYC to increase the expression of so-called house-keeping genes such as those involved in metabolism and ribosome biogenesis. It is surmised that these genes are required for virtually all nucleated cells to maintain biosynthesis and homeostasis, and hence, their chromatin structures are opened and accessible to transcription factors. Intriguingly, canonical MYC E-boxes are enriched in these genes, suggesting that MYC's affinity for these DNA-binding sites drives the expression of metabolic and ribosome biogenesis genes at physiological MYC levels, which favor high-affinity binding sites (van Riggelen et al. 2010; Dang et al. 2008). As MYC protein levels increased upon stimulation of cell growth, other genes are also bound such as those encoding other transcription factors, for example, E2F, microRNAs—particularly the miR-17-92a cluster, or long noncoding RNAs or lncRNAs to fine-tune the transcriptional program induced by MYC (Sears et al. 1997; Hung et al. 2014; O'Donnell et al. 2005; Aguda et al. 2008).

Transcriptional positive and negative feed-forward loops (Alon 2007) allow for delayed MYC transcriptional regulation of genes involved in S-phase when MYC is activated early in G1 phase of the cell cycle. For example, MYC is well documented to induce the expression of the transcription factor E2F1 as part of a program for cell entry into S-phase once cell growth has reached a critical point (Wong et al. 2011). MYC also potently induces the miR-17-92 microRNA cluster that attenuates E2F expression, resulting in a biphasic expression of E2F1 that is increased with physiological MYC and diminished with high levels of MYC through induction of miR-17-92 (O'Donnell et al. 2005; Aguda et al. 2008). Abrogation of miR-17-92 together with high MYC expression caused DNA replication stress, suggesting that feed-forward loops are required to attenuate MYC's overactivity and allow cells to enter into S-phase that had been properly prepared through adequate stimulation of nucleotide biosynthetic genes by MYC (Pickering et al. 2009; Liu et al. 2008). Upon completion of S-phase, cells appear to require MYC to traverse the G2-M phase of the cell cycle to complete cell replication.

In early G1 phase, when ribosome biogenesis dominates for cell growth, MYC appears to stimulate glycolytic enzyme genes (van Riggelen et al. 2010; Dang et al. 2008). Early on after MYC was documented to function as a transcription factor, efforts toward identifying MYC targets revealed through subtraction cloning that MYC could bind to the promoter and activate the expression of lactate dehydrogenase A (LDHA) among several dozens of genes that responded to MYC overexpression (Zeller et al. 2006). In subsequent years, more detailed studies revealed that MYC regulates glucose transporters and hexokinase 2 (HK2) prominently, as well as virtually all genes involved in glycolysis, except for a few such as PGAM, a target of p53 (Osthus et al. 2000; Berkers et al. 2013). In fact, examination of the promoter regions of these glycolytic genes revealed phylogenetically conserved

MYC E-boxes that are bound by MYC and presumably through which MYC activates expression of the corresponding genes (Kim et al. 2004). Further, MYC-binding sites were found to overlap with the binding sites of the hypoxia inducible factor 1 (HIF-1), which would increase the expression of glycolytic genes under oxygen-deprived conditions to increase anaerobic glycolytic as a means to adapt and survive hypoxia (Semenza et al. 1996; Iyer et al. 1998; Seagroves et al. 2001). This observation also suggests that a collaboration between deregulated MYC and HIF-1 in stimulating glycolytic gene expression could provide an advantageous edge for cancer cells to survive in the hypoxic tumor microenvironments (Kim et al. 2006).

Beyond glycolysis, MYC was also shown to increase iron metabolism through its induction of the transferrin receptor, which was recently exploited for MYC-driven prostate cancer imaging directed at this receptor (O'Donnell et al. 2006; Holland et al. 2012). MYC also stimulates glutaminolysis, which was hinted by the fact that deregulated MYC not only sensitizes cells to glucose-deprivation-induced death, but also renders cells addicted to glutamine (Yuneva et al. 2007). Two independent studies documented the role of MYC in regulating glutaminolysis (Wise et al. 2008; Gao et al. 2009). One study was driven by this earlier observation of glutamine dependency of MYC-driven cells; it documented the increase in expression of genes involved in glutamine metabolism (Wise et al. 2008). The other study set out to determine whether MYC could alter the mitochondrial proteome, particularly since MYC was shown to induce mitochondrial biogenesis (Gao et al. 2009). Using purified mitochondria from a MYC-inducible human B cell line, proteomic analysis revealed that a dozen or more protein spots on high-resolution 2-D PAGE gels were increased in the mitochondrial proteome upon MYC induction. One such protein was found to be glutaminase by mass spectrometry. These studies led to the concept that MYC not only induces glycolysis, but it also induces glutaminolysis in growing, proliferating cells. In fact, MYC could induce the Warburg effect concurrently with increased TCA cycling and oxygen consumption. This obviates the misconception that the Warburg effect occurs in cancer and growing cells exclusive of oxidative phosphorylation.

As mentioned, several studies documented that MYC can induce mitochondrial biogenesis by directly regulating genes involved in mitochondria structure and function as well as those required for mitochondrial biogenesis, such as PGC1β (Li et al. 2005; Zhang et al. 2007). Knockdown of PGC1β significantly diminished the ability of MYC to increase mitochondrial mass in the P493 MYC-inducible Burkitt's lymphoma model cell line. Mitochondria are critically important for glucose and glutamine metabolism, which contributes to ATP production and the generation of amino acids, lipids, and nucleotides. Indeed two early studies documented in many systems that MYC could directly induce genes involved in nucleotide metabolism, involving both purine and pyrimidine biosyntheses (Liu et al. 2008; Mannava et al. 2008). The TCA cycle is critically important for de novo purine and pyrimidine synthesis to support growing cells. Glucose provides a source for glycine, which is required for purine synthesis. Further glucose through its conversion to serine contributes to single-carbon metabolism involving tetrahydrofolate cycling and nucleotide synthesis (Amelio et al. 2014). Glucose is also the source for ribose through the pentose phosphate pathway. Nitrogen donation by glutamine is essential for purine synthesis that is highly dependent on folate-mediated single-carbon metabolism. Further glutamine is converted through the TCA cycle to oxaloacetate, which is converted to aspartate by glutamate–oxaloacetate transaminase (GOT). Aspartate is used for de novo pyrimidine synthesis for growing cells. Intriguingly, GOT2 appears to be critical for nucleotide biosynthesis in breast cancer cells in a MYC-dependent fashion (Korangath et al. 2015). Genes involved in these pathways behave as MYC targets.

It is particularly notable the MYC was first documented to directly induce the expression of carbamoyl phosphate synthetase (CAD), the first gene involved in pyrimidine synthesis (Miltenberger et al. 1995). CAD is intriguing in that it is also activated via phosphorylation by mTOR (Ben-Sahra et al. 2013; Robitaille et al. 2013). Hence, nutrient sensing and growth factor stimulation converges at a key enzyme involved in nucleotide synthesis. Another MYC target key enzyme, dihydroorotate dehydrogenase (DHODH) that is involved in pyrimidine synthesis, requires coupling with the mitochondrial electron transport chain for the conversion of dihydroxyorotate to orotic acid (Liu et al. 2008). Orotate is a precursor for pyrimidine synthesis.

In addition to nucleotide synthesis, the mitochondrion is instrumental for fatty acid and cholesterol synthesis, because citrate produced by the TCA cycle is exported into the cytosol and then converted to acetyl-CoA and oxaloacetate by acetyl-CoA lyase (ACLY). MYC affects lipid metabolism and also induces the key enzymes, fatty acid synthase (FASN), and stearoyl-CoA desaturase (SCD), which are vital for fatty acid chain elongation and monosaturation to produce palmitate, for example, and oleate, respectively (Zeller et al. 2006; Carroll et al. 2015; Eberlin et al. 2014; Zirath et al. 2013). Monosaturation of fatty acids is critical for limiting lipotoxicity that is associated with unsaturated fatty acids such as palmitate, which induces the unfolded protein response (Hatanaka et al. 2014). In this regard, SREBP plays an important role in the transactivation of SCD such that loss of SREBP function results in loss of cancer cellular viability (Williams et al. 2013). Because MYC can drive glucose and glutamine into the production of long-chain fatty acids through FAS, the corresponding increase in SCD activity is required to diminish lipotoxicity. Thus, it is surmised that an interplay between MYC and SREBP is critical for growing cells. This notion further underscores the complexity of MYC-mediated transcription, which does not simply result in an overall nondiscriminating amplification of gene expression. From many gene expression profiles in response to MYC overexpression, it could be gleaned that some of the genes involved in cholesterol synthesis also respond to MYC; however, whether this results from a collaboration between MYC and the SREBPs remains to be established.

4 MYC, Protein Synthesis, and Ribosomal Biogenesis

One of the MYC target genes identified early on was eIF4E, which is critically important for translational initiation, thereby linking MYC to protein synthesis (Jones et al. 1996). Subsequent studies further revealed that enzymes involved in amino acid transport and synthesis, such as proline, and aminoacyl-tRNA synthetases are robust targets of MYC (Wise et al. 2008; Gao et al. 2009; Liu et al. 2012; Coller et al. 2000). Over the last decade, the role of MYC in protein synthesis also became more apparent when *Drosophila* dMyc was linked to ribosomal biogenesis (Gallant et al. 1996; Schreiber-Agus et al. 1997; Johnston et al. 1999). Specifically, the small fly body size associated with the genetic complementation group termed Minutes contains genotypes that comprise both hypomorphic alleles of dMyc and mutations in ribosomal protein genes (Kongsuwan et al. 1985). The group of flies with the same phenotype of having a small body size resulting from small cell size implies that dMyc is linked to ribosomal biogenesis. Studies in mammalian B cells and subsequently in mouse livers acutely transduced with adenovirally mediated human MYC expression indicate that MYC can induce ribosomal protein genes (Kim et al. 2000). Later, MYC was documented to be able to induce transcription from genes driven by RNA polymerases I and III in addition to its previously recognized role in RNA polymerase II-mediated transcription (van Riggelen et al. 2010; Grandori et al. 2005; Gomez-Roman et al. 2003). Importantly, all three RNA polymerases are required for the production of ribosomes, comprising ribosomal proteins and RNA Pol I-induced rRNAs and Pol III-induced 5S RNA. Thus, the ribosomes lie at the nexus of all three RNA polymerases as well as being produced from both glucose and glutamine, coupling protein and nucleotide synthesis. In this regard, it is notable that MYC was shown to induce PRPS2, which initiates purine synthesis and is regulated by both transcription and protein syntheses (Cunningham et al. 2014).

Intriguingly, ribosome biogenesis also appears to play a nutrient-sensing role (Mayer and Grummt 2006). rRNA synthesis is sensitive to nutrient availability. Further, a putative imbalance in rRNA synthesis compared to ribosomal protein synthesis could result in excess ribosomal subunits, such as RPL11 and RPL5, which can bind to MDM2 and sequester it away from p53, resulting in activation of p53 (Lohrum et al. 2003; Dai and Lu 2004). In this manner, imbalances in ribosome biogenesis resulting from nutrient stress induced by overexpression of MYC, for example, could induce p53 function to halt cells from progressing in the cell cycle unless nutrients are adequate. Overall, the ability of MYC to drive glycolysis and glutaminolysis, and in turn drive protein and nucleic acid synthesis, is important to ensure adequate building supplies for ribosomal biogenesis. Concurrent mitochondrial biogenesis ensures that the growing cell is adequately powered by both ATP production and increased biosynthetic flux through mitochondria (Li et al. 2005). Thus, ribosome biogenesis is central to MYC function as gleaned from a MYC signature derived from rigorous overlap of many MYC target gene studies. This MYC signature comprises 51 genes, which through gene set enrichment

analysis revealed ribosome biogenesis as the most enrich set (Ji et al. 2011) Biologically, the role of MYC in ribosome biogenesis is important both for the
physiology and for the pathophysiology of cancer. A recent study of hypomorphic
Myc mice carrying one allele of Myc revealed that they live longer than their
wild-type counterparts (Hofmann et al. 2015). Gene expression profiles uncover a
decrease in ribosomal protein and biogenesis genes, suggesting that reduced protein
synthesis and energy demand may underlie the prolonged longevity phenotype of
these Myc hypomorphic mice. The prolonged longevity of these mice is reminis-
cent of the effects of metformin which inhibits mitochondrial function and rapamycin that inhibits mTORC1 (Harrison et al. 2009; Martin-Montalvo et al. 2013).
Intriguingly, loss of one copy of RPL24 significantly delayed MYC-induced
lymphoma in transgenic mice, further attesting to the role of MYC-induced ribosome biogenesis in cell growth and tumorigenesis (Barna et al. 2008). Inhibition of
RNA Pol I activity has also been shown to delay tumorigenesis, further adding to
the overarching view that MYC-induced ribosomal biogenesis is part-and-parcel of
cell growth and tumorigenesis (Hein et al. 2013).

5 MYC Metabolic Synthetic Lethal Interactions and Therapy

Constitutive activation of MYC in certain types of cancers could result in deregulated biomass accumulation that renders cancer cells addicted to nutrients. In this
regard, early studies revealed that MYC-overexpressing cells caused them to
become addicted to glucose or glutamine, such that nutrient deprivation led to cell
death (Yuneva et al. 2007; Shim et al. 1998). Similarly, when nutrient sensing is
uncoupled from biomass accumulation in yeast by genetic manipulation, mutant
yeast cells that are constitutive for ribosome biogenesis become nutrient-addicted
and are nonviable with glucose or glutamine deprivation (Dang 2012a). As such,
overexpression of MYC in mammalian cells is surmised to render them addicted to
certain metabolic pathways (Fig. 2a).

As mentioned, oncogenic overexpression of MYC results in a qualitatively
different transcriptome than that induced by normal physiological MYC levels
found in the growing, normal cells. The basis for this difference is depicted in
Fig. 1, which illustrates the nonlinear increase in gene expression as MYC
expression exceeds the normal range, in which MYC begins to bind low-affinity
binding sites and invade enhancer in a manner that selective amplifies the
expression of enhancer-regulated genes over those that are unassociated with an
enhancer. The imbalance in expressions of specific target genes results in nonsto-
ichiometric production of components of a multi-subunit structure, for example, the
ribosome or multienzyme complexes. The nonstoichiometry may result in protein
misfolding that triggers the stress-unfolded protein response (UPR). The imbalanced expression of RNA splicing factors can result in an imbalance in the
expression of alternative spliced mRNA variants and imbalanced in the mixture of
the resultant proteins (Koh et al. 2015). Hence, along with MYC-driven nutrient

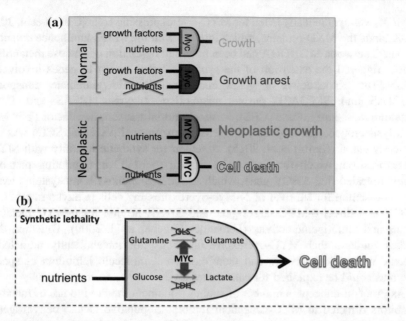

Fig. 2 Schematic of nutrient addiction and synthetic lethality in Myc-overexpressing neoplastic cells. Normal cells need both growth factors and nutrients to initiate growth. In the absence of nutrient, normal cells enter a state of arrest. High level of Myc converts cancer cells to become addicted to nutrients, so that withdrawal of nutrients leads to cell death. Similar to nutrient withdrawal, blocking key metabolic pathways, such as lactate dehydrogenase (LDHA) and glutaminase (GLS), is synthetic lethal for Myc-driven neoplastic cells

addiction to increase biosynthesis, the imbalance metabolic state due to nonlinearity of gene expression amplification could be exploited to uncover synthetic lethal interactions (Fig. 2b).

A number of synthetic lethality screens have been performed with MYC or MYCN overexpression, revealing numerous targets that seem to be cell system dependent (Carroll et al. 2015; Toyoshima et al. 2012; Cermelli et al. 2014). While the overlap between the different synthetic lethal screens is minimal or sometimes nonexistent, several themes have emerged. For example, the nonmetabolic targets such as Aurora kinases have appeared in multiple systems (Yang et al. 2010; den Hollander et al. 2010). Although metabolic targets have appeared in these screens, there is a lack of consistency between the systems that may result from the lack of robustness of the screening (i.e., ineffective knock down), off-target effects of interference RNAs, or from cell type specificities. Nonetheless, a number of interesting metabolic targets have emerged. Using human fibroblasts, a screen revealed that loss of glucose metabolism genes (aldolase A (ALDOA) and pyruvate dehydrogenase kinase 1 (PDK1)), nucleotide metabolism gene (CTPS), or nutrient transporters (SLC1A4 and SLC25A6) had synthetically lethal interactions with MYC overexpression (Toyoshima et al. 2012). A screen directed at the extended MYC transcription factor network revealed that knockdown of MONDOA, a partner of MLX, was synthetically lethal for MYC-overexpressing cells (Carroll et al. 2015). MLX binds the MXD proteins, which interact with the MAX, and hence extending the MYC network. MONDOA had been linked to regulation of glucose metabolism, which triggered the extension of the synthetic lethal screen to genes involved in metabolism. Knockdown of genes encoding glutamine/glutamate transporters (SLC1A5 and SLC3A2), purine metabolism enzymes (PFAS and CAD), cystathionine-beta-synthase (CBS), a mitochondrial transcription factor (TFAM), a glycolysis enzyme (ENO3), and lipogenesis enzymes (FASN and SCD) was synthetically lethal (Carroll et al. 2015). A screen for synthetic lethality with MYCN overexpression, which functions quite similar to MYC in regulating metabolic genes, revealed that AHCY gene, which encodes S-adenosylhomocysteine hydrolase, is essential for survival of N-Myc-overexpressing cells (Chayka et al. 2015). AHCY is a direct c-MYC target gene and SAHH is vital for c-MYC metabolic effects and tumorigenic activity (Fernandez-Sanchez et al. 2009). Together, these studies indicate that MYC-overexpressing cells are metabolically addicted to specific metabolic pathways and therefore, small molecule inhibitors of specific enzymes could be exploited for cancer therapy.

As proof-of-concept, a number of studies document the in vivo use of metabolic inhibitors directed at MYC-dependent tumors. A putative lactate dehydrogenase inhibitor, FX11, was shown to reduce a MYC-dependent lymphomagenesis in a xenograft model (Le et al. 2010). This study also illustrates the tumor-reducing activity of FK866, an inhibitor of NAMPT that is also a MYC target (Le et al. 2010). Further, glutaminase inhibitors, 986 and BPTES, have inhibitory activities in this model (Le et al. 2012; Xiang et al. 2015). BPTES, a potent allosteric inhibitor of glutaminase, also has an in vivo effect that prolongs the survival of an aggressive model of MYC-inducible liver cancer and renal cell carcinoma (Xiang et al. 2015; Shroff et al. 2015). An inhibitor of the MYC target lactate transporter, MCT1, also displayed in vivo activity (Doherty et al. 2014; Sonveaux et al. 2008). Collectively, these studies provide evidence that targeting metabolism is feasible, but the findings also suggest that understanding metabolic rewiring is required for strategic combination therapies to have a clinical benefit.

6 Concluding Remarks

Although a deeper understanding of MYC's molecular function in physiologic versus pathologic conditions requires additional investigations, the collective knowledge points to a transcription factor that can amplify the expression of thousands of genes, which are involved in many cellular processes. Key metabolic pathways driving cell growth are regulated by MYC so that biomass accumulation for cell proliferation is coupled with nutrient supplies. High oncogenic levels of MYC result in a nonlinear amplification of the normal transcriptome that causes imbalanced metabolism, cellular stress, and elicits checkpoints. Loss of checkpoints and adaptation to stress unleashes MYC's oncogenic potential to drive uncontrolled cell growth and neoplasia. Oncogenic MYC also renders cells vulnerable to metabolic perturbations as revealed by synthetic lethality screens, which have generated new opportunities for cancer therapy in addition to the key metabolic pathways that have been identified through hypothesis-driven research.

References

Aguda BD, Kim Y, Piper-Hunter MG, Friedman A, Marsh CB (2008) MicroRNA regulation of a cancer network: consequences of the feedback loops involving miR-17-92, E2F, and Myc. Proc Natl Acad Sci USA 105(50):19678–19683. doi:10.1073/pnas.0811166106

Alon U (2007) Network motifs: theory and experimental approaches. Nat Rev Genet 8(6):450–461. doi:10.1038/nrg2102

Altman BJ, Hsieh AL, Sengupta A, Krishnanaiah SY, Stine ZE, Walton ZE, Gouw AM, Venkataraman A, Li B, Goraksha-Hicks P, Diskin SJ, Bellovin DI, Simon MC, Rathmell JC, Lazar MA, Maris JM, Felsher DW, Hogenesch JB, Weljie AM, Dang CV (2015) MYC disrupts the circadian clock and metabolism in cancer cells. Cell Metab. doi:10.1016/j.cmet. 2015.09.003

Amelio I, Cutruzzola F, Antonov A, Agostini M, Melino G (2014) Serine and glycine metabolism in cancer. Trends Biochem Sci 39(4):191–198. doi:10.1016/j.tibs.2014.02.004

Barna M, Pusic A, Zollo O, Costa M, Kondrashov N, Rego E, Rao PH, Ruggero D (2008) Suppression of Myc oncogenic activity by ribosomal protein haploinsufficiency. Nature 456 (7224):971–975. doi:10.1038/nature07449

Ben-Sahra I, Howell JJ, Asara JM, Manning BD (2013) Stimulation of de novo pyrimidine synthesis by growth signaling through mTOR and S6K1. Science 339(6125):1323–1328. doi:10.1126/science.1228792

Berkers CR, Maddocks OD, Cheung EC, Mor I, Vousden KH (2013) Metabolic regulation by p53 family members. Cell Metab 18(5):617–633. doi:10.1016/j.cmet.2013.06.019

Blackwell TK, Huang J, Ma A, Kretzner L, Alt FW, Eisenman RN, Weintraub H (1993) Binding of myc proteins to canonical and noncanonical DNA sequences. Mol Cell Biol 13(9):5216–5224

Carroll PA, Diolaiti D, McFerrin L, Gu H, Djukovic D, Du J, Cheng PF, Anderson S, Ulrich M, Hurley JB, Raftery D, Ayer DE, Eisenman RN (2015) Deregulated Myc requires mondoA/Mlx for metabolic reprogramming and tumorigenesis. Cancer Cell 27(2):271–285. doi:10.1016/j. ccell.2014.11.024

Cermelli S, Jang IS, Bernard B, Grandori C (2014) Synthetic lethal screens as a means to understand and treat MYC-driven cancers. Cold Spring Harbor Perspect Med 4(3). doi:10. 1101/cshperspect.a014209

Chayka O, D'Acunto CW, Middleton O, Arab M, Sala A (2015) Identification and pharmacological inactivation of the MYCN gene network as a therapeutic strategy for neuroblastic tumor cells. J Biol Chem 290(4):2198–2212. doi:10.1074/jbc.M114.624056

Chen X, Xu H, Yuan P, Fang F, Huss M, Vega VB, Wong E, Orlov YL, Zhang W, Jiang J, Loh YH, Yeo HC, Yeo ZX, Narang V, Govindarajan KR, Leong B, Shahab A, Ruan Y, Bourque G, Sung WK, Clarke ND, Wei CL, Ng HH (2008) Integration of external signaling pathways with the core transcriptional network in embryonic stem cells. Cell 133(6):1106–1117. doi:10.1016/j.cell.2008.04.043

Ciriello G, Miller ML, Aksoy BA, Senbabaoglu Y, Schultz N, Sander C (2013) Emerging landscape of oncogenic signatures across human cancers. Nat Genet 45(10):1127–1133. doi:10.1038/ng.2762

Coller HA, Grandori C, Tamayo P, Colbert T, Lander ES, Eisenman RN, Golub TR (2000) Expression analysis with oligonucleotide microarrays reveals that MYC regulates genes involved in growth, cell cycle, signaling, and adhesion. Proc Natl Acad Sci USA 97(7):3260–3265

Cunningham JT, Moreno MV, Lodi A, Ronen SM, Ruggero D (2014) Protein and nucleotide biosynthesis are coupled by a single rate-limiting enzyme, PRPS2, to drive cancer. Cell 157 (5):1088–1103. doi:10.1016/j.cell.2014.03.052

Dai MS, Lu H (2004) Inhibition of MDM2-mediated p53 ubiquitination and degradation by ribosomal protein L5. J Biol Chem 279(43):44475–44482. doi:10.1074/jbc.M403722200

Dang CV (2012a) Links between metabolism and cancer. Genes Dev 26(9):877–890. doi:10.1101/gad.189365.112

Dang CV (2012b) MYC on the path to cancer. Cell 149(1):22–35. doi:10.1016/j.cell.2012.03.003

Dang CV (2013) MYC, metabolism, cell growth, and tumorigenesis. Cold Spring Harb Perspect Med 3 (8). doi:10.1101/cshperspect.a014217

Dang CV, Kim JW, Gao P, Yustein J (2008) The interplay between MYC and HIF in cancer. Nat Rev Cancer 8(1):51–56. doi:10.1038/nrc2274

den Hollander J, Rimpi S, Doherty JR, Rudelius M, Buck A, Hoellein A, Kremer M, Graf N, Scheerer M, Hall MA, Goga A, von Bubnoff N, Duyster J, Peschel C, Cleveland JL, Nilsson JA, Keller U (2010) Aurora kinases A and B are up-regulated by Myc and are essential for maintenance of the malignant state. Blood 116(9):1498–1505. doi:10.1182/blood-2009-11-251074

Doherty JR, Yang C, Scott KE, Cameron MD, Fallahi M, Li W, Hall MA, Amelio AL, Mishra JK, Li F, Tortosa M, Genau HM, Rounbehler RJ, Lu Y, Dang CV, Kumar KG, Butler AA, Bannister TD, Hooper AT, Unsal-Kacmaz K, Roush WR, Cleveland JL (2014) Blocking lactate export by inhibiting the Myc target MCT1 Disables glycolysis and glutathione synthesis. Cancer Res 74(3):908–920. doi:10.1158/0008-5472.CAN-13-2034

Eberlin LS, Gabay M, Fan AC, Gouw AM, Tibshirani RJ, Felsher DW, Zare RN (2014) Alteration of the lipid profile in lymphomas induced by MYC overexpression. Proc Natl Acad Sci USA 111(29):10450–10455. doi:10.1073/pnas.1409778111

Efeyan A, Zoncu R, Sabatini DM (2012) Amino acids and mTORC1: from lysosomes to disease. Trends Mol Med 18(9):524–533. doi:10.1016/j.molmed.2012.05.007

Eischen CM, Weber JD, Roussel MF, Sherr CJ, Cleveland JL (1999) Disruption of the ARF-Mdm2-p53 tumor suppressor pathway in Myc-induced lymphomagenesis. Genes Dev 13 (20):2658–2669

Elkon R, Zeller KI, Linhart C, Dang CV, Shamir R, Shiloh Y (2004) In silico identification of transcriptional regulators associated with c-Myc. Nucleic Acids Res 32(17):4955–4961. doi:10.1093/nar/gkh816

Farrell PJ, Allan GJ, Shanahan F, Vousden KH, Crook T (1991) p53 is frequently mutated in Burkitt's lymphoma cell lines. The EMBO journal 10(10):2879–2887

Fernandez PC, Frank SR, Wang L, Schroeder M, Liu S, Greene J, Cocito A, Amati B (2003) Genomic targets of the human c-Myc protein. Genes Dev 17(9):1115–1129. doi:10.1101/gad.1067003

Fernandez-Sanchez ME, Gonatopoulos-Pournatzis T, Preston G, Lawlor MA, Cowling VH (2009) S-adenosyl homocysteine hydrolase is required for Myc-induced mRNA cap methylation, protein synthesis, and cell proliferation. Mol Cell Biol 29(23):6182–6191. doi:10.1128/MCB.00973-09

Gallant P, Shiio Y, Cheng PF, Parkhurst SM, Eisenman RN (1996) Myc and Max homologs in Drosophila. Science 274(5292):1523–1527

Gao P, Tchernyshyov I, Chang TC, Lee YS, Kita K, Ochi T, Zeller KI, De Marzo AM, Van Eyk JE, Mendell JT, Dang CV (2009) c-Myc suppression of miR-23a/b enhances mitochondrial glutaminase expression and glutamine metabolism. Nature 458(7239):762–765. doi:10.1038/nature07823

Gomez-Roman N, Grandori C, Eisenman RN, White RJ (2003) Direct activation of RNA polymerase III transcription by c-Myc. Nature 421(6920):290–294. doi:10.1038/nature01327

Grandori C, Gomez-Roman N, Felton-Edkins ZA, Ngouenet C, Galloway DA, Eisenman RN, White RJ (2005) c-Myc binds to human ribosomal DNA and stimulates transcription of rRNA genes by RNA polymerase I. Nat Cell Biol 7(3):311–318. doi:10.1038/ncb1224

Gregory MA, Qi Y, Hann SR (2003) Phosphorylation by glycogen synthase kinase-3 controls c-myc proteolysis and subnuclear localization. J Biol Chem 278(51):51606–51612. doi:10.1074/jbc.M310722200

Harrison DE, Strong R, Sharp ZD, Nelson JF, Astle CM, Flurkey K, Nadon NL, Wilkinson JE, Frenkel K, Carter CS, Pahor M, Javors MA, Fernandez E, Miller RA (2009) Rapamycin fed late in life extends lifespan in genetically heterogeneous mice. Nature 460(7253):392–395. doi:10.1038/nature08221

Hart LS, Cunningham JT, Datta T, Dey S, Tameire F, Lehman SL, Qiu B, Zhang H, Cerniglia G, Bi M, Li Y, Gao Y, Liu H, Li C, Maity A, Thomas-Tikhonenko A, Perl AE, Koong A, Fuchs SY, Diehl JA, Mills IG, Ruggero D, Koumenis C (2012) ER stress-mediated autophagy promotes Myc-dependent transformation and tumor growth. J Clin Investig 122(12):4621–4634. doi:10.1172/JCI62973

Hatanaka M, Maier B, Sims EK, Templin AT, Kulkarni RN, Evans-Molina C, Mirmira RG (2014) Palmitate induces mRNA translation and increases ER protein load in islet beta-cells via activation of the mammalian target of rapamycin pathway. Diabetes 63(10):3404–3415. doi:10.2337/db14-0105

Hein N, Hannan KM, George AJ, Sanij E, Hannan RD (2013) The nucleolus: an emerging target for cancer therapy. Trends Mol Med 19(11):643–654. doi:10.1016/j.molmed.2013.07.005

Hofmann JW, Zhao X, De Cecco M, Peterson AL, Pagliaroli L, Manivannan J, Hubbard GB, Ikeno Y, Zhang Y, Feng B, Li X, Serre T, Qi W, Van Remmen H, Miller RA, Bath KG, de Cabo R, Xu H, Neretti N, Sedivy JM (2015) Reduced expression of MYC increases longevity and enhances healthspan. Cell. doi:10.1016/j.cell.2014.12.016

Holland JP, Evans MJ, Rice SL, Wongvipat J, Sawyers CL, Lewis JS (2012) Annotating MYC status with 89Zr-transferrin imaging. Nat Med 18(10):1586–1591. doi:10.1038/nm.2935

Howell JJ, Manning BD (2011) mTOR couples cellular nutrient sensing to organismal metabolic homeostasis. Trends Endocrinol Metab TEM 22(3):94–102. doi:10.1016/j.tem.2010.12.003

Hung CL, Wang LY, Yu YL, Chen HW, Srivastava S, Petrovics G, Kung HJ (2014) A long noncoding RNA connects c-Myc to tumor metabolism. Proc Natl Acad Sci USA 111(52):18697–18702. doi:10.1073/pnas.1415669112

Iyer NV, Leung SW, Semenza GL (1998) The human hypoxia-inducible factor 1alpha gene: HIF1A structure and evolutionary conservation. Genomics 52(2):159–165. doi:10.1006/geno.1998.5416

Ji H, Wu G, Zhan X, Nolan A, Koh C, De Marzo A, Doan HM, Fan J, Cheadle C, Fallahi M, Cleveland JL, Dang CV, Zeller KI (2011) Cell-type independent MYC target genes reveal a primordial signature involved in biomass accumulation. PLoS ONE 6(10):e26057. doi:10.1371/journal.pone.0026057

Johnston LA, Prober DA, Edgar BA, Eisenman RN, Gallant P (1999) Drosophila myc regulates cellular growth during development. Cell 98(6):779–790

Jones RM, Branda J, Johnston KA, Polymenis M, Gadd M, Rustgi A, Callanan L, Schmidt EV (1996) An essential E box in the promoter of the gene encoding the mRNA cap-binding protein (eukaryotic initiation factor 4E) is a target for activation by c-myc. Mol Cell Biol 16(9):4754–4764

Jorgensen P, Tyers M (2004) How cells coordinate growth and division. Curr Biol CB 14(23):R1014–R1027. doi:10.1016/j.cub.2004.11.027

Kim S, Li Q, Dang CV, Lee LA (2000) Induction of ribosomal genes and hepatocyte hypertrophy by adenovirus-mediated expression of c-Myc in vivo. Proc Natl Acad Sci USA 97(21):11198–11202. doi:10.1073/pnas.200372597

Kim JW, Zeller KI, Wang Y, Jegga AG, Aronow BJ, O'Donnell KA, Dang CV (2004) Evaluation of myc E-box phylogenetic footprints in glycolytic genes by chromatin immunoprecipitation assays. Mol Cell Biol 24(13):5923–5936. doi:10.1128/MCB.24.13.5923-5936.2004

Kim JW, Tchernyshyov I, Semenza GL, Dang CV (2006) HIF-1-mediated expression of pyruvate dehydrogenase kinase: a metabolic switch required for cellular adaptation to hypoxia. Cell Metab 3(3):177–185. doi:10.1016/j.cmet.2006.02.002

Kim J, Roh M, Doubinskaia I, Algarroba GN, Eltoum IE, Abdulkadir SA (2012) A mouse model of heterogeneous, c-MYC-initiated prostate cancer with loss of Pten and p53. Oncogene 31 (3):322–332. doi:10.1038/onc.2011.236

Koh CM, Bezzi M, Low DH, Ang WX, Teo SX, Gay FP, Al-Haddawi M, Tan SY, Osato M, Sabo A, Amati B, Wee KB, Guccione E (2015) MYC regulates the core pre-mRNA splicing machinery as an essential step in lymphomagenesis. Nature. doi:10.1038/nature14351

Kongsuwan K, Yu Q, Vincent A, Frisardi MC, Rosbash M, Lengyel JA, Merriam J (1985) A Drosophila Minute gene encodes a ribosomal protein. Nature 317(6037):555–558

Korangath P, Teo WW, Sadik H, Han L, Mori N, Huijts CM, Wildes F, Bharti S, Zhang Z, Santa-Maria CA, Tsai HL, Dang CV, Stearns V, Bhujwalla Z, Sukumar S (2015) Targeting glutamine metabolism in breast cancer with aminooxyacetate. Clin Cancer Res Official J Am Assoc Cancer Res. doi:10.1158/1078-0432.CCR-14-1200

Kroemer G, Pouyssegur J (2008) Tumor cell metabolism: cancer's Achilles' heel. Cancer Cell 13 (6):472–482. doi:10.1016/j.ccr.2008.05.005

Le A, Cooper CR, Gouw AM, Dinavahi R, Maitra A, Deck LM, Royer RE, Vander Jagt DL, Semenza GL, Dang CV (2010) Inhibition of lactate dehydrogenase A induces oxidative stress and inhibits tumor progression. Proc Natl Acad Sci USA 107(5):2037–2042. doi:10.1073/pnas. 0914433107

Le A, Lane AN, Hamaker M, Bose S, Gouw A, Barbi J, Tsukamoto T, Rojas CJ, Slusher BS, Zhang H, Zimmerman LJ, Liebler DC, Slebos RJ, Lorkiewicz PK, Higashi RM, Fan TW, Dang CV (2012) Glucose-independent glutamine metabolism via TCA cycling for proliferation and survival in B cells. Cell Metab 15(1):110–121. doi:10.1016/j.cmet.2011.12.009

Li F, Wang Y, Zeller KI, Potter JJ, Wonsey DR, O'Donnell KA, Kim JW, Yustein JT, Lee LA, Dang CV (2005) Myc stimulates nuclearly encoded mitochondrial genes and mitochondrial biogenesis. Mol Cell Biol 25(14):6225–6234. doi:10.1128/MCB.25.14.6225-6234.2005

Lin CY, Loven J, Rahl PB, Paranal RM, Burge CB, Bradner JE, Lee TI, Young RA (2012) Transcriptional amplification in tumor cells with elevated c-Myc. Cell 151(1):56–67. doi:10. 1016/j.cell.2012.08.026

Lippman SI, Broach JR (2009) Protein kinase A and TORC1 activate genes for ribosomal biogenesis by inactivating repressors encoded by Dot6 and its homolog Tod6. Proc Natl Acad Sci USA 106(47):19928–19933. doi:10.1073/pnas.0907027106

Liu YC, Li F, Handler J, Huang CR, Xiang Y, Neretti N, Sedivy JM, Zeller KI, Dang CV (2008) Global regulation of nucleotide biosynthetic genes by c-Myc. PLoS ONE 3(7):e2722. doi:10. 1371/journal.pone.0002722

Liu W, Le A, Hancock C, Lane AN, Dang CV, Fan TW, Phang JM (2012) Reprogramming of proline and glutamine metabolism contributes to the proliferative and metabolic responses regulated by oncogenic transcription factor c-MYC. Proc Natl Acad Sci USA 109(23):8983–8988. doi:10.1073/pnas.1203244109

Lohrum MA, Ludwig RL, Kubbutat MH, Hanlon M, Vousden KH (2003) Regulation of HDM2 activity by the ribosomal protein L11. Cancer Cell 3(6):577–587

Mannava S, Grachtchouk V, Wheeler LJ, Im M, Zhuang D, Slavina EG, Mathews CK, Shewach DS, Nikiforov MA (2008) Direct role of nucleotide metabolism in C-MYC-dependent proliferation of melanoma cells. Cell Cycle 7(15):2392–2400

Martin-Montalvo A, Mercken EM, Mitchell SJ, Palacios HH, Mote PL, Scheibye-Knudsen M, Gomes AP, Ward TM, Minor RK, Blouin MJ, Schwab M, Pollak M, Zhang Y, Yu Y, Becker KG, Bohr VA, Ingram DK, Sinclair DA, Wolf NS, Spindler SR, Bernier M, de Cabo R (2013) Metformin improves healthspan and lifespan in mice. Nat Commun 4:2192. doi:10. 1038/ncomms3192

Mayer C, Grummt I (2006) Ribosome biogenesis and cell growth: mTOR coordinates transcription by all three classes of nuclear RNA polymerases. Oncogene 25(48):6384–6391. doi:10.1038/sj. onc.1209883

McCubrey JA, Steelman LS, Chappell WH, Abrams SL, Wong EW, Chang F, Lehmann B, Terrian DM, Milella M, Tafuri A, Stivala F, Libra M, Basecke J, Evangelisti C, Martelli AM, Franklin RA (2007) Roles of the Raf/MEK/ERK pathway in cell growth, malignant transformation and drug resistance. Biochim Biophys Acta 1773(8):1263–1284. doi:10.1016/j.bbamcr.2006.10.001

Miltenberger RJ, Sukow KA, Farnham PJ (1995) An E-box-mediated increase in cad transcription at the G1/S-phase boundary is suppressed by inhibitory c-Myc mutants. Mol Cell Biol 15 (5):2527–2535

Nie Z, Hu G, Wei G, Cui K, Yamane A, Resch W, Wang R, Green DR, Tessarollo L, Casellas R, Zhao K, Levens D (2012) c-Myc is a universal amplifier of expressed genes in lymphocytes and embryonic stem cells. Cell 151(1):68–79. doi:10.1016/j.cell.2012.08.033

O'Donnell KA, Wentzel EA, Zeller KI, Dang CV, Mendell JT (2005) c-Myc-regulated microRNAs modulate E2F1 expression. Nature 435(7043):839–843. doi:10.1038/nature03677

O'Donnell KA, Yu D, Zeller KI, Kim JW, Racke F, Thomas-Tikhonenko A, Dang CV (2006) Activation of transferrin receptor 1 by c-Myc enhances cellular proliferation and tumorigenesis. Mol Cell Biol 26(6):2373–2386. doi:10.1128/MCB.26.6.2373-2386.2006

Osthus RC, Shim H, Kim S, Li Q, Reddy R, Mukherjee M, Xu Y, Wonsey D, Lee LA, Dang CV (2000) Deregulation of glucose transporter 1 and glycolytic gene expression by c-Myc. J Biol Chem 275(29):21797–21800. doi:10.1074/jbc.C000023200

Pickering MT, Stadler BM, Kowalik TF (2009) miR-17 and miR-20a temper an E2F1-induced G1 checkpoint to regulate cell cycle progression. Oncogene 28(1):140–145. doi:10.1038/onc.2008. 372

Rahl PB, Lin CY, Seila AC, Flynn RA, McCuine S, Burge CB, Sharp PA, Young RA (2010) c-Myc regulates transcriptional pause release. Cell 141(3):432–445. doi:10.1016/j.cell.2010. 03.030

Robitaille AM, Christen S, Shimobayashi M, Cornu M, Fava LL, Moes S, Prescianotto-Baschong C, Sauer U, Jenoe P, Hall MN (2013) Quantitative phosphoproteomics reveal mTORC1 activates de novo pyrimidine synthesis. Science 339(6125):1320–1323. doi:10.1126/science. 1228771

Sabo A, Kress TR, Pelizzola M, de Pretis S, Gorski MM, Tesi A, Morelli MJ, Bora P, Doni M, Verrecchia A, Tonelli C, Faga G, Bianchi V, Ronchi A, Low D, Muller H, Guccione E, Campaner S, Amati B (2014) Selective transcriptional regulation by Myc in cellular growth control and lymphomagenesis. Nature. doi:10.1038/nature13537

Schreiber-Agus N, Stein D, Chen K, Goltz JS, Stevens L, DePinho RA (1997) Drosophila Myc is oncogenic in mammalian cells and plays a role in the diminutive phenotype. Proc Natl Acad Sci USA 94(4):1235–1240

Seagroves TN, Ryan HE, Lu H, Wouters BG, Knapp M, Thibault P, Laderoute K, Johnson RS (2001) Transcription factor HIF-1 is a necessary mediator of the pasteur effect in mammalian cells. Mol Cell Biol 21(10):3436–3444. doi:10.1128/MCB.21.10.3436-3444.2001

Sears R, Ohtani K, Nevins JR (1997) Identification of positively and negatively acting elements regulating expression of the E2F2 gene in response to cell growth signals. Mol Cell Biol 17 (9):5227–5235

Semenza GL, Jiang BH, Leung SW, Passantino R, Concordet JP, Maire P, Giallongo A (1996) Hypoxia response elements in the aldolase A, enolase 1, and lactate dehydrogenase A gene promoters contain essential binding sites for hypoxia-inducible factor 1. J Biol Chem 271 (51):32529–32537

Shim H, Chun YS, Lewis BC, Dang CV (1998) A unique glucose-dependent apoptotic pathway induced by c-Myc. Proc Natl Acad Sci USA 95(4):1511–1516

Shroff EH, Eberlin LS, Dang VM, Gouw AM, Gabay M, Adam SJ, Bellovin DI, Tran PT, Philbrick WM, Garcia-Ocana A, Casey SC, Li Y, Dang CV, Zare RN, Felsher DW (2015) MYC oncogene overexpression drives renal cell carcinoma in a mouse model through glutamine metabolism. Proc Natl Acad Sci USA. doi:10.1073/pnas.1507228112

Sonveaux P, Vegran F, Schroeder T, Wergin MC, Verrax J, Rabbani ZN, De Saedeleer CJ, Kennedy KM, Diepart C, Jordan BF, Kelley MJ, Gallez B, Wahl ML, Feron O, Dewhirst MW (2008) Targeting lactate-fueled respiration selectively kills hypoxic tumor cells in mice. J Clin Investig 118(12):3930–3942. doi:10.1172/JCI36843

Thai M, Graham NA, Braas D, Nehil M, Komisopoulou E, Kurdistani SK, McCormick F, Graeber TG, Christofk HR (2014) Adenovirus E4ORF1-induced MYC activation promotes host cell anabolic glucose metabolism and virus replication. Cell Metab 19(4):694–701. doi:10. 1016/j.cmet.2014.03.009

Thomas LR, Wang Q, Grieb BC, Phan J, Foshage AM, Sun Q, Olejniczak ET, Clark T, Dey S, Lorey S, Alicie B, Howard GC, Cawthon B, Ess KC, Eischen CM, Zhao Z, Fesik SW, Tansey WP (2015) Interaction with WDR5 promotes target gene recognition and tumorigenesis by MYC. Mol Cell 58(3):440–452. doi:10.1016/j.molcel.2015.02.028

Toyoshima M, Howie HL, Imakura M, Walsh RM, Annis JE, Chang AN, Frazier J, Chau BN, Loboda A, Linsley PS, Cleary MA, Park JR, Grandori C (2012) Functional genomics identifies therapeutic targets for MYC-driven cancer. Proc Natl Acad Sci USA 109(24):9545–9550. doi:10.1073/pnas.1121119109

van Riggelen J, Yetil A, Felsher DW (2010) MYC as a regulator of ribosome biogenesis and protein synthesis. Nat Rev Cancer 10(4):301–309. doi:10.1038/nrc2819

Walz S, Lorenzin F, Morton J, Wiese KE, von Eyss B, Herold S, Rycak L, Dumay-Odelot H, Karim S, Bartkuhn M, Roels F, Wustefeld T, Fischer M, Teichmann M, Zender L, Wei CL, Sansom O, Wolf E, Eilers M (2014) Activation and repression by oncogenic MYC shape tumour-specific gene expression profiles. Nature. doi:10.1038/nature13473

Warburg O (1956a) On the origin of cancer cells. Science 123(3191):309–314

Warburg O (1956b) On respiratory impairment in cancer cells. Science 124(3215):269–270

Williams KJ, Argus JP, Zhu Y, Wilks MQ, Marbois BN, York AG, Kidani Y, Pourzia AL, Akhavan D, Lisiero DN, Komisopoulou E, Henkin AH, Soto H, Chamberlain BT, Vergnes L, Jung ME, Torres JZ, Liau LM, Christofk HR, Prins RM, Mischel PS, Reue K, Graeber TG, Bensinger SJ (2013) An essential requirement for the SCAP/SREBP signaling axis to protect cancer cells from lipotoxicity. Cancer Res 73(9):2850–2862. doi:10.1158/0008-5472.CAN-13-0382-T

Wise DR, DeBerardinis RJ, Mancuso A, Sayed N, Zhang XY, Pfeiffer HK, Nissim I, Daikhin E, Yudkoff M, McMahon SB, Thompson CB (2008) Myc regulates a transcriptional program that stimulates mitochondrial glutaminolysis and leads to glutamine addiction. Proc Natl Acad Sci USA 105(48):18782–18787. doi:10.1073/pnas.0810199105

Wolf E, Lin CY, Eilers M, Levens DL (2014) Taming of the beast: shaping Myc-dependent amplification. Trends Cell Biol. doi:10.1016/j.tcb.2014.10.006

Wong JV, Yao G, Nevins JR, You L (2011) Viral-mediated noisy gene expression reveals biphasic E2f1 response to MYC. Mol Cell 41(3):275–285. doi:10.1016/j.molcel.2011.01.014

Xiang Y, Stine ZE, Xia J, Lu Y, O'Connor RS, Altman BJ, Hsieh AL, Gouw AM, Thomas AG, Gao P, Sun L, Song L, Yan B, Slusher BS, Zhuo J, Ooi LL, Lee CG, Mancuso A, McCallion AS, Le A, Milone MC, Rayport S, Felsher DW, Dang CV (2015) Targeted inhibition of tumor-specific glutaminase diminishes cell-autonomous tumorigenesis. J Clin Investig. doi:10.1172/JCI75836

Yang D, Liu H, Goga A, Kim S, Yuneva M, Bishop JM (2010) Therapeutic potential of a synthetic lethal interaction between the MYC proto-oncogene and inhibition of aurora-B kinase. Proc Natl Acad Sci USA 107(31):13836–13841. doi:10.1073/pnas.1008366107

Yuneva M, Zamboni N, Oefner P, Sachidanandam R, Lazebnik Y (2007) Deficiency in glutamine but not glucose induces MYC-dependent apoptosis in human cells. J Cell Biol 178(1):93–105. doi:10.1083/jcb.200703099

Zeller KI, Zhao X, Lee CW, Chiu KP, Yao F, Yustein JT, Ooi HS, Orlov YL, Shahab A, Yong HC, Fu Y, Weng Z, Kuznetsov VA, Sung WK, Ruan Y, Dang CV, Wei CL (2006) Global mapping of c-Myc binding sites and target gene networks in human B cells. Proc Natl Acad Sci USA 103(47):17834–17839. doi:10.1073/pnas.0604129103

Zhang H, Gao P, Fukuda R, Kumar G, Krishnamachary B, Zeller KI, Dang CV, Semenza GL (2007) HIF-1 inhibits mitochondrial biogenesis and cellular respiration in VHL-deficient renal cell carcinoma by repression of C-MYC activity. Cancer Cell 11(5):407–420. doi:10.1016/j.ccr.2007.04.001

Zirath H, Frenzel A, Oliynyk G, Segerstrom L, Westermark UK, Larsson K, Munksgaard Persson M, Hultenby K, Lehtio J, Einvik C, Pahlman S, Kogner P, Jakobsson PJ, Arsenian Henriksson M (2013) MYC inhibition induces metabolic changes leading to accumulation of lipid droplets in tumor cells. Proc Natl Acad Sci USA. doi:10.1073/pnas.1222404110

The Role of pH Regulation in Cancer Progression

Alan McIntyre and Adrian L. Harris

Abstract

Frequently observed phenotypes of tumours include high metabolic activity, hypoxia and poor perfusion; these act to produce an acidic microenvironment. Cellular function depends on pH homoeostasis, and thus, tumours become dependent on pH regulatory mechanisms. Many of the proteins involved in pH regulation are highly expressed in tumours, and their expression is often of prognostic significance. The more acidic tumour microenvironment also has important implications with regard to chemotherapeutic and radiotherapeutic interventions. In addition, we review pH-sensing mechanisms, the role of pH regulation in tumour phenotype and the use of pH regulatory mechanisms as therapeutic targets.

Keywords

pH regulation · Hypoxia · Tumour · Carbonic anhydrase 9 · acidic microenvironment

A. McIntyre · A.L. Harris (✉)
Molecular Oncology Laboratories, Department of Medical Oncology,
Weatherall Institute of Molecular Medicine, University of Oxford, Oxford, UK
e-mail: adrian.harris@oncology.ox.ac.uk

© Springer International Publishing Switzerland 2016
T. Cramer and C.A. Schmitt (eds.), *Metabolism in Cancer*,
Recent Results in Cancer Research 207, DOI 10.1007/978-3-319-42118-6_5

1   Introduction

1.1   Measurements of Intracellular and Extracellular Tumour pH

For many years, it was rightly assumed that the increased conversion of glucose to lactic acid (the Warburg effect) seen in tumours would result in more acidic tumour microenvironment. It was also assumed that this would result in a more acidic intracellular pH. Indeed, initial measurements of tumour pH using pH electrodes inserted into tumours showed a more acidic pH (range 5.6–7.6) than normal tissues (range 6.9–7.6) (Griffiths 1991). However, the pH probes used in these studies were often large compared to single tumour cells and are now understood to have mostly measured extracellular pH (Griffiths 1991). Tumour pH has been measured by a number of imaging techniques including positron emission tomography (PET), magnetic resonance spectroscopy (MRS) and magnetic resonance imaging (MRI) (Zhang et al. 2010). Measurement of pH by ^{31}P-MRS tends to reflect the intracellular pH of tumours which is much more neutral than the extracellular pH in tumour cells (range 6.8–7.4) (Griffiths 1991; Zhang et al. 2010; Gerweck and Seetharaman 1996). In normal tissue, the intracellular pH tends to be more similar to, or more acidic than, the extracellular pH (range 6.8–7.3) (Griffiths 1991; Gerweck and Seetharaman 1996). In addition, the intracellular pH can be more alkaline in tumour cells compared to normal cells; for example, assessment of pH in malignant gliomas revealed a more alkaline steady-state intracellular pH (7.31–7.48) than was identified in normal astrocytes (6.98) (McLean et al. 2000). Therefore, there is a shift in the intracellular/extracellular balance of pH in tumour cells compared to normal cells, producing at least a reduced gradient and in many cases reversal in pH gradients for cancer cells. This has also been shown experimentally in animal studies (Stubbs et al. 1994).

More recently, research has identified higher resolution, less toxic and more clinically useful approaches for in vivo measurement of pH. These will give a greater idea of the pH of the tumour microenvironment in individual patients and enable a more tailored treatment regimen. These include the use of hyperpolarized ^{13}C-labelled bicarbonate imaging by MRS to measure the interstitial pH of tumours, ^{99m}Technetium-labelled pH low-insertion peptide (pHLIP) to measure extracellular pH by CT/SPECT imaging (Macholl et al. 2012; Fendos and Engelman 2012) and diamagnetic and paramagnetic chemical exchange saturation transfer (CEST)-MRI pH-responsive contrast agents to asses pH measurements in vivo (Longo et al. 2012; Liu et al. 2012; Sheth et al. 2012; Aime et al. 2002).

2 Why Are Tumours Acidic?

It is clear that tumours produce a more acidic environment and the Warburg effect first described in the 1920s identified increased glycolysis, increased glucose uptake and lactate production even in normoxic conditions seen in most cancer cells, termed aerobic glycolysis (Warburg 1956; Hanahan and Weinberg 2011). The conversion of pyruvate to lactic acid provides a source of increased H^+ production in cancer cells. However, in addition to this, acidity can arise from other steps in the metabolic processes. For example in the pentose phosphate pathway, conversion of 1 glucose-6-phosphate to 1 fructose 6-phophate produces $3CO_2$ and $6H^+$ (6 NADPH + H^+). The pentose phosphate pathway is up-regulated in cancer, and stabilization of $HIF1\alpha$ increases expression of genes involved in the pentose phosphate pathway (Riganti et al. 2012). The Krebs cycle, which still takes place in tumours, produces $6CO_2$ for every molecule of glucose metabolized. Fatty acid synthesis of 1 molecule of palmitate from citrate produces 7 CO_2 and $7H^+$ and requires 7 molecules of HCO_3^-. An additional $2H^+$ (which are NADPH + H^+) are used in the process (Salway 2000). Fatty acid synthesis is increased in many tumour types and is associated with malignant transformation and HIF1 stabilization (Kuhajda et al. 1994; Kuhajda 2000; Yang et al. 2002; Menendez and Lupu 2007).

CO_2 forms a weak acid, carbonic acid (H_2CO_3) in solution. It can also be hydrated to form HCO_3^- and an H^+, a reaction that is catalysed by members of the family of carbonic anhydrases (Supuran 2008). HCO_3^- is important in regulation of both intracellular and extracellular pH through its titration of H^+ (Hulikova et al. 2012a, b); therefore, HCO_3^- use in fatty acid synthesis will increase intracellular acidification.

In addition, the lack of adequate functional vascularization, also a common feature of tumours, affects the acidification of tumours (Vaupel et al. 1989). The lack of vasculature reduces oxygen supply and removal of waste products including acid resulting in accumulation of H^+ in the poorly perfused microenvironment (Vaupel et al. 1989). Therefore, the increased CO_2 and H^+ production from up-regulated metabolic processes coupled with poor perfusion result in a more acidic tumour microenvironment.

2.1 pH Sensing

Despite the increased production of acidity, tumour cells maintain a relatively more alkali intracellular pH than normal cells, suggesting an increase in activity of mechanisms to regulate pH and therefore sense the pH changes occurring. The transcription factors such as $HIF1\alpha$ and $HIF2\alpha$, which are stabilized in many types of tumours and which affect many metabolic pathways (Schulze and Harris 2012), are stabilized by acidic extracellular pH, in addition to the well-recognized low oxygen levels (Mekhail et al. 2004). Regulation of $HIF1\alpha$ and $HIF2\alpha$ stabilization under acidic extracellular conditions is due to nucleolar sequestration of VHL (which degrades HIF under normoxic conditions) (Mekhail et al. 2004). Further to this, a later study identified increased normoxic expression of carbonic anhydrase IX (CAIX) in response to extracellular acidosis (Ihnatko et al. 2006). *CA9* is a HIF1α target, which regulates intracellular and extracellular pH (this is discussed in detail later) (Swietach et al. 2009). Increased expression of *CA9* transcription in response to extracellular acidity is regulated both by HIF and also independently of HIF (Ihnatko et al. 2006). Further investigation revealed that inhibition of the MAPK and PI3K pathways resulted in complete suppression of CAIX induction by acidosis (Ihnatko et al. 2006).

Proton-sensing G-protein-coupled receptors sense extracellular pH (Ludwig et al. 2003). Ovarian cancer G-protein-coupled receptor 1 (ORG1) was identified to be completely activated at pH 6.8 but was inactive at pH 7.8 (Ludwig et al. 2003). This study identified, by mutation studies, that the histidines in the extracellular domain of ORG1 were involved in pH sensing (Ludwig et al. 2003). ORG1 stimulated inositol phosphate formation in response to acidic extracellular pH (Ludwig et al. 2003). GPR4 and TDAG8, two additional G-protein coupled receptors, increased cyclic AMP formation in response to pH changes (Ludwig et al. 2003; Wang et al. 2004; Ishii et al. 2005). Inositol phosphate formation resulted in increased phospholipase IP_3 signalling, which can induce Ca^{2+} release from intracellular stores (Huang et al. 2008; Rahman 2012). Extracellular acidification resulted in phospholipase C activation, IP_3 formation and Ca^{2+} release, leading to increased phosphorylation of ERK (Huang et al. 2008). ERK forms part of the MAPK cascade, which is required for acid-induced CAIX expression described above (Ihnatko et al. 2006). Differentially localized Ca^{2+} concentrations activate many different processes (Petersen and Tepikin 2008). Further to this membrane potential changes induced by Ca^{2+} release have many critical roles including regulation of cell cycle, proliferation and intracellular pH (Kunzelmann 2005). OGR1 is strongly expressed in medulloblastoma patient samples (Huang et al. 2008).

Four acid-sensing ion channel genes (ASIC1, ASIC2, ASIC3 and ASIC4) have been identified, which open in response to extracellular H^+ (Bassler et al. 2001; Glitsch 2011). These allow non-selective transport of cations (K^+, Na^+, Ca^{2+}, etc.) across the membrane (Glitsch 2011). ASIC1 is more highly expressed in gliomas than in normal astrocytes (Kapoor et al. 2009). ASIC1 knockdown reduced proliferation and cell migration of glioblastoma cells in two separate studies (Kapoor et al. 2009; Rooj et al. 2012), identifying a role for ASICs in tumour cells.

In addition to the mechanisms described above, H^+ can inhibit or potentiate additional ion channels and functions through allosteric binding including K^+ channels and ATP-gated ion channels (Glitsch 2011).

2.2 Mechanisms of pH Regulation

Regulation of cellular cytoplasmic pH is mediated through 5 main families of proteins (a diagram of these is shown in Fig. 1). Three of these directly transport H^+ across the membrane, namely the monocarboxylate transporters, the sodium hydrogen ion exchangers and the vacuolar-type H^+-adenosine triphosphatases.

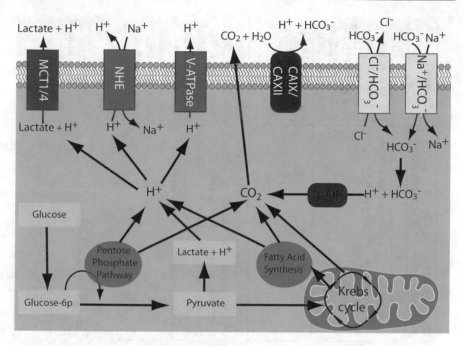

Fig. 1 A model of metabolically produced acid and the mechanisms by which it is removed from tumour cells. MCT1/4, monocarboxylate transporters 1 and 4; NHE, Na^+/H^+ exchangers; V-ATPase, vacuolar-type H^+-adenosine triphosphatases; CAIX/CAXII, carbonic anhydrases 9 and 12; Cl^-/HCO_3^-, chloride bicarbonate anion exchangers; Na^+/HCO_3^-, sodium-dependent bicarbonate cotransporters

Another mechanism of pH regulation is through the uptake of bicarbonate, which is used to titrate intracellular H^+. The bicarbonate transporters of the SLC4 and SLC26 family of genes facilitate this. Finally, there is the family of carbonic anhydrases, which regulate pH through their ability to catalyse the reversible hydration of CO_2. Each of these is covered in more detail below, including their mechanism of action, expression and prognostic associations in cancer. This information and additional information, covered later in this chapter, regarding the role for each of these pH-regulating proteins on tumour growth/proliferation and metastasis, have been tabulated for quick reference in Table 1.

Table 1 pH regulators

Protein	Function	Hypoxic regulation	Expression in tumours	Prognostic tumour marker	Effect on tumour growth/proliferation	Effect on migration/invasion	References
Monocarboxylate transporters							
MCT1		1/3 studies	High levels in GISTs, colorectal and breast cancer. Associated with the basal-like subtype of breast cancer	–	Knockdown or inhibition reduces xenograft growth rate	Knockdown reduced invasion in vitro	Pinheiro et al. (2008), Pinheiro et al. (2010b), Izumi et al. (2011), Le Floch et al. (2011), Boidot et al. (2012), de Oliviera et al. (2012)
MCT2	Proton-coupled moncarboxylate transporters (lactate, pyruvate and ketone)	–	Increased expression in colorectal and prostate				Pinheiro et al. (2008), Pertega-Gomes et al. (2011)
MCT4		Yes	High expression in colorectal cancer, breast cancer non-small cell lung cancer and prostate cancer	High expression is associated with; relapse-free survival in ccRCC and poorer prognosis in prostate cancer and colon cancer	Knockdown reduces xenograft growth rates. Overexpression in RAS-transformed fibroblasts increased growth in xenografts	Knockdown reduced migration in ARPE-19, MDCK cells and MDA-MB 231 in vitro. Knockdown also reduced invasion in vitro	Ullah et al. (2006), Gallagher et al. (2007), Pinheiro et al. (2008), Pinheiro et al. (2010b), Izumi et al. (2011), Le Floch et al. (2011), Pertega-Gomes et al. (2011), Rademakers et al. (2011), Chiche et al. (2012), Gerlinger et al. (2012), Meijer et al. (2012), Nakayama et al. (2012)

(continued)

Table 1 (continued)

Protein	Function	Hypoxic regulation	Expression in tumours	Prognostic tumour marker	Effect on tumour growth/proliferation	Effect on migration/invasion	References
Sodium hydrogen ion exchangers							
NHE1	Electroneutral exchange of Na^+ and H^+	Yes	Expressed in breast cancer DCIS and a colon cancer cell line	–	Knockdown in a small cell lung cancer cell line reduced proliferation and increased the number of cells in vitro. EIPA, a NHE inhibitor, reduced proliferation in a gastric cancer cell line in vitro and induced Gi arrest	Overexpression increased invasion of a breast cancer cell line in vitro. EIPA inhibition of NHE reduced hypoxia-induced migration of HEPG2 cells in vitro	Rios et al. (2005), Gatenby et al. (2007), Beltran et al. (2008), Li et al. (2009), Hosogi et al. (2012), Onishi et al. (2012)
NHE2			Expressed in a colon cancer cell line				Beltran et al. (2008)
NHE4			Expressed in a colon cancer cell line				Beltran et al. (2008)
Vacuolar-type H^+ ATPase							
V-ATPase	Actively transport H^+ across membranes, 2H$^+$ for every ATP consumed		Expression of the a4 subunit in 11/32 gangliomas. Expression in 42/46 invasive ductal pancreatic carcinomas, where increased intensity is associated with increasing pancreatic cancer stage			Knockdown of subunits a3 or a4 reduced invasion in vitro in a breast cancer cell line. Inhibition with an a3 subunit-specific inhibitor reduced bone metastasis of B16 melaoma cells in vivo	Ohta et al. (1996), Hinton et al. (2009), Chung et al. (2011), Glieze et al. (2012)

(continued)

Table 1 (continued)

Protein	Function	Hypoxic regulation	Expression in tumours	Prognostic tumour marker	Effect on tumour growth/proliferation	Effect on migration/invasion	References
Extracellular membrane-tethered carbonic anhydrases							
CAIX	Catalyses the extracellular reversible hydration of CO_2 to form HCO_3^- and H^+	yes	Increased expression in many tumour types including ccRCC, colon and breast cancer	Expression is associated with poor prognosis in most tumour types including breast and colorectal cancer	Knockdown reduced spheroid growth rate in vitro in 2 colon cancer cell lines. Knockdown or inhibition reduces xenograft growth rate *in vivo* in 1 glioblastoma, 2 colon and 2 breast cancer models	Inhibition reduced migration of HeLa cells in vitro. Knockdown or inhibition reduced metastasis in a mouse breast cancer model in vivo	Liao et al. (1997), Wykoff et al. (2000), Chia et al. (2001), Wykoff et al. (2001), Kivela et al. (2005), Cleven et al. (2008), Chiche et al. (2009), Lou et al. (2011), McIntyre et al. (2012), Svastova et al. (2012)
CAXII		yes	Expressed in breast cancer where it is associated with estrogen receptor-positive tumours	Expression is a marker of good prognosis in breast cancer	Knockdown in combination with CAIX knockdown further reduced xenograft growth rates in vivo in 1 colon cancer model		Wykoff et al. (2000), Watson et al. (2003), Barnett et al. (2008), Chiche et al. (2009)
Bicarbonate transporters							
AE1-3, CLD, Pendri PAT1, SLC26A7, SLC26A9	Chloride–bicarbonate anion exchanger		AE2 is expressed in 68 % of hepatocellular carcinomas			Inhibtion with DIDS inhibited migration in vitro	Wu et al. (2006), Klein et al. (2000)

(continued)

Table 1 (continued)

Protein	Function	Hypoxic regulation	Expression in tumours	Prognostic tumour marker	Effect on tumour growth/proliferation	Effect on migration/invasion	References
NBCe1, NBCe2, NBCn1 AE4, NBCn2	Sodium-dependent bicarbonate cotransporters					Inhibition with DIDS inhibited migration in vitro	Klein et al. (2000)
NDCBE	Sodium-dependent chloride–bicarbonate anion exchanger					Inhibtion with DIDS inhibited migration in vitro	Klein et al. (2000)

A table summarizing the published data on pH regulators in cancer, including function, evidence of hypoxic regulation, expression, prognostic significance in tumours and effects on tumour growth, proliferation, migration and invasion. Gastrointestinal stromal tumours (GISTs), ductal carcinoma in situ (DCIS), clear cell renal cell carcinoma (ccRCC), ethyl-isopropyl amiloride (EIPA) and 4,4′-diisothiocyanate-stilbene-2,2′-disulfonic acid (DIDS). ARPE-19 is a retinal pigment epithelia cell line. Madin–Darby canine kidney (MDCK). MDA-MB-231 is a triple receptor negative breast cancer cell line. HEPG2 is a hepatocellular carcinoma cell line. HeLa is a cervical cancer cell line—denotes no significant published data

3 Lactate Acid Secretion (Monocarboxylate Transporters 1-4)

Monocarboxylate transporters (MCT) 1–4 are members of the solute carrier (SLC) 16A family (Halestrap and Meredith 2004). There are 14 SLC16A family members (Halestrap and Meredith 2004). MCT1–MCT4 are proton-coupled monocarboxylate (lactate, pyruvate and ketone) transporters (Halestrap and Wilson 2012). These are electroneutral 1:1 monocarboxylate:H^+ (Halestrap and Wilson 2012). MCT1 has the ability for both lactate and H^+ efflux or influx, the direction of which is controlled by lactate and H^+ gradients (Boidot et al. 2012). Whilst studies on MCT4 have characterized it as facilitating lactate and H^+ efflux (Halestrap and Wilson 2012; Manning Fox et al. 2000), MCT4 expression increases the gradient between intracellular and extracellular pH in RAS-transformed hamster fibroblast deficient in NHE-1 (Na^+/H^+ exchanger-1), grown as xenografts (Chiche et al. 2012). This work highlights the important role for MCT4 in regulation of intracellular pH (Chiche et al. 2012). Further to this, siRNA knockdown of MCT4 produced intracellular acidosis and Warburg effect reversion (Gerlinger et al. 2012).

MCT4 (*SLC16A3*) is hypoxia regulated in a HIF1α-dependent manner (Ullah et al. 2006), and this initial study did not identify up-regulation of MCT1 in response to hypoxia (Ullah et al. 2006). However, a further study identified increased expression of MCT1 in response to hypoxia, which was p53 dependent and NFκB regulated (Boidot et al. 2012). A separate study investigating the expression of MCT1 and 4 showed that MCT4 but not MCT1 staining had a hypoxic pattern of expression in head and neck cancer biopsies (Rademakers et al. 2011). A symbiosis between hypoxic cells and normoxic cells has been identified with regard to the role of lactate. Cancer cells switch to a much greater role of glycolysis in the hypoxic microenvironment. Lactate is extruded by MCT4 from cells in these regions of tumours and is then taken up by oxygenated cancer cells and used as a fuel in respiration (Sonveaux et al. 2008). This symbiosis would also act to remove H^+ from poorly perfused regions of tumour.

MCT1 is expressed ubiquitously in normal tissues with high expression in the heart and red muscle. MCT4 is also widely expressed (Halestrap and Wilson 2012; Halestrap and Price 1999). CD147/Basigin is a chaperone protein whose expression correlates with MCT1 and MCT4 (Kennedy and Dewhirst 2010). CD147 is required for proper folding and membrane expression of MCT1 and MCT4 (Kirk et al. 2000). High expression of MCT4 was associated with poorer relapse-free survival in ccRCC patients (Gerlinger et al. 2012). MCT1, MCT2 and MCT4 have high levels of expression in gastrointestinal stromal tumours (GISTs) (de Oliveira et al. 2012). Co-expression of MCT1 and CD147 was associated with lower GIST patient survival (de Oliveira et al. 2012). MCT2 and MCT4 had higher expression in prostate cancer samples compared with normal prostate material (Pertega-Gomes et al. 2011). MCT4 and CD147 expression correlated with poorer prognosis in prostate cancer (Pertega-Gomes et al. 2011). Increased expression of MCT1, MCT2 and MCT4 was identified in a large study of 126 colorectal cancer patient samples by immunohistochemistry (Pinheiro et al. 2008). However, the increased expression of MCT2 was also associated with a decrease in membrane-associated expression of the protein in cancer cells (Pinheiro et al. 2008). Another study of colorectal cancer identified MCT4 expression in 50 % of colorectal cancer patients where it was significantly associated with poorer prognosis (Nakayama et al. 2012). High levels of MCT1 and MCT4 were identified in breast carcinoma compared with normal tissue (Pinheiro et al. 2010a). A further study identified that MCT1 and CD147 expressions were associated with variables associated with poorer prognosis including higher-grade tumours and the basal-like subtype (Pinheiro et al. 2010b). A high level of MCT4 in addition to high GLUT1 levels was associated with poor survival in non-small cell lung adenocarcinomas (Meijer et al. 2012).

4 Sodium Hydrogen Ion Exchangers (NHE)

There are 9 Na^+/H^+ exchangers (NHE1–NHE9) (Slepkov et al. 2007). These are members of the solute carrier (SLC) 9A family, member 1–9 (Slepkov et al. 2007). These have heterogeneous patterns of expression (Slepkov et al. 2007). NHE proteins facilitate pH regulation via electroneutral, 1:1, exchange of Na^+ and H^+ along their respective gradients (Slepkov et al. 2007; Orlowski and Grinstein 2004). Establishing Na^+ gradients achieved by Na^+/K^+-ATPase pumps can result in excess extrusion of H^+ as a method of removal of metabolically derived acid (Orlowski and Grinstein 2004).

The role of NHE in pH regulation of tumour cells has been investigated. These studies include analysis of the effect of NHE1 inhibition, in the absence of CO_2/HCO_3^- buffering solution. This resulted in no effect on normal astrocytes whilst the malignant gliomas investigated displayed significant intracellular acidification (McLean et al. 2000). NHE1 antisense treatment of a small cell lung cancer cell line acidified the intracellular pH of these cells (Li et al. 2009). Expression of NHE1, NHE2 and NHE4 regulated pH in a colon cancer cell line (T84) (Beltran et al. 2008). Interestingly, NHE7 can shuttle between the plasma membrane, endosomes and the trans-Golgi network, regulating the intracellular pH of these organelles and cytoplasm (Orlowski and Grinstein 2004; Onishi et al. 2012).

The expression of NHE has been investigated in tumours in a small number of studies detailed below. NHE1 is the most studied and understood of the NHE family members, and is ubiquitously expressed in normal tissues (Slepkov et al. 2007). Chronic hypoxia increased NHE1 expression and NHE activity in isolated mouse pulmonary arterial smooth muscle cells (Rios et al. 2005). In addition, NHE1 expression increased in response to increases in intracellular superoxide (O_2^-) concentrations (Akram et al. 2006). NHE1 expression in breast cancer ductal carcinoma in situ (DCIS) had a pattern indicating hypoxic regulation (Gatenby et al. 2007). NHE1, NHE2 and NHE4 expressions were all identified in a colon cancer cell line (T84) (Beltran et al. 2008). Interestingly, inhibition of NHE reduced VEGF expression levels although the mechanism for this was unclear (He et al. 2010).

5 Vacuolar-Type H⁺ ATPase

Vacuolar-type H^+-adenosine triphosphatases (V-ATPases) actively transport H^+ across membranes (Beyenbach and Wieczorek 2006; Sennoune et al. 2004a; Toei et al. 2010; Perez-Sayans et al. 2009). Figure 2 shows the structure and subunits of the V-ATPase. The ratio of H^+ transported to ATP consumed is 2:1, respectively (Tomashek and Brusilow 2000). The functions associated with the V-ATPase include the acidification of intracellular compartments including lysosomes, regulation of intracellular pH and uptake of cations (Na^+, Ca^{2+}, etc.) (Beyenbach and Wieczorek 2006; Dietz et al. 2001). The V-ATPase is a multi-subunit protein containing up to 14 subunits, which make up two separate complexes (Fig. 2) (Beyenbach and Wieczorek 2006; Sennoune et al. 2004a; Perez-Sayans et al. 2009). The first is the V_1 complex, which is situated on the cytoplasmic side of the membrane and interacts with ATP and is between 400 and 600 kDa in size

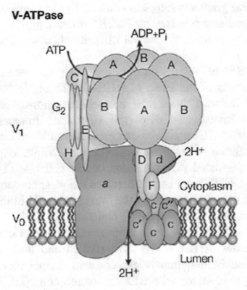

Fig. 2 Structure of V-ATPase. Hydrolysis of ATP by the V_1 domain (shown in *yellow*) drives proton transport through the membrane integral V_0 domain (shown in *green*). The V_1 domain comprises of 8 subunits (*A–H*). Subunit *A* contains the catalytic site for ATP hydrolysis. The V_0 domain comprises of 6 subunits. Subunits *a*, *c*, *c′* and *c″* make up the H^+ transport site. Reprinted by permission from Macmillan Publishers Ltd: (Nature reviews molecular cell biology) (Nishi and Forgac 2002), © 2002 www.nature.com/reviews/molcellbio

(Beyenbach and Wieczorek 2006; Sennoune et al. 2004a; Perez-Sayans et al. 2009). The second is the V_0 complex, which is transmembrane and pumps the H^+ across the membranes and is between 150 and 350 kDa. The variation in size depends on the specific subunit isoforms (Beyenbach and Wieczorek 2006; Sennoune et al. 2004a; Perez-Sayans et al. 2009).

The functional importance of V-ATPases was analyzed in human tumour cell lines using Bafilomycin, a V-ATPase inhibitor. This identified a role for V-ATPase in pH regulation on the plasma membrane of 5/13 of the cell lines tested (Martinez-Zaguilan et al. 1993). Cytoplasmic pH was acidified in the breast cancer cell line MDA-MB-231 by isoform-specific knockdown of V-ATPase subunit a3 (Hinton et al. 2009).

The expression of V-ATPase is ubiquitous in mammalian cells with expression localized to the plasma membrane and intracellular membranes including, Golgi vesicles, endosomes and lysosomes (Stevens and Forgac 1997). MTORC regulates V-ATPase expression via regulation of phosphorylation and nuclear localization of the transcription factor TFEB (Pena-Llopis et al. 2011). A study of glioma biopsies identified expression of the a4 subunit (normally expressed in kidney and epididymis) in 11/32 grade III oligodendrogliomas, 19/28 astrocytomas and 6/16 gangliomas (Gleize et al. 2012). An initial study in pancreatic carcinoma identified 42/46 invasive ductal pancreatic carcinomas that expressed cytoplasmic V-ATPase compared with no non-invasive ductal pancreatic cancers or benign pancreatic cysts (Ohta et al. 1996). A later immunohistochemical study identified increased intensity of V-ATPase expression corresponding to increasing grade of pancreatic cancer (Chung et al. 2011).

6 Extracellular Membrane-tethered Carbonic Anhydrases (CAIX/CAXII)

Carbonic anhydrases are zinc metalloproteins that catalyse the reversible hydration of CO_2 to form HCO_3^- and H^+ (Sly and Hu 1995; Swietach et al. 2008). The carbonic anhydrase family has 14 members that can be categorized by cellular localization. There are the mitochondrial CAs (CAV A, CAV B), the cytoplasmic CAs (CAI, CAII, CAIII, CAVII, CAX, CAXI and CAXIII), the membrane-bound CA (CAIV), the transmembrane CAs (CAIX, CAXII and CAIV) and finally the secreted CA (CAVI) (Supuran 2008; Pastorek et al. 1994). CAIX and CAXII are both transmembrane carbonic anhydrase with an extracellular catalytic domain (Opavsky et al. 1996; Tureci et al. 1998). Both CAIX and CAXII are regulated by hypoxia in a HIF1α-dependent manner (Wykoff et al. 2000).

CAIX is a key enzyme in pH regulation; this was identified in work done in three-dimensional spheroid cultures. Spheroids mimic the three-dimensional gradients achieved within tumours for oxygen, lactate, ATP (Hirschhaeuser et al. 2010), etc. Analysis of pH gradients using spheroids also reveals clear intracellular and extracellular gradients (Swietach et al. 2009). These gradients are regulated by CAIX expression. In experiments comparing a colon cancer cell line in which CAIX had been overexpressed compared to empty vector controls, CAIX expression resulted in more alkali intracellular pH (~ 6.6 compared to ~ 6.2 empty vector controls) and more acidic extracellular pH (~ 6.6 compared to ~ 6.9 empty vector controls)(the pH maps of spheroids from these experiments can be seen in Fig. 3). CO_2 forms carbonic acid, a weak acid in solution. It is therefore preferential for the cell to extrude it. In hydrating CO_2 to form HCO_3^- and H^+, CAIX acts to maintain an efflux gradient of CO_2. This enables continued passive diffusion of CO_2 across the membrane despite in an inability to remove the CO_2 from the hypoxic milieu (Hulikova et al. 2011, 2012b; Swietach et al. 2009). The crystal structure of CAIX has been resolved, and this work suggests that CAIX functions as a dimer (Alterio et al. 2009).

CAIX expression is up-regulated in many cancer types including breast (Wykoff et al. 2001), colon (Kivela et al. 2005), renal cell cancer (ccRCC) (Liao et al. 1997; Murakami et al. 1999), etc. In most tumour types, CAIX expression is associated with poor prognosis including breast (Chia et al. 2001), non-small cell lung cancer (Giatromanolaki et al. 2001), colon (Cleven et al. 2008), carcinoma of the cervix

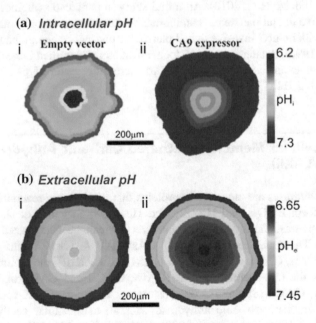

Fig. 3 CAIX expression maintains a more neutral intracellular pH in the hypoxic spheroid core whilst producing a more acidic extracellular pH. **a** A map of spheroid intracellular pH, measured using carboxy SNARF-1. **b** A map of spheroid extracellular pH, measured using Fluorescein-5-(and-6-)-sulfonic acid. pH maps are shown for spheroids of HCT116 (empty vector) which has low levels of hypoxia-induced CAIX and HCT116 with constitutive CAIX expression (CA9 expressor) ($n = 20$, spheroid radius of 222.0 ± 10.6 µm (*panel i*) or of 299.0 ± 2.7 µm)(*panel ii*). This figure is modified from Fig. 3a and b of Swietach et al. (2009). Modified with permission © 2009 The Journal of Biological Chemistry

(Loncaster et al. 2001) and soft tissue sarcoma (Maseide et al. 2004). CAIX expression is associated with better prognosis in ccRCC (Phuoc et al. 2008; Muriel Lopez et al. 2012) in contrast to other tumour types; however, non-papillary or clear cell RCC has von Hippel–Lindau (VHL) mutations, promoter hyper-methylation or loss of gene in nearly all cases (Gossage and Eisen 2010). The function of VHL protein is to regulate stability of HIF1α and HIF2α based on O_2 tension. Under normoxic conditions, HIF1α and HIF2α are hydroxylated on conserved proline residues by the prolyl hydroxylases (PHDs) 1-3. This modification enables VHL to bind HIF1α and HIF2α with the E3 ubiquitin ligase complex which polyubiquiti-nates and targets them for degradation by the proteasome (Gossage and Eisen 2010). One explanation for the association of CAIX with good prognosis in the VHL mutant ccRCC could be that the CAIX expression needs to be more tightly regulated in response to O_2 levels. High constitutive CAIX expression increased the levels of histological necrosis and also apoptosis in a colon cancer xenograft study (McIntyre et al. 2012). Also, HIF2 expression is associated with a more aggressive phenotype and does not regulate CAIX, so a switch to a HIF2-driven cancer would indirectly be associated with poor prognosis (Holmquist-Mengelbier et al. 2006). This evidence suggests that the loss of CAIX expression regulation in response to O_2 levels may have a negative impact on tumour survival. Interestingly, a CAIX knockout mouse was functionally normal apart from gastric hyperplasia, although this knockout has not been confirmed and may target another gene (Gut et al. 2002; Leppilampi et al. 2005).

A study of 103 breast cancer samples identified CAXII expression as a marker for good prognosis and positive estrogen receptor expression (Watson et al. 2003). The correlation between CAXII expression and estrogen receptor alpha expression was the focus of another study in breast cancer, which showed that CAXII expression was regulated by estrogen in breast cancer (Barnett et al. 2008). A study in renal cell carcinoma identified a correlation between lower CAIX and CAXII expression and worse prognosis (Kim et al. 2005). High CAXII expression was associated with a lower risk of metastasis in primary cervical cancer (Kim et al. 2006). High expression of CAXII in non-small cell lung cancer (NSCLC) was associated with better prognosis lower grade and with the squamous cell carcinoma type (Ilie et al. 2011). The shed extracellular domain of CAXII has been identified as a blood serum marker for diagnosis of NSCLC (Kobayashi et al. 2012).

7 Bicarbonate Transporters

Bicarbonate transporter proteins can be broadly subdivided into two main categories. These are the sodium-coupled bicarbonate transporters and the chloride–bicarbonate exchangers (Alper 2006; Romero et al. 2004; Cordat and Casey 2009). There are 14 genes encoding bicarbonate transporters, and these are split between 2 gene families that have distinct evolution and little amino acid homology (Alper 2006; Romero et al. 2004; Cordat and Casey 2009). These are the SLC4 and SLC26 families (Alper

2006; Romero et al. 2004; Cordat and Casey 2009). Within these families, there are two main subtypes of bicarbonate transporter. These are the chloride–bicarbonate anion exchangers and the sodium-dependent bicarbonate cotransporters. One of the bicarbonate transporters, NBCDE, fits into both categories (Alper 2006; Romero et al. 2004; Cordat and Casey 2009). In normal cells, bicarbonate transport is important mostly in the regulation of HCO_3^- removal as a by-product of the Krebs cycle (Alper 2006; Romero et al. 2004; Cordat and Casey 2009).

There are eight chloride–bicarbonate anion exchangers. These transporters exchange Cl^- for HCO_3^- and consist of (*gene name*(protein name)): *SLC4A1* (AE1), *SLC4A2* (AE2), *SLC4A3* (AE3), *SLC26A3* (chloride anion exchanger (CLD)), *SLC26A4* (pendrin), *SLC26A6* (PAT-1), *SLC26A7* (SLC26A7) and *SLC26A9* (SLC26A9) (Alper 2006; Romero et al. 2004; Cordat and Casey 2009; Kopito and Lodish 1985; Alper et al. 1988; Kudrycki et al. 1990; Walker et al. 2009; Scott et al. 1999; Waldegger et al. 2001; Lohi et al. 2002; Chernova et al. 2005). Within this grouping, there is also some difference in the electrogenicity. AE1-3, CLD, pendrin and PAT-1 are electroneutral and therefore result in no exchange of charge, and one Cl^- is exchanged for one HCO_3^- (Kopito and Lodish 1985; Alper et al. 1988; Kudrycki et al. 1990; Walker et al. 2009; Scott et al. 1999; Waldegger et al. 2001; Lohi et al. 2002; Alper 2006; Romero et al. 2004; Cordat and Casey 2009; Chernova et al. 2005). SLC26A7 and SLC26A9 are electrogenic, although there is some disagreement within the literature (Cordat and Casey 2009; Lohi et al. 2002). Diseases associated with defects in these transporters include haemolytic anaemia (AE1) (Southgate et al. 1996; Bouhassira et al. 1992; Jarolim et al. 1992), idiopathic epilepsy (AE3) (Sander et al. 2002) and Pendred syndrome (pendrin) (Everett et al. 1997).

Six sodium-dependent bicarbonate cotransporters consist of (*gene name* (protein name)): *SLC4A4* (NBCe1), *SLC4A5* (NBCe2), *SLC4A7* (NBCn1), *SLC4A8* (NDCBE), *SLC4A9* (AE4) and *SLC4A10* (NBCn2). These transporters cotransport Na^+ and HCO_3^- across the membrane (Romero et al. 1997; Virkki et al. 2002; Choi et al. 2001; Grichtchenko et al. 2001; Parker et al. 2008a, b). NBCe1 and NBCe2 are electrogenic, whilst NBCn1, NDCBE, AE4 and NBCn2 result in no exchange of charge across the membrane (Romero et al. 1997; Virkki et al. 2002; Choi et al. 2001; Grichtchenko et al. 2001; Parker et al. 2008a, b). In addition, NBCe1 can also cotransport CO_3^{2-} (Grichtchenko et al. 2001) although the physiological significance is unclear due to the greater than 100-fold higher levels of HCO_3^- in the cell than CO_3^{2-} (Cordat and Casey 2009). NDCBE is a Na^+-dependent Cl^-/HCO_3^- exchanger (Parks et al. 2011).

AE1–AE3 bicarbonate transporters are most likely restricted to bicarbonate efflux as theses anion exchangers are driven by Cl^- and HCO_3^- gradients except in the case of red blood cells (Cordat and Casey 2009), although if the Cl^- gradients were reversed, the anion exchangers can equally take up bicarbonate (Sterling and Casey 1999). The sodium-dependent bicarbonate transporters similarly are dependent on Na^+ and HCO_3^- gradients, the levels of which are higher extracellular than intracellular (Cordat and Casey 2009).

Bicarbonate transporters have roles in many normal physiological processes including gastric acid neutralization as it enters the intestine (Kuijpers and De Pont 1987) and multiple roles in the nervous system (Majumdar and Bevensee 2010). Bicarbonate transporters have three main functions: efflux of CO_2 and HCO_3^- produced by respiration, pH regulation and regulation of cell volume (Cordat and Casey 2009).

HCO_3^- is membrane impermeant and, therefore, unlike CO_2, needs to be transported across the membrane (Cordat and Casey 2009). However, bicarbonate transport via the bicarbonate transporters does not require ATP. This means that in situations of reduced ATP production such as stress induced by hypoxia, the lack of energy requirement should increase the importance of bicarbonate transport over other methods of pH regulation, and this has been shown in (Hulikova et al. 2011, 2012a). Investigation into the role of bicarbonate transport in normoxia versus hypoxia revealed a greater importance for bicarbonate transport in hypoxia (Hulikova et al. 2012a). Modelling the three-dimensional microenvironment of tumours in vitro using spheroids highlighted the dual importance of both bicarbonate transport and NHE activity. This work used the NHE inhibitor 5-(N,N-dimethly) amiloride (DMA),and also the SLC4 family bicarbonate transport inhibitor 4,4'-diisothiocyanatostilbene-2,2'-disulphonic acid (DIDS). The dependence on the bicarbonate transport and NHE activity for pH regulation was cell line dependent and therefore possibly also cancer-type dependent (Hulikova et al. 2011). This work also highlighted the importance of carbonic anhydrase function in cellular use of bicarbonate as a mobile buffer. Finally, the dominance of bicarbonate transport and NHE activity in pH regulation depends on regional variation in spheroids. In general, the further from the spheroid periphery, the greater the reliance on bicarbonate transport (Hulikova et al. 2011). Figure 4 shows data from Hulikova et al. 2011 identifying the importance of bicarbonate transport on intracellular pH regulation in the hypoxic core of spheroids, in two colorectal cancer cell lines.

Increased levels of bicarbonate transporters have been reported in tumours. For example, *SLC4A1* levels are raised in gastric cancer and its knockdown reduced tumour progression (Suo et al. 2012). *SLC4A2* expression is increased in hepatocellular carcinoma and colorectal cancer (Gorbatenko et al. 2014). *SLC4A4* expression is increased in chronic myeloid leukaemia (Gorbatenko et al. 2014), and knockdown in a colorectal tumour cell line increased sensitivity to methotrexate (Gorbatenko et al. 2014). *SLC4A7* expression is regulated by ErbB1, ErbB2 and ErbB3 in breast cancer cell lines via AKT, ERK, Src and KLF4 (Gorbatenko et al. 2014). Furthermore, recent analysis of bicarbonate transporter gene expression from TCGA data sets for lung colorectal adenocarcinoma and breast cancer revealed the heterogeneous patterns of expression across and within the tumour types (Gorbatenko et al. 2014). High levels of *SLC26A9* were associated with high levels of *CA9* in breast cancer, but the opposing relationship was seen in lung cancer

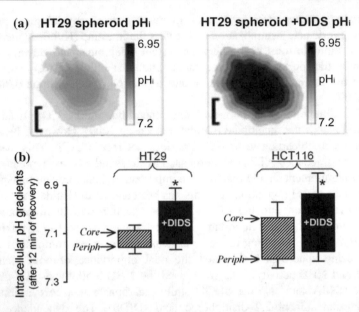

Fig. 4 Bicarbonate transport is key in pH regulation in the hypoxic area of spheroids. A 20 mM ammonium prepulse was performed to impose an acid load on HT29 and HCT116 spheroids. Spheroids were allowed to recover for 12 min, and the data shown here are of the end point analysis of pH_i gradients for spheroids untreated or treated with 150 uM of the bicarbonate transport inhibitor DIDS (4,4′-diisothiocyanate-stilbene-2,2′-disulfonic acid). The superfusate used in these experiments was buffered by 5 % CO_2/22 mM HCO_3^-, pH 7.4. **a** pH_i map of HT29 spheroids with and without DIDS treatment. pH_i gradients for untreated HT29 spheroids = 0.071 ± 0.023. pH_i gradients for DIDS treated HT29 spheroids = 0.180 ± 0.037 (mean spheroid radius = 182.6 ± 15.3um)(bar = 100um). **b** Histograms of the pH_i gradients between the spheroid core and periphery (periph) for HT29 and HCT116 spheroids untreated or treated with DIDS. DIDS treatment increased the pH_i gradient within spheroids due to reducing pH_i in the core of spheroids (*$p < 0.05$, Student's t test). This figure is modified from Figs. 3a and b and 6a of Hulikova *et al.*, 2011. Modified with permission © 2011 The Journal of Biological chemistry

(Gorbatenko et al. 2014). In breast cancer, expression of *SLC4A5* tended to be raised, whilst in colon cancer, increased expression of *SLC4A2* and *SLC4A5* was identified. In lung cancer, there was an increased in expression of *SLC4A3*, *SLC4A7* and *SLC26A6* (Gorbatenko et al. 2014). In colon cancer, *SLC4A4* expression was increased in adenocarcinoma compared to squamous cell carcinoma which had increased expression of *SLC4A3* and also tended to have higher expression of *CA9* and *CA12* (Gorbatenko et al. 2014). Furthermore, hypoxia increases *SLC4A4* expression in the Ls174T colorectal cancer cell line (Parks and Pouyssegur 2015).

8 Metabolons

Interactions between members of the families of proteins involved in pH regulation have been identified, and the formation of pH regulation metabolons proposed, examples of these interactions are described below. Carbonic anhydrase II (CAII) increases the transport activity of MCT1 and MCT4 in a catalytic and non-catalytic way (Becker et al. 2011). CAII function is facilitated by a H^+ shuttle, which was key for this function. The authors hypothesize that the mechanism by which CAII positively regulates MCT1 and MCT4 transport is though removal of high local H^+ around MCT1 and MCT4 (Becker et al. 2011). Similarly, the transport activity of MCT2 was enhanced by the extracellular carbonic anhydrase CAIV (Klier et al. 2011). This effect was only slightly reduced in response to mutation of the inter-molecular H^+ shuttle of CAIV (Klier et al. 2011). It could be hypothesized that CAIX may also potentiate extracellular transport by MCTs dependent on localized H^+ concentrations.

Interactions have also been identified between members of the bicarbonate transporter and carbonic anhydrase families. The bicarbonate transporter SLC26a6 binds directly to CAII through a CAII binding (CAB) site, mutation of which greatly reduced SLC26A6 activity. This highlights the role of metabolon formation in bicarbonate transport and could be a possible therapeutic target (Alvarez et al. 2005). Binding of SLC26A6 to CAII was negatively regulated by protein kinase C (PKC) (Alvarez et al. 2005). CAIX co-immunoprecipitated with bicarbonate transporters such as AE1, AE2 and AE3 but not with SLC26A7 (Morgan et al. 2007). Co-expression of CAIX with AE1, AE2 or AE3 increased the transport rate of these bicarbonate transporters by 32, 28 and 37 %, respectively (Morgan et al. 2007). Further to this, interactions between CAIX and two bicarbonate transporters, AE2 and NBCe1, were identified by a proximity ligation assay (Svastova et al. 2012).

8.1 pH and Cell Phenotype

8.1.1 Proliferation and Growth

An intracellular pH of approximately higher than 7.2 has been identified as a threshold for growth factor-induced DNA synthesis to reinitiate G-/S-phase transition (Pouyssegur et al. 1985). This work compared wild-type and Na^+/H^+ mutant fibroblasts (Pouyssegur et al. 1985). Similarly, an acidic microenvironment (pH 6.6) suppressed G_2-/M-phase transition, after G_2 arrest induced by radiation, by modulating the kinase activity of the cyclin B1–Cdc2 complex (Park et al. 2000). Regulation of intracellular pH by NHE1 activity affected the entry of G_2/M. This work, carried out using NHE1 mutant and wild-type fibroblasts, identified an increase in intracellular pH just before G_2/M which if attenuated in the NHE1 mutant, delays S phase and inhibits G_2/M entry (Putney and Barber 2003). Further work also showed a reduction in the kinase activity of the cyclin B1–Cdc2 complex in response to the more acidic intracellular pH and showed an increase in the expression of Wee1 kinase (Putney and Barber 2003). Knockdown of NHE1 in a small cell lung cancer cell line reduced proliferation and resulted in increased number of cells in G_1 in cell cycle analysis (Li et al. 2009). Ethyl-isopropyl amiloride (EIPA) is an inhibitor of NHE (Hosogi et al. 2012) and EIPA treatment of a gastric cancer cell line reduced proliferation and induced a G_0/G_1 arrest; however, it did not affect intracellular pH (Hosogi et al. 2012). In addition, NHE activity regulates cell volume (Grinstein et al. 1992).

Knockdown of CD147, which acts as a chaperone protein and is required for the correct membrane localization of MCT1 and MCT4, reduced in vitro and in vivo growth of pancreatic cancer cells (Schneiderhan et al. 2009). A further study in a colon cancer cell line found that inhibition or knockdown of MCT1 or MCT4 by shRNA or by targeting CD147 reduced xenograft growth rate (Le Floch et al. 2011). Conversely, constitutive MCT4 expression in RAS-transformed fibroblasts increased growth in xenografts (Chiche et al. 2012). Knockdown and inhibition of MCT4 and MCT1, respectively, increased cellular dependence on oxidative phosphorylation (Marchiq et al. 2015). MCT1 and MCT4 deficiency sensitized cells to a mitochondrial complex 1 inhibitor, phenformin, the combination of which reduced cellular ATP levels and inhibited xenograft growth (Marchiq et al. 2015).

CAIX expression increased 3D spheroid culture growth rate in 2 colon cancer cell lines (McIntyre et al. 2012). CAIX knockdown, investigated in 3 separate studies, reduced the growth rate of xenografts in two human colon, one human breast, one human glioblastoma and one mouse breast cancer cell line model in vivo (McIntyre et al. 2012; Chiche et al. 2009; Lou et al. 2011). Further to this, the reduced xenograft growth rate effect with CAIX knockdown was increased in combination with Bevacizumab, an anti-VEGF inhibitor, which increased the hypoxic fraction of cells within the xenograft (McIntyre et al. 2012). The effects of CAIX knockdown or inhibition with and without Bevacizumab treatment in colon carcinoma and glioblastoma xenografts from McIntyre et al. 2012 are shown in Fig. 5. A role for CAIX in the cancer stem cell niche has been identified. The stem cell niche increases in hypoxia (Lock et al. 2013). mTORC1 signalling that regulates cell metabolism, proliferation and invasion is inhibited by acidic intracellular pH (Balgi et al. 2011). CAIX regulates the expression of markers of cancer stem cells such as *Notch* and *Jagged1* via maintenance of mTORC1 signalling in the hypoxic tumour microenvironment (Lock et al. 2013). Knockdown of the bicarbonate transporter *SLC4A4* in the colorectal cancer cell line Ls174T reduced 2D growth and increased cell mortality in acidic conditions in vitro (Parks and Pouyssegur 2015).

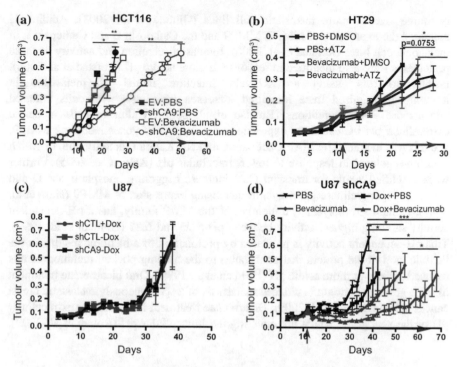

Fig. 5 CAIX knockdown reduces growth rate and enhances Bevacizumab treatment in colon cancer and glioblastoma xenografts. **a** Xenograft growth curves of HCT116 clones ± Bevacizumab treatment. shCA9 (*CA9* knockdown by shRNA) xenografts grow slower than EV (empty vector)(*$p < 0.05$, $n = 5$). Bevacizumab treatment reduced EV growth rate (**$p < 0.01$, $n = 5$). shCA9 xenografts treated with Bevacizumab grow slower than EV treated with Bevacizumab (**$p < 0.01$, $n = 5$). B. Xenograft growth curves of HT29 wt ± acetazolamide (ATZ) ± Bevacizumab treatment. Acetazolamide and Bevacizumab combination treatment reduced xenografts growth rate compared to the untreated xenografts (*$p < 0.05$, $n = 5$) and Bevacizumab treatment alone (*$p < 0.05$, $n = 5$). **c** Xenograft growth curves of U87 doxycycline (Dox) inducible shCTL (control shRNA) ± Dox and shCA9—Dox xenografts. These xenografts grow equally. **d** Xenograft growth curves of U87 doxycycline (Dox) inducible shCA9 ± Dox ± Bevacizumab. Dox-induced shCA9 (*$p < 0.05$, $n = 5$) and Bevacizumab treatment (*p, 0.05, $n = 5$) alone significantly reduced growth rate. Dox-induced shCA9 and Bevacizumab in combination significantly reduced xenografts growth rate compared to the untreated xenografts (***$p < 0.001$, $n = 5$) and Dox (*$p < 0.05$, $n = 5$) or Bevacizumab treatment (*$p < 0.05$, $n = 5$) alone. *Red arrows* denote the start of Bevacizumab treatment (10 mg/kg/3 times a week) and/or acetazolamide treatment (20 mg/kg/day) at 150 mm³ xenograft volume. This figure is reproduced from Fig. 6 of McIntyre et al. (2012). Reproduced with permission © 2012 The AACR

9 The Effect of pH on Migration and Metastasis

Culturing of two tumour cell lines long term in more acidic media (pH 6.8) increased migration and invasion in both (Martinez-Zaguilan et al. 1996). Acidic extracellular pH altered lysosome location to the periphery of the cell and increased lysosome size in more metastatic cell lines (Glunde et al. 2003). Acidic pH increased the expression of MMP2, MMP9 and the Cathepsin B and Cathepsin L in melanoma cell lines (Rofstad et al. 2006). Increased secretion and activity of these proteinases were also identified in response to more acidic pH (Rofstad et al. 2006). Further to this, using a tail vein injection model of metastasis in immune-compromised mice identified increased metastasis with cells cultured under more acidic conditions (Rofstad et al. 2006). Culturing cells in acidic extracellular pH increased the expression of a number of genes including MMP9, and this was mediated by MAPK activation and NF-Kappa B activity both of which were up-regulated in response to low extracellular pH (Kato et al. 2005). Further work identified a role for transient Ca^{2+} increase triggering phospholipase D and acidic sphingomyelinase pathways, in increasing expression of MMP9 (Kato et al. 2007). Human stromelysin-1, a member of the MMP family, has a pH-dependent activity showing highest activity between pH 5.75 and 6.25 (Holman et al. 1999). This pH-dependent activity is regulated by protonation of a histidine residue in the flexible loop of the protein that contributes to the binding pocket, mutation of this reduced activity at more acidic pH (Holman et al. 1999). Oral bicarbonate treatment of mice growing xenografts reduced metastasis of a spontaneously metastatic breast cancer model (Robey et al. 2009). Bicarbonate treatment increased the extracellular pH of the xenografts whilst not affecting the intracellular pH (Robey et al. 2009).

10   The Expression of PH Regulators and Metastasis

V-ATPases were more highly expressed at the membrane of a highly metastatic breast cancer cell line compared with a less metastatic line. This was also associated with greater V-ATPase activity in isolated membranes in the highly metastatic line an effect that was decreased by V-ATPase inhibitors (Sennoune et al. 2004b). Migrating cells have an extracellular pH gradient which is more acid pH at the leading edge and less acid environment at the rear of the cell (Stock et al. 2007). This gradient was related to NHE activity which was increased at the leading edge compared to the rear of the cell (Stock et al. 2007). In a separate study in Madin–Darby canine kidney (MDCK) cells, NHE1 and AE2 localized to the leading edge of lamellipodia (Klein et al. 2000). NHE1 localized to the leading edge of human melanoma cells (Stock et al. 2007). A more recent study identified CAIX localization to the leading edge of lamellipodia upon hepatocyte growth factor (HGF) stimulation (Svastova et al. 2012). At the leading edge of lamellipodia, CAIX colocalizes with 2 bicarbonate transporters, anion exchanger 2 (AE2) and NBCe1 (Svastova et al. 2012). A further study identified that CAIX localizes in focal adhesion structures of lamellipodia and plays a role in focal contact regulation enabling enhanced migration (Csaderova et al. 2013). MCT4 is also localized in the leading edge of migrating cells (Gallagher et al. 2009). Immunohistochemistry analysis of tumour material has identified a correlation with levels of MCT1 and MCT4 and tumour invasiveness (Izumi et al. 2011). In addition, co-expression of

MCT1 and CD147 was associated with metastasis in cervical adenocarcinoma (Pinheiro et al. 2009).

11 Functional Studies of pH Regulators and Their Effects on Migration and Metastasis

MCT4 knockdown reduced migration in ARPE-19 and MDCK cells (Gallagher et al. 2009). MCT4 knockdown in the breast cancer cell line MDA-MB-231 also reduced migration (Gallagher et al. 2007). In addition, another study showed reduced invasion but not migration in response to MCT1 or MCT4 knockdown by siRNA in vitro (Izumi et al. 2011).

Inhibition of NHE reduced migration and acidified Madin–Darby canine Kidney (MCDK-F) cells (Klein et al. 2000). Application of the NHE inhibitors just to the lamellipodium reduced migration, whereas when the inhibitors were directed to the cell body, no effect was seen (Klein et al. 2000). Serum starvation increased cell migration and invasion via increased NHE1 activity (Paradiso et al. 2004). A further study showed that NHE inhibition reduced serum starvation increased cell motility and invasiveness in breast cancer cell lines (Reshkin et al. 2000). In this study, inhibition of PI3-Kinase signalling NHE activity and reduced invasion suggests that PI3-Kinase regulates NHE in serum starvation conditions (Reshkin et al. 2000). Inhibition of NHE1 (using EIPA) suppressed hypoxia-induced migration of HepG2 cells (Yang et al. 2010). Overexpression of NHE1 or NHE7 in a breast cancer cell line increased invasion (Onishi et al. 2012). Lysosomal exocytosis hypothesized to be required for protease secretion and was inhibited by NHE inhibition with EIPA (Steffan et al. 2009). Further to this, NHE inhibition with EIPA or combination of NHE inhibition with cariporide and NHE3 inhibition with s3266 reduced Cathepsin B secretion and HGF-induced invasion in prostate cancer cells in vitro (Steffan and Cardelli 2010). Additionally, NHE1 inhibition also reduced MMP2 and MMP9 expressions via an ERK1/2-dependent mechanism (Yang et al. 2010).

Invasion was reduced in the breast cancer cell line MDA-MB-231 by isoform-specific knockdown of V-ATPase subunit a3 (Hinton et al. 2009). Knockdown of the a4 subunit also reduced invasion but did not affect intracellular pH (Hinton et al. 2009). MMP-9 activity was reduced in vitro in response to V-ATPase inhibition or knockdown with siRNA in pancreatic cell lines. In addition, V-ATPase colocalized with cortactin a factor directing MMP release (Chung et al. 2011). Further to this, FR167356 a specific V-ATPase a3 subunit inhibitor reduced bone metastasis of B16 melanoma cells in vivo (Nishisho et al. 2011). TM9SF4 is a recently identified V-ATPase-associated protein that interacts with the VATPase subunit ATP6V1H, resulting in aberrant constitutive activation of VATPase (Lozupone et al. 2015). Knockdown of TM9SF4 acidified cytosolic pH, alkalinised the extracellular pH and reduced the invasive capacity of colon cancer cell lines (Lozupone et al. 2015).

CAIX overexpression in fibroblasts increased wound healing (migration) at a pH of 6.5 where no difference was seen at a pH of 7.4 (Chiche et al. 2010). Inhibition of CAIX also reduced migration in HeLa cells (Svastova et al. 2012). Knockdown or inhibition of CAIX with small molecular inhibitors reduced metastasis in a mouse breast cancer cell line model in vivo (Lou et al. 2011). CAIX is required for the NF-κB-regulated expression of granulocyte colony-stimulating factor (G-CSF) which drives the mobilization of granulocyte myeloid-derived suppressor cells to the lung metastatic niche of breast cancer (Chafe et al. 2015). This colonization of bone-marrow-derived cells to the premetastatic niche enables metastasis (Chafe et al. 2015). Reactive oxygen species-driven expression of CAIX in prostate cancer-associated fibroblasts (CAF) was identified in normoxia and was found to increase extracellular acidity and expression of matrix metalloproteinase expression by CAF which enabled increased invasion in vitro and metastasis in vivo (Fiaschi et al. 2013). Similarly, inhibition of bicarbonate transport using the inhibitor 4,4'-diisothiocyanate-stilbene-2,2'-disulfonic acid (DIDS) inhibited cell migration in cell culture (Klein et al. 2000). Furthermore, knockdown of the bicarbonate transporter *SLC4A4* in the breast cancer cell line MDA-MB-231 reduced migration and invasion in vitro (Parks and Pouyssegur 2015).

12 Therapeutic Considerations

12.1 The Effects of pH on Chemotherapy and Radiotherapy

Protonation of weakly basic drugs in the acidic extracellular environment is proposed to inhibit uptake of these drugs (Gerweck and Seetharaman 1996). Doxorubicin, a weakly basic drug, increased in concentration and toxicity with a corresponding increase in extracellular pH (Gerweck et al. 1999). Similarly, the toxicity of doxorubicin treatment on a breast cancer cell line in vitro was reduced by reducing the pH from 7.4 to 6.8 (Raghunand et al. 1999). Bicarbonate treatment of mice growing xenografts reduced extracellular pH and increased toxicity of doxorubicin to significantly further reduce xenograft growth rate (Raghunand et al. 1999). A further study of three weakly basic chemotherapeutic compounds (mitoxantrone, doxorubicin and danorubicin) and three weakly acidic chemotherapeutic compounds (cyclophosphamide, 5-flourouracil and chlorambucil) identified a significant difference in IC_{50} response to these compounds in a breast cancer cell line in vitro in media of pH 7.4 versus pH 6.8 (Mahoney et al. 2003). More acidic pH reduced the toxicity of weakly basic drugs whilst increasing toxicity of weakly acidic drugs (Mahoney et al. 2003). This study also identified reduced uptake of weakly basic drugs in more acidic medium, identifying ion trapping as a physiological and microenvironmental explanation for drug resistance (Mahoney et al. 2003). An in vivo study altering the extracellular pH of xenografts by administration of glucose (resulting in a 0.2 pH increase in pH gradient) changed the ratio of the growth delay effect of a weakly acidic drug (Chlorambucil) versus a weakly basic drug (doxorubicin) (Gerweck et al. 2006).

Increased doxorubicin resistance was associated with increased expression and activity of NHE. Inhibition of NHE with EIPA increased doxorubicin uptake and sensitized the resistant line to doxorubicin treatment (Miraglia et al. 2005). The authors hypothesized that NHE resulted in a higher intracellular pH, which reduced doxorubicin accumulation in the resistant cells (Miraglia et al. 2005). NHE1 knockdown in a drug-resistant SCLC cell line significantly increased sensitivity to a number of chemotherapeutic drugs including cisplatin, etoposide and vincristine (Li et al. 2009).

Expression of V-ATPase was increased in 3 cisplatin-resistant cell lines, which had significantly more alkali intracellular pH than the parental cells (Murakami et al. 2001). Knockdown of the V-ATPase subunit ATP6L, identified in this previous study to be overexpressed in cisplatin-resistant cell lines, sensitized a drug-resistant breast cancer cell line to basic chemotherapeutics including doxorubicin and vincristine (You et al. 2009).

In addition to effecting chemotherapeutic response, pH modulation has also been displayed to effect response to radiation therapy. CAIX knockdown and inhibition enhanced the effect of tumour irradiation in a colon cancer xenograft model (Dubois et al. 2011). Similar experiments in a second colon cancer xenograft model with knockdown of CAIX showed a similar effect where CAIX knockdown enhanced radiotherapeutic treatment (Doyen et al. 2012). Furthermore, MCT1 inhibition or siRNA knockdown resulted in increased tumour sensitivity to irradiation (Sonveaux et al. 2008).

13 pH Modulators as Therapeutic Targets

The effects of pH regulation on tumour growth, proliferation and survival in addition to their role in chemotherapy and radiotherapy resistance described above make the pH regulators appealing targets for cancer therapy. Therefore, a number of inhibitors for these compounds have been designed. AR-C155858, an inhibitor of MCT1 and MCT2, has been identified (Ovens et al. 2010). However, it remains to be investigated whether this has any effect on pH regulation in tumours. However, MCT1 may not be a good target for cancer therapy given its role in the rapid proliferation of T cells upon activation that plays an important in immune response (Murray et al. 2005). There is a phase I clinical trial for the MCT1 inhibitor AZD3965 (AstraZeneca) (Table 2).

A number of NHE inhibitors have been developed (reviewed in (Masereel et al. 2003)). Cariporide, a NHE inhibitor, was used in preclinical studies (Masereel et al. 2003; Karmazyn 2000) and in a phase III clinical study investigating the inhibition of NHE on death or myocardial infarction in patients undergoing coronary artery bypass graft surgery (Mentzer et al. 2008). The results showed a decrease in myocardial infarction whilst increasing mortality due to an increase in cerebrovascular events possibly due to cariporide (Mentzer et al. 2008).

Table 2 Clinical trials targeting pH regulators

Target	Compound	Generic name	Study title	Indications	Company/sponsor	Phase	Clinical trials identifier	Status
MCT1	AZD3965		A Phase I Trial of AZD3965 in Patients With Advanced Cancer	Solid tumours	AstraZeneca	Phase I	NCT01791595	Active recruiting
V-ATPase		Omeprazole	Docetaxel and Cisplatin Chemotherapy With or Without High Dose Proton Pump Inhibitor in Metastatic Breast Cancer	Metatatic breast cancer	Fudan University	Phase II	NCT01069081	Completed
V-ATPase		Omeprazole	A Study to Examine the Effects of Esomeprazole on The Pharmacokinetics of Orally Administered Lapatinib in Subjects With Metastatic ErbB2 Positive Breast Cancer	Metastatic ErbB2 positive breast cancer	GlaxoSmithKline	phase I	NCT00849329	Completed
V-ATPase		Pantoprazole	Pantoprazole and Docetaxel for Men With Metastatic Castration-Resistant Prostate Cancer	Metastatic castration-resistant prostate cancer	University Health Network, Toronto	Phase II	NCT01748500	Active
V-ATPase		Pantoprazole	Study Evaluating Pantoprazole With Doxorubicin for Advanced Cancer Patients With Extension Cohort of Patients With Solid Tumours	Solid tumours	University Health Network, Toronto	Phase I	NCT01163903	Unknown
CAIX	cG250	Girentuximab	Monoclonal Antibody Therapy (Rencarex®) in Treating Patients Who Have Undergone Surgery for Non-metastatic Kidney Cancer	Kidney cancer	Wilex	Phase III	NCT00087022	Completed

(continued)

Table 2 (continued)

Target	Compound	Generic name	Study title	Indications	Company/sponsor	Phase	Clinical trials identifier	Status
CAIX	^{124}I-cG250	Girentuximab	Pre-surgical Detection of Clear Cell Renal Cell Carcinoma Using Radiolabeled G250-Antibody	Cancer imaging	Wilex	Phase III	NCT00606632	Completed
CAIX	3ee9-MMAE	BAY-79-4620	cG250 coupled to the cytotoxic drug auristatin	Solid tumours	Bayer	Phase I	NCT01028755	Completed
CAIX	SLC-0111		Safety Study of SLC-0111 in Subjects With Advanced Solid Tumours	Solid tumours	SignalChem Lifesciences Corporation	Phase I	NCT02215850	Active Recruiting
CAIX	DTP348		Trial to Determine Optimal Phase II Dose of the Oral Dual CAIX Inhibitor/Radiosensitizer	Solid tumours	Maastricht Radiation Oncology	Phase I	NCT02216669	Not yet open to participant recruitment

A table detailing some of the clinical trials targeting proteins which regulate pH. Information obtained from clinicaltrials.gov (a service of the US National Institutes of Health), which is a database of worldwide publicly and privately supported human clinical studies. Clinical trials targeting MCT1, V-ATPase and CAIX have been started and in some cases completed. The clinical trials utilising a CAIX-specific chimeric monoclonal antibody (cG250) to target CAIX are not inhibiting CAIX CO_2 hydration capacity. The CAIX-targeting trials use cG250 to induce antibody-dependent cell mediated cytotoxicity, image CAIX-expressing tumours using a radiolabelled version of this antibody and specifically target CAIX-positive tumour cells using a cytotoxic drug–antibody conjugate

Inhibition of V-ATPase can be achieved with proton pump inhibitors (Spugnini et al. 2010). Proton pump inhibitors including esomeprazole have been used for many years clinically for the treatment of gastric acid disorders with few clinical side effects (Der 2003). Treatment of melanoma xenografts with 2.5 mg/kg of esomeprazole reduced tumour tissue pH gradients and tumour growth (De Milito et al. 2010). In addition, NiK-12192, a V-ATPase inhibitor, induced cell death in colon carcinoma cell lines (Supino et al. 2009). This work and other have provided a strong rationale for more testing of V-ATPase inhibitors in the treatment of tumours (Spugnini et al. 2010; De Milito et al. 2012). There are multiple clinical trials using the proton pump inhibitors, omeprazole or pantoprazole, in combination therapy (examples are shown in Table 2).

Many carbonic anhydrase inhibitors have been developed (see (Supuran 2008) for a review). Much work has been done in recent years to identify CAIX-/CAXII-specific inhibitors (Neri and Supuran 2011). CAIX/CAXII inhibitors are derivatives of sulphonamides, sulphamates, sulphamides, coumarins and thio-coumarins (Supuran 2010; Vullo et al. 2005; Pastorekova et al. 2004; Cecchi et al. 2005). Acetazolamide (Diamox) a general carbonic anhydrase inhibitor (Supuran 2008) is used clinically in a number of non-cancer treatment schedules including glaucoma (Kaur et al. 2002). Acetazolamide has also been used in CAIX inhibition in in vivo tumour models (McIntyre et al. 2012; Dubois et al. 2011). However, many drug studies have used much higher concentrations than needed to inhibit CA activity, suggesting off target effects. In addition to chemical inhibition, antibodies targeting CAIX and also CAXII function have been developed (Murri-Plesko et al. 2011; Battke et al. 2011). The CAXII inhibitory antibody (6A10) inhibits CAXII very effectively at low concentrations (Battke et al. 2011). However, the CAIX inhibitory antibody can inhibit a maximum of 76 % of CAIX CO_2 hydration capacity (Murri-Plesko et al. 2011). The CAXII inhibitory antibody (6A10) reduced spheroid growth in vitro and significantly reduced the growth rate of xenografts (Gondi et al. 2013). An antibody against CAIX, which does not inhibit CAIX activity, has also been developed (cG250) and is in clinical trials for clear cell renal cell carcinoma, where it induces natural killer cell interactions via antibody-dependent cellular toxicity (Surfus et al. 1996; Siebels et al. 2011) (Table 2). cG250 is also in 2 additional clinical trials. A phase III study for presurgical detection of clear cell renal cell carcinoma uses a radiolabelled $^{124-}$I-cG250. Additionally in a phase I study, G250 is coupled to the cytotoxic drug auristatin for the treatment of solid tumours (Table 2). A dual targeting bioreductive nitroimidazole-based anti-CAIX sulfamide drug (DH348) has been developed (Dubois et al. 2013). DH348 reduced xenograft growth and sensitized CAIX-positive tumours to radiotherapy (Dubois et al. 2013). A phase I clinical trial to optimize dosing has been initiated (Table 2).

Bicarbonate transport by the SLC4A and SLC26A families is largely inhibited by 4,4′-diisothiocyanatostilbene-2,2′-disulfonic acid (DIDS) (Hulikova et al. 2011). There are also two more specific bicarbonate transport inhibitors. S0859, a specific Na^+-dependent bicarbonate transport inhibitor, has been identified (Ch'en et al. 2008; Larsen et al. 2012). In addition, NDCBE is specifically inhibited by S3705

(Wong et al. 2002). S3705 treatment reduced the intracellular pH of cholangio-carcinoma cells, reduced proliferation, increased apoptosis and inhibited activation of the ERK and AKT pathways upon serum stimulation (Di Sario et al. 2007).

14 Conclusion

The identified phenotypic roles of proteins involved in pH regulation, in proliferation, invasion and metastasis, make these good therapeutic targets. This highlights the importance of further developing drugs against these for clinical assessment. With regard to the development of therapeutic targeting agents, it is worth also considering the numerous normal physiological roles of the monocarboxylate transporters, NHE, the V-ATPases and the bicarbonate transporters. These normal roles make these a less cancer-specific target than CAIX. CAIX is more tightly regulated by hypoxia and tends to be tumour associated, and a CAIX knockout mouse shows little effect on pathology (Gut et al. 2002; Leppilampi et al. 2005). Development of compounds, which inhibit the protein interactions, required for formation of metabolons, or which inhibit membrane localization of the pH regulation proteins, would also be worth pursuing. Future research directions should include the effect of deregulating tumour pH controls on metabolism (Parks et al. 2013) given that this is the source of the acidity and also on signalling cascades following up the effects of pH on signalling described earlier. It is clear that the increased acidity of the tumour microenvironment has many effects and adaptation to this, and other microenvironmental stresses are key in cancer development and progression. Furthermore, understanding this heterogeneous tumour microenvironment will be crucial in developing effective therapeutic strategies.

Acknowledgments This work was supported by funds from Cancer Research UK, Breast Cancer Research Foundation, EU Framework 7 Metoxia and the Oxford Cancer Imaging Centre.

References

Aime S, Barge A, Delli Castelli D, Fedeli F, Mortillaro A, Nielsen FU, Terreno E (2002) Paramagnetic lanthanide(III) complexes as pH-sensitive chemical exchange saturation transfer (CEST) contrast agents for MRI applications. Magn Reson Med Official J Soc Magn Reson Med/Soc Magn Reson Med 47(4):639–648

Akram S, Teong HF, Fliegel L, Pervaiz S, Clement MV (2006) Reactive oxygen species-mediated regulation of the Na+−H+ exchanger 1 gene expression connects intracellular redox status with cells' sensitivity to death triggers. Cell Death Differ 13(4):628–641. doi:10.1038/sj.cdd.4401775

Alper SL (2006) Molecular physiology of SLC4 anion exchangers. Exp Physiol 91(1):153–161. doi:10.1113/expphysiol.2005.031765

Alper SL, Kopito RR, Libresco SM, Lodish HF (1988) Cloning and characterization of a murine band 3-related cDNA from kidney and from a lymphoid cell line. J Biol Chem 263(32):17092–17099

Alterio V, Hilvo M, Di Fiore A, Supuran CT, Pan P, Parkkila S, Scaloni A, Pastorek J, Pastorekova S, Pedone C, Scozzafava A, Monti SM, De Simone G (2009) Crystal structure of the catalytic domain of the tumor-associated human carbonic anhydrase IX. Proc Natl Acad Sci USA 106(38):16233–16238. doi:10.1073/pnas.0908301106

Alvarez BV, Vilas GL, Casey JR (2005) Metabolon disruption: a mechanism that regulates bicarbonate transport. EMBO J 24(14):2499–2511. doi:10.1038/sj.emboj.7600736

Balgi AD, Diering GH, Donohue E, Lam KK, Fonseca BD, Zimmerman C, Numata M, Roberge M (2011) Regulation of mTORC1 signaling by pH. PLoS ONE 6(6):e21549. doi:10.1371/journal.pone.0021549

Barnett DH, Sheng S, Charn TH, Waheed A, Sly WS, Lin CY, Liu ET, Katzenellenbogen BS (2008) Estrogen receptor regulation of carbonic anhydrase XII through a distal enhancer in breast cancer. Cancer Res 68(9):3505–3515. doi:10.1158/0008-5472.CAN-07-6151

Bassler EL, Ngo-Anh TJ, Geisler HS, Ruppersberg JP, Grunder S (2001) Molecular and functional characterization of acid-sensing ion channel (ASIC) 1b. J Biol Chem 276(36):33782–33787. doi:10.1074/jbc.M104030200

Battke C, Kremmer E, Mysliwietz J, Gondi G, Dumitru C, Brandau S, Lang S, Vullo D, Supuran C, Zeidler R (2011) Generation and characterization of the first inhibitory antibody targeting tumour-associated carbonic anhydrase XII. Cancer Immunol Immunother 60(5):649–658. doi:10.1007/s00262-011-0980-z

Becker HM, Klier M, Schuler C, McKenna R, Deitmer JW (2011) Intramolecular proton shuttle supports not only catalytic but also noncatalytic function of carbonic anhydrase II. Proc Natl Acad Sci USA 108(7):3071–3076. doi:10.1073/pnas.1014293108

Beltran AR, Ramirez MA, Carraro-Lacroix LR, Hiraki Y, Reboucas NA, Malnic G (2008) NHE1, NHE2, and NHE4 contribute to regulation of cell pH in T84 colon cancer cells. Pflugers Arch 455(5):799–810. doi:10.1007/s00424-007-0333-0

Beyenbach KW, Wieczorek H (2006) The V-type H+ ATPase: molecular structure and function, physiological roles and regulation. J Exp Biol 209(Pt 4):577–589. doi:10.1242/jeb.02014

Boidot R, Vegran F, Meulle A, Le Breton A, Dessy C, Sonveaux P, Lizard-Nacol S, Feron O (2012) Regulation of monocarboxylate transporter MCT1 expression by p53 mediates inward and outward lactate fluxes in tumors. Cancer Res 72(4):939–948. doi:10.1158/0008-5472.CAN-11-2474

Bouhassira EE, Schwartz RS, Yawata Y, Ata K, Kanzaki A, Qiu JJ, Nagel RL, Rybicki AC (1992) An alanine-to-threonine substitution in protein 4.2 cDNA is associated with a Japanese form of hereditary hemolytic anemia (protein 4.2NIPPON). Blood 79(7):1846–1854

Cecchi A, Hulikova A, Pastorek J, Pastorekova S, Scozzafava A, Winum JY, Montero JL, Supuran CT (2005) Carbonic anhydrase inhibitors. Design of fluorescent sulfonamides as probes of tumor-associated carbonic anhydrase IX that inhibit isozyme IX-mediated acidification of hypoxic tumors. J Med Chem 48(15):4834–4841. doi:10.1021/jm0501073

Chafe SC, Lou Y, Sceneay J, Vallejo M, Hamilton MJ, McDonald PC, Bennewith KL, Moller A, Dedhar S (2015) Carbonic anhydrase IX promotes myeloid-derived suppressor cell mobilization and establishment of a metastatic niche by stimulating G-CSF production. Cancer Res 75(6):996–1008. doi:10.1158/0008-5472.CAN-14-3000

Ch'en FF, Villafuerte FC, Swietach P, Cobden PM, Vaughan-Jones RD (2008) S0859, an N-cyanosulphonamide inhibitor of sodium-bicarbonate cotransport in the heart. Br J Pharmacol 153(5):972–982. doi:10.1038/sj.bjp.0707667

Chernova MN, Jiang L, Friedman DJ, Darman RB, Lohi H, Kere J, Vandorpe DH, Alper SL (2005) Functional comparison of mouse slc26a6 anion exchanger with human SLC26A6 polypeptide variants: differences in anion selectivity, regulation, and electrogenicity. J Biol Chem 280(9):8564–8580. doi:10.1074/jbc.M411703200

Chia SK, Wykoff CC, Watson PH, Han C, Leek RD, Pastorek J, Gatter KC, Ratcliffe P, Harris AL (2001) Prognostic significance of a novel hypoxia-regulated marker, carbonic anhydrase IX, in invasive breast carcinoma. J Clin Oncol Official J Am Soc Clin Oncol 19(16):3660–3668

Chiche J, Ilc K, Laferriere J, Trottier E, Dayan F, Mazure NM, Brahimi-Horn MC, Pouyssegur J (2009) Hypoxia-inducible carbonic anhydrase IX and XII promote tumor cell growth by counteracting acidosis through the regulation of the intracellular pH. Cancer Res 69(1):358–368. doi:10.1158/0008-5472.CAN-08-2470

Chiche J, Ilc K, Brahimi-Horn MC, Pouyssegur J (2010) Membrane-bound carbonic anhydrases are key pH regulators controlling tumor growth and cell migration. Adv Enzyme Regul 50 (1):20–33. doi:10.1016/j.advenzreg.2009.10.005

Chiche J, Le Fur Y, Vilmen C, Frassineti F, Daniel L, Halestrap AP, Cozzone PJ, Pouyssegur J, Lutz NW (2012) In vivo pH in metabolic-defective Ras-transformed fibroblast tumors: key role of the monocarboxylate transporter, MCT4, for inducing an alkaline intracellular pH. Int J Cancer J Int Cancer 130(7):1511–1520. doi:10.1002/ijc.26125

Choi JY, Muallem D, Kiselyov K, Lee MG, Thomas PJ, Muallem S (2001) Aberrant CFTR-dependent HCO3− transport in mutations associated with cystic fibrosis. Nature 410 (6824):94–97. doi:10.1038/35065099

Chung C, Mader CC, Schmitz JC, Atladottir J, Fitchev P, Cornwell ML, Koleske AJ, Crawford SE, Gorelick F (2011) The vacuolar-ATPase modulates matrix metalloproteinase isoforms in human pancreatic cancer. Lab Invest 91(5):732–743. doi:10.1038/labinvest.2011.8

Cleven AH, Wouters BG, Schutte B, Spiertz AJ, van Engeland M, de Bruine AP (2008) Poorer outcome in stromal HIF-2 alpha- and CA9-positive colorectal adenocarcinomas is associated with wild-type TP53 but not with BNIP3 promoter hypermethylation or apoptosis. Br J Cancer 99(5):727–733. doi:10.1038/sj.bjc.6604547

Cordat E, Casey JR (2009) Bicarbonate transport in cell physiology and disease. Biochem J 417 (2):423–439. doi:10.1042/BJ20081634

Csaderova L, Debreova M, Radvak P, Stano M, Vrestiakova M, Kopacek J, Pastorekova S, Svastova E (2013) The effect of carbonic anhydrase IX on focal contacts during cell spreading and migration. Front Physiol 4:271. doi:10.3389/fphys.2013.00271

De Milito A, Canese R, Marino ML, Borghi M, Iero M, Villa A, Venturi G, Lozupone F, Iessi E, Logozzi M, Della Mina P, Santinami M, Rodolfo M, Podo F, Rivoltini L, Fais S (2010) pH-dependent antitumor activity of proton pump inhibitors against human melanoma is mediated by inhibition of tumor acidity. Int J Cancer J Int Cancer 127(1):207–219. doi:10. 1002/ijc.25009

De Milito A, Marino ML, Fais S (2012) A rationale for the use of proton pump inhibitors as antineoplastic agents. Curr Pharm Des 18(10):1395–1406

de Oliveira AT, Pinheiro C, Longatto-Filho A, Brito MJ, Martinho O, Matos D, Carvalho AL, Vazquez VL, Silva TB, Scapulatempo C, Saad SS, Reis RM, Baltazar F (2012) Co-expression of monocarboxylate transporter 1 (MCT1) and its chaperone (CD147) is associated with low survival in patients with gastrointestinal stromal tumors (GISTs). J Bioenerg Biomembr 44 (1):171–178. doi:10.1007/s10863-012-9408-5

Der G (2003) An overview of proton pump inhibitors. Gastroenterol Nurs 26(5):182–190

Di Sario A, Bendia E, Omenetti A, De Minicis S, Marzioni M, Kleemann HW, Candelaresi C, Saccomanno S, Alpini G, Benedetti A (2007) Selective inhibition of ion transport mechanisms regulating intracellular pH reduces proliferation and induces apoptosis in cholangiocarcinoma cells. Dig Liver Dis 39(1):60–69. doi:10.1016/j.dld.2006.07.013

Dietz KJ, Tavakoli N, Kluge C, Mimura T, Sharma SS, Harris GC, Chardonnens AN, Golldack D (2001) Significance of the V-type ATPase for the adaptation to stressful growth conditions and its regulation on the molecular and biochemical level. J Exp Bot 52(363):1969–1980

Doyen J, Parks SK, Marcié S, Pouysségur J, Chiche J (2012) Knock-down of hypoxia-induced carbonic anhydrases IX and XII radiosensitizes tumor cells by increasing intracellular acidosis. Front Mol Cellular Oncol 2 (199) doi:10.3389/fonc.2012.00199

Dubois L, Peeters S, Lieuwes NG, Geusens N, Thiry A, Wigfield S, Carta F, McIntyre A, Scozzafava A, Dogne JM, Supuran CT, Harris AL, Masereel B, Lambin P (2011) Specific inhibition of carbonic anhydrase IX activity enhances the in vivo therapeutic effect of tumor irradiation. Radiother Oncol 99(3):424–431. doi:10.1016/j.radonc.2011.05.045

Dubois L, Peeters SG, van Kuijk SJ, Yaromina A, Lieuwes NG, Saraya R, Biemans R, Rami M, Parvathaneni NK, Vullo D, Vooijs M, Supuran CT, Winum JY, Lambin P (2013) Targeting carbonic anhydrase IX by nitroimidazole based sulfamides enhances the therapeutic effect of tumor irradiation: a new concept of dual targeting drugs. Radiother Oncol J Eur Soc Ther Radiol Oncol 108(3):523–528. doi:10.1016/j.radonc.2013.06.018

Everett LA, Glaser B, Beck JC, Idol JR, Buchs A, Heyman M, Adawi F, Hazani E, Nassir E, Baxevanis AD, Sheffield VC, Green ED (1997) Pendred syndrome is caused by mutations in a putative sulphate transporter gene (PDS). Nat Genet 17(4):411–422. doi:10.1038/ng1297-411

Fendos J, Engelman D (2012) pHLIP and acidity as a universal biomarker for cancer. Yale J Biol Med 85(1):29–35

Fiaschi T, Giannoni E, Taddei ML, Cirri P, Marini A, Pintus G, Nativi C, Richichi B, Scozzafava A, Carta F, Torre E, Supuran CT, Chiarugi P (2013) Carbonic anhydrase IX from cancer-associated fibroblasts drives epithelial-mesenchymal transition in prostate carcinoma cells. Cell Cycle 12(11):1791–1801. doi:10.4161/cc.24902

Gallagher SM, Castorino JJ, Wang D, Philp NJ (2007) Monocarboxylate transporter 4 regulates maturation and trafficking of CD147 to the plasma membrane in the metastatic breast cancer cell line MDA-MB-231. Cancer Res 67(9):4182–4189. doi:10.1158/0008-5472.CAN-06-3184

Gallagher SM, Castorino JJ, Philp NJ (2009) Interaction of monocarboxylate transporter 4 with beta1-integrin and its role in cell migration. Am J Physiol Cell Physiol 296(3):C414–C421. doi:10.1152/ajpcell.00430.2008

Gatenby RA, Smallbone K, Maini PK, Rose F, Averill J, Nagle RB, Worrall L, Gillies RJ (2007) Cellular adaptations to hypoxia and acidosis during somatic evolution of breast cancer. Br J Cancer 97(5):646–653. doi:10.1038/sj.bjc.6603922

Gerlinger M, Santos CR, Spencer-Dene B, Martinez P, Endesfelder D, Burrell RA, Vetter M, Jiang M, Saunders RE, Kelly G, Dykema K, Rioux-Leclercq N, Stamp G, Patard JJ, Larkin J, Howell M, Swanton C (2012) Genome-wide RNA interference analysis of renal carcinoma survival regulators identifies MCT4 as a Warburg effect metabolic target. J Pathol 227(2):146–156. doi:10.1002/path.4006

Gerweck LE, Seetharaman K (1996) Cellular pH gradient in tumor versus normal tissue: potential exploitation for the treatment of cancer. Cancer Res 56(6):1194–1198

Gerweck LE, Kozin SV, Stocks SJ (1999) The pH partition theory predicts the accumulation and toxicity of doxorubicin in normal and low-pH-adapted cells. Br J Cancer 79(5–6):838–842. doi:10.1038/sj.bjc.6690134

Gerweck LE, Vijayappa S, Kozin S (2006) Tumor pH controls the in vivo efficacy of weak acid and base chemotherapeutics. Mol Cancer Ther 5(5):1275–1279. doi:10.1158/1535-7163.MCT-06-0024

Giatromanolaki A, Koukourakis MI, Sivridis E, Pastorek J, Wykoff CC, Gatter KC, Harris AL (2001) Expression of hypoxia-inducible carbonic anhydrase-9 relates to angiogenic pathways and independently to poor outcome in non-small cell lung cancer. Cancer Res 61(21):7992–7998

Gleize V, Boisselier B, Marie Y, Poea-Guyon S, Sanson M, Morel N (2012) The renal v-ATPase a4 subunit is expressed in specific subtypes of human gliomas. Glia 60(6):1004–1012. doi:10.1002/glia.22332

Glitsch M (2011) Protons and Ca2 + : ionic allies in tumor progression? Physiology (Bethesda) 26 (4):252–265. doi:10.1152/physiol.00005.2011

Glunde K, Guggino SE, Solaiyappan M, Pathak AP, Ichikawa Y, Bhujwalla ZM (2003) Extracellular acidification alters lysosomal trafficking in human breast cancer cells. Neoplasia 5 (6):533–545

Gondi G, Mysliwietz J, Hulikova A, Jen JP, Swietach P, Kremmer E, Zeidler R (2013) Antitumor efficacy of a monoclonal antibody that inhibits the activity of cancer-associated carbonic anhydrase XII. Cancer Res 73(21):6494–6503. doi:10.1158/0008-5472.CAN-13-1110

Gorbatenko A, Olesen CW, Boedtkjer E, Pedersen SF (2014) Regulation and roles of bicarbonate transporters in cancer. Front Physiol 5:130. doi:10.3389/fphys.2014.00130

Gossage L, Eisen T (2010) Alterations in VHL as potential biomarkers in renal-cell carcinoma. Nat Rev Clin Oncol 7(5):277–288. doi:10.1038/nrclinonc.2010.42

Grichtchenko II, Choi I, Zhong X, Bray-Ward P, Russell JM, Boron WF (2001) Cloning, characterization, and chromosomal mapping of a human electroneutral Na(+)-driven Cl-HCO3 exchanger. J Biol Chem 276(11):8358–8363. doi:10.1074/jbc.C000716200

Griffiths JR (1991) Are cancer cells acidic? Br J Cancer 64(3):425–427

Grinstein S, Woodside M, Sardet C, Pouyssegur J, Rotin D (1992) Activation of the Na+/H + antiporter during cell volume regulation. Evidence for a phosphorylation-independent mechanism. J Biol Chem 267(33):23823–23828

Gut MO, Parkkila S, Vernerova Z, Rohde E, Zavada J, Hocker M, Pastorek J, Karttunen T, Gibadulinova A, Zavadova Z, Knobeloch KP, Wiedenmann B, Svoboda J, Horak I, Pastorekova S (2002) Gastric hyperplasia in mice with targeted disruption of the carbonic anhydrase gene Car9. Gastroenterology 123(6):1889–1903. doi:10.1053/gast.2002.37052

Halestrap AP, Meredith D (2004) The SLC16 gene family-from monocarboxylate transporters (MCTs) to aromatic amino acid transporters and beyond. Pflugers Arch 447(5):619–628. doi:10.1007/s00424-003-1067-2

Halestrap AP, Price NT (1999) The proton-linked monocarboxylate transporter (MCT) family: structure, function and regulation. Biochem J 343(Pt 2):281–299

Halestrap AP, Wilson MC (2012) The monocarboxylate transporter family—role and regulation. IUBMB Life 64(2):109–119. doi:10.1002/iub.572

Hanahan D, Weinberg RA (2011) Hallmarks of cancer: the next generation. Cell 144(5):646–674. doi:10.1016/j.cell.2011.02.013

He B, Zhang M, Zhu R (2010) Na+/H+ exchanger blockade inhibits the expression of vascular endothelial growth factor in SGC7901 cells. Oncol Rep 23(1):79–87

Hinton A, Sennoune SR, Bond S, Fang M, Reuveni M, Sahagian GG, Jay D, Martinez-Zaguilan R, Forgac M (2009) Function of a subunit isoforms of the V-ATPase in pH homeostasis and in vitro invasion of MDA-MB231 human breast cancer cells. J Biol Chem 284(24):16400–16408. doi:10.1074/jbc.M901201200

Hirschhaeuser F, Menne H, Dittfeld C, West J, Mueller-Klieser W, Kunz-Schughart LA (2010) Multicellular tumor spheroids: an underestimated tool is catching up again. J Biotechnol 148 (1):3–15. doi:10.1016/j.jbiotec.2010.01.012

Holman CM, Kan CC, Gehring MR, Van Wart HE (1999) Role of His-224 in the anomalous pH dependence of human stromelysin-1. Biochemistry 38(2):677–681. doi:10.1021/bi9822170

Holmquist-Mengelbier L, Fredlund E, Lofstedt T, Noguera R, Navarro S, Nilsson H, Pietras A, Vallon-Christersson J, Borg A, Gradin K, Poellinger L, Pahlman S (2006) Recruitment of HIF-1alpha and HIF-2alpha to common target genes is differentially regulated in neuroblastoma: HIF-2alpha promotes an aggressive phenotype. Cancer Cell 10(5):413–423. doi:10. 1016/j.ccr.2006.08.026

Hosogi S, Miyazaki H, Nakajima KI, Ashihara E, Niisato N, Kusuzaki K, Marunaka Y (2012) An inhibitor of Na/H(+) exchanger (NHE), ethyl-isopropyl amiloride (EIPA), diminishes proliferation of MKN28 human gastric cancer cells by decreasing the cytosolic Cl(−) concentration via DIDS-sensitive pathways. Cell Physiol Biochem 30(5):1241–1253. doi:10. 1159/000343315

Huang WC, Swietach P, Vaughan-Jones RD, Ansorge O, Glitsch MD (2008) Extracellular acidification elicits spatially and temporally distinct Ca2+ signals. Curr Biol 18(10):781–785. doi:10.1016/j.cub.2008.04.049

Hulikova A, Vaughan-Jones RD, Swietach P (2011) Dual role of CO2/HCO3(-) buffer in the regulation of intracellular pH of three-dimensional tumor growths. J Biol Chem 286 (16):13815–13826. doi:10.1074/jbc.M111.219899

Hulikova A, Harris AL, Vaughan-Jones RD, Swietach P (2012a) Regulation of intracellular pH in cancer cell lines under normoxia and hypoxia. J Cell Physiol. doi:10.1002/jcp.24221

Hulikova A, Harris AL, Vaughan-Jones RD, Swietach P (2012b) Acid-extrusion from tissue: the interplay between membrane transporters and pH buffers. Curr Pharm Des 18(10):1331–1337

Ihnatko R, Kubes M, Takacova M, Sedlakova O, Sedlak J, Pastorek J, Kopacek J, Pastorekova S (2006) Extracellular acidosis elevates carbonic anhydrase IX in human glioblastoma cells via transcriptional modulation that does not depend on hypoxia. Int J Oncol 29(4):1025–1033

Ilie MI, Hofman V, Ortholan C, Ammadi RE, Bonnetaud C, Havet K, Venissac N, Mouroux J, Mazure NM, Pouyssegur J, Hofman P (2011) Overexpression of carbonic anhydrase XII in tissues from resectable non-small cell lung cancers is a biomarker of good prognosis. Int J Cancer J Int Cancer 128(7):1614–1623. doi:10.1002/ijc.25491

Ishii S, Kihara Y, Shimizu T (2005) Identification of T cell death-associated gene 8 (TDAG8) as a novel acid sensing G-protein-coupled receptor. J Biol Chem 280(10):9083–9087. doi:10.1074/jbc.M407832200

Izumi H, Takahashi M, Uramoto H, Nakayama Y, Oyama T, Wang KY, Sasaguri Y, Nishizawa S, Kohno K (2011) Monocarboxylate transporters 1 and 4 are involved in the invasion activity of human lung cancer cells. Cancer Sci 102(5):1007–1013. doi:10.1111/j.1349-7006.2011.01908.x

Jarolim P, Palek J, Rubin HL, Prchal JT, Korsgren C, Cohen CM (1992) Band 3 Tuscaloosa: Pro327—Arg327 substitution in the cytoplasmic domain of erythrocyte band 3 protein associated with spherocytic hemolytic anemia and partial deficiency of protein 4.2. Blood 80 (2):523–529

Kapoor N, Bartoszewski R, Qadri YJ, Bebok Z, Bubien JK, Fuller CM, Benos DJ (2009) Knockdown of ASIC1 and epithelial sodium channel subunits inhibits glioblastoma whole cell current and cell migration. J Biol Chem 284(36):24526–24541. doi:10.1074/jbc.M109.037390

Karmazyn M (2000) Pharmacology and clinical assessment of cariporide for the treatment coronary artery diseases. Expert Opin Investig Drugs 9(5):1099–1108. doi:10.1517/13543784.9.5.1099

Kato Y, Lambert CA, Colige AC, Mineur P, Noel A, Frankenne F, Foidart JM, Baba M, Hata R, Miyazaki K, Tsukuda M (2005) Acidic extracellular pH induces matrix metalloproteinase-9 expression in mouse metastatic melanoma cells through the phospholipase D-mitogen-activated protein kinase signaling. J Biol Chem 280(12):10938–10944. doi:10.1074/jbc.M411313200

Kato Y, Ozawa S, Tsukuda M, Kubota E, Miyazaki K, St-Pierre Y, Hata R (2007) Acidic extracellular pH increases calcium influx-triggered phospholipase D activity along with acidic sphingomyelinase activation to induce matrix metalloproteinase-9 expression in mouse metastatic melanoma. FEBS J 274(12):3171–3183. doi:10.1111/j.1742-4658.2007.05848.x

Kaur IP, Smitha R, Aggarwal D, Kapil M (2002) Acetazolamide: future perspective in topical glaucoma therapeutics. Int J Pharm 248(1–2):1–14

Kennedy KM, Dewhirst MW (2010) Tumor metabolism of lactate: the influence and therapeutic potential for MCT and CD147 regulation. Future Oncol 6(1):127–148. doi:10.2217/fon.09.145

Kim HL, Seligson D, Liu X, Janzen N, Bui MH, Yu H, Shi T, Belldegrun AS, Horvath S, Figlin RA (2005) Using tumor markers to predict the survival of patients with metastatic renal cell carcinoma. J Urol 173(5):1496–1501. doi:10.1097/01.ju.0000154351.37249.f0

Kim JY, Shin HJ, Kim TH, Cho KH, Shin KH, Kim BK, Roh JW, Lee S, Park SY, Hwang YJ, Han IO (2006) Tumor-associated carbonic anhydrases are linked to metastases in primary cervical cancer. J Cancer Res Clin Oncol 132(5):302–308. doi:10.1007/s00432-005-0068-2

Kirk P, Wilson MC, Heddle C, Brown MH, Barclay AN, Halestrap AP (2000) CD147 is tightly associated with lactate transporters MCT1 and MCT4 and facilitates their cell surface expression. EMBO J 19(15):3896–3904. doi:10.1093/emboj/19.15.3896

Kivela AJ, Parkkila S, Saarnio J, Karttunen TJ, Kivela J, Parkkila AK, Bartosova M, Mucha V, Novak M, Waheed A, Sly WS, Rajaniemi H, Pastorekova S, Pastorek J (2005) Expression of von Hippel-Lindau tumor suppressor and tumor-associated carbonic anhydrases IX and XII in normal and neoplastic colorectal mucosa. World J Gastroenterol 11(17):2616–2625

Klein M, Seeger P, Schuricht B, Alper SL, Schwab A (2000) Polarization of Na(+)/H(+) and Cl (−)/HCO (3)(−) exchangers in migrating renal epithelial cells. J Gen Physiol 115(5):599–608

Klier M, Schuler C, Halestrap AP, Sly WS, Deitmer JW, Becker HM (2011) Transport activity of the high-affinity monocarboxylate transporter MCT2 is enhanced by extracellular carbonic anhydrase IV but not by intracellular carbonic anhydrase II. J Biol Chem 286(31):27781–27791. doi:10.1074/jbc.M111.255331

Kobayashi M, Matsumoto T, Ryuge S, Yanagita K, Nagashio R, Kawakami Y, Goshima N, Jiang SX, Saegusa M, Iyoda A, Satoh Y, Masuda N, Sato Y (2012) CAXII Is a sero-diagnostic marker for lung cancer. PLoS ONE 7(3):e33952. doi:10.1371/journal.pone.0033952

Kopito RR, Lodish HF (1985) Primary structure and transmembrane orientation of the murine anion exchange protein. Nature 316(6025):234–238

Kudrycki KE, Newman PR, Shull GE (1990) cDNA cloning and tissue distribution of mRNAs for two proteins that are related to the band 3 Cl−/HCO3− exchanger. J Biol Chem 265(1):462–471

Kuhajda FP (2000) Fatty-acid synthase and human cancer: new perspectives on its role in tumor biology. Nutrition 16(3):202–208

Kuhajda FP, Jenner K, Wood FD, Hennigar RA, Jacobs LB, Dick JD, Pasternack GR (1994) Fatty acid synthesis: a potential selective target for antineoplastic therapy. Proc Natl Acad Sci USA 91(14):6379–6383

Kuijpers GA, De Pont JJ (1987) Role of proton and bicarbonate transport in pancreatic cell function. Annu Rev Physiol 49:87–103. doi:10.1146/annurev.ph.49.030187.000511

Kunzelmann K (2005) Ion channels and cancer. J Membr Biol 205(3):159–173. doi:10.1007/s00232-005-0781-4

Larsen AM, Krogsgaard-Larsen N, Lauritzen G, Olesen CW, Honore Hansen S, Boedtkjer E, Pedersen SF, Bunch L (2012) Gram-scale solution-phase synthesis of selective sodium bicarbonate co-transport inhibitor S0859: in vitro efficacy studies in breast cancer cells. ChemMedChem 7(10):1808–1814. doi:10.1002/cmdc.201200335

Le Floch R, Chiche J, Marchiq I, Naiken T, Ilk K, Murray CM, Critchlow SE, Roux D, Simon MP, Pouyssegur J (2011) CD147 subunit of lactate/H+ symporters MCT1 and hypoxia-inducible MCT4 is critical for energetics and growth of glycolytic tumors. Proc Natl Acad Sci USA 108(40):16663–16668. doi:10.1073/pnas.1106123108

Leppilampi M, Karttunen TJ, Kivela J, Gut MO, Pastorekova S, Pastorek J, Parkkila S (2005) Gastric pit cell hyperplasia and glandular atrophy in carbonic anhydrase IX knockout mice: studies on two strains C57/BL6 and BALB/C. Transgenic Res 14(5):655–663. doi:10.1007/s11248-005-7215-z

Li S, Bao P, Li Z, Ouyang H, Wu C, Qian G (2009) Inhibition of proliferation and apoptosis induced by a Na+/H+ exchanger-1 (NHE-1) antisense gene on drug-resistant human small cell lung cancer cells. Oncol Rep 21(5):1243–1249

Liao SY, Aurelio ON, Jan K, Zavada J, Stanbridge EJ (1997) Identification of the MN/CA9 protein as a reliable diagnostic biomarker of clear cell carcinoma of the kidney. Cancer Res 57(14):2827–2831

Liu G, Li Y, Sheth VR, Pagel MD (2012) Imaging in vivo extracellular pH with a single paramagnetic chemical exchange saturation transfer magnetic resonance imaging contrast agent. Mol Imaging 11(1):47–57

Lock FE, McDonald PC, Lou Y, Serrano I, Chafe SC, Ostlund C, Aparicio S, Winum JY, Supuran CT, Dedhar S (2013) Targeting carbonic anhydrase IX depletes breast cancer stem cells within the hypoxic niche. Oncogene 32(44):5210–5219. doi:10.1038/onc.2012.550

Lohi H, Kujala M, Makela S, Lehtonen E, Kestila M, Saarialho-Kere U, Markovich D, Kere J (2002) Functional characterization of three novel tissue-specific anion exchangers SLC26A7, −A8, and −A9. J Biol Chem 277(16):14246–14254. doi:10.1074/jbc.M111802200

Loncaster JA, Harris AL, Davidson SE, Logue JP, Hunter RD, Wycoff CC, Pastorek J, Ratcliffe PJ, Stratford IJ, West CM (2001) Carbonic anhydrase (CA IX) expression, a potential new intrinsic marker of hypoxia: correlations with tumor oxygen measurements and prognosis in locally advanced carcinoma of the cervix. Cancer Res 61(17):6394–6399

Longo DL, Busato A, Lanzardo S, Antico F, Aime S (2012) Imaging the pH evolution of an acute kidney injury model by means of iopamidol, a MRI-CEST pH-responsive contrast agent. Magn Reson Med Official J Soc Mag Reson Med/Soc Mag Reson Med. doi:10.1002/mrm.24513

Lou Y, McDonald PC, Oloumi A, Chia S, Ostlund C, Ahmadi A, Kyle A, Auf dem Keller U, Leung S, Huntsman D, Clarke B, Sutherland BW, Waterhouse D, Bally M, Roskelley C, Overall CM, Minchinton A, Pacchiano F, Carta F, Scozzafava A, Touisni N, Winum JY, Supuran CT, Dedhar S (2011) Targeting tumor hypoxia: suppression of breast tumor growth and metastasis by novel carbonic anhydrase IX inhibitors. Cancer Res 71(9):3364–3376. doi:10.1158/0008-5472.CAN-10-4261

Lozupone F, Borghi M, Marzoli F, Azzarito T, Matarrese P, Iessi E, Venturi G, Meschini S, Canitano A, Bona R, Cara A, Fais S (2015) TM9SF4 is a novel V-ATPase-interacting protein that modulates tumor pH alterations associated with drug resistance and invasiveness of colon cancer cells. Oncogene. doi:10.1038/onc.2014.437

Ludwig MG, Vanek M, Guerini D, Gasser JA, Jones CE, Junker U, Hofstetter H, Wolf RM, Seuwen K (2003) Proton-sensing G-protein-coupled receptors. Nature 425(6953):93–98. doi:10.1038/nature01905

Macholl S, Morrison MS, Iveson P, Arbo BE, Andreev OA, Reshetnyak YK, Engelman DM, Johannesen E (2012) In vivo pH imaging with (99 m)Tc-pHLIP. Mol Imaging Biol 14(6):725–734. doi:10.1007/s11307-012-0549-z

Mahoney BP, Raghunand N, Baggett B, Gillies RJ (2003) Tumor acidity, ion trapping and chemotherapeutics. I. Acid pH affects the distribution of chemotherapeutic agents in vitro. Biochem Pharmacol 66(7):1207–1218

Majumdar D, Bevensee MO (2010) Na-coupled bicarbonate transporters of the solute carrier 4 family in the nervous system: function, localization, and relevance to neurologic function. Neuroscience 171(4):951–972. doi:10.1016/j.neuroscience.2010.09.037

Manning Fox JE, Meredith D, Halestrap AP (2000) Characterisation of human monocarboxylate transporter 4 substantiates its role in lactic acid efflux from skeletal muscle. J Physiol 529(Pt 2):285–293

Marchiq I, Le Floch R, Roux D, Simon MP, Pouyssegur J (2015) Genetic disruption of lactate/H+ symporters (MCTs) and their subunit CD147/BASIGIN sensitizes glycolytic tumor cells to phenformin. Cancer Res 75(1):171–180. doi:10.1158/0008-5472.CAN-14-2260

Martinez-Zaguilan R, Lynch RM, Martinez GM, Gillies RJ (1993) Vacuolar-type H(+)-ATPases are functionally expressed in plasma membranes of human tumor cells. Am J Physiol 265(4 Pt 1):C1015–C1029

Martinez-Zaguilan R, Seftor EA, Seftor RE, Chu YW, Gillies RJ, Hendrix MJ (1996) Acidic pH enhances the invasive behavior of human melanoma cells. Clin Exp Metastasis 14(2):176–186

Maseide K, Kandel RA, Bell RS, Catton CN, O'Sullivan B, Wunder JS, Pintilie M, Hedley D, Hill RP (2004) Carbonic anhydrase IX as a marker for poor prognosis in soft tissue sarcoma. Clin Cancer Res Official J Am Assoc Cancer Res 10(13):4464–4471. doi:10.1158/1078-0432. CCR-03-0541

Masereel B, Pochet L, Laeckmann D (2003) An overview of inhibitors of Na(+)/H(+) exchanger. Eur J Med Chem 38(6):547–554

McIntyre A, Patiar S, Wigfield S, Li JL, Ledaki I, Turley H, Leek R, Snell C, Gatter K, Sly WS, Vaughan-Jones RD, Swietach P, Harris AL (2012) Carbonic anhydrase IX promotes tumor growth and necrosis in vivo and inhibition enhances anti-VEGF therapy. Clin Cancer Res Official J Am Assoc Cancer Res 18(11):3100–3111. doi:10.1158/1078-0432.CCR-11-1877

McLean LA, Roscoe J, Jorgensen NK, Gorin FA, Cala PM (2000) Malignant gliomas display altered pH regulation by NHE1 compared with nontransformed astrocytes. Am J Physiol Cell Physiol 278(4):C676–C688

Meijer TW, Schuurbiers OC, Kaanders JH, Looijen-Salamon MG, de Geus-Oei LF, Verhagen AF, Lok J, van der Heijden HF, Rademakers SE, Span PN, Bussink J (2012) Differences in metabolism between adeno- and squamous cell non-small cell lung carcinomas: spatial distribution and prognostic value of GLUT1 and MCT4. Lung Cancer 76(3):316–323. doi:10.1016/j.lungcan.2011.11.006

Mekhail K, Gunaratnam L, Bonicalzi ME, Lee S (2004) HIF activation by pH-dependent nucleolar sequestration of VHL. Nat Cell Biol 6(7):642–647. doi:10.1038/ncb1144

Menendez JA, Lupu R (2007) Fatty acid synthase and the lipogenic phenotype in cancer pathogenesis. Nat Rev Cancer 7(10):763–777. doi:10.1038/nrc2222

Mentzer RM Jr, Bartels C, Bolli R, Boyce S, Buckberg GD, Chaitman B, Haverich A, Knight J, Menasche P, Myers ML, Nicolau J, Simoons M, Thulin L, Weisel RD (2008) Sodium-hydrogen exchange inhibition by cariporide to reduce the risk of ischemic cardiac events in patients undergoing coronary artery bypass grafting: results of the EXPEDITION study. Ann Thorac Surg 85(4):1261–1270. doi:10.1016/j.athoracsur.2007.10.054

Miraglia E, Viarisio D, Riganti C, Costamagna C, Ghigo D, Bosia A (2005) Na +/H + exchanger activity is increased in doxorubicin-resistant human colon cancer cells and its modulation modifies the sensitivity of the cells to doxorubicin. Int J Cancer J Int Cancer 115(6):924–929. doi:10.1002/ijc.20959

Morgan PE, Pastorekova S, Stuart-Tilley AK, Alper SL, Casey JR (2007) Interactions of transmembrane carbonic anhydrase, CAIX, with bicarbonate transporters. Am J Physiol Cell Physiol 293(2):C738–C748. doi:10.1152/ajpcell.00157.2007

Murakami Y, Kanda K, Tsuji M, Kanayama H, Kagawa S (1999) MN/CA9 gene expression as a potential biomarker in renal cell carcinoma. BJU Int 83(7):743–747

Murakami T, Shibuya I, Ise T, Chen ZS, Akiyama S, Nakagawa M, Izumi H, Nakamura T, Matsuo K, Yamada Y, Kohno K (2001) Elevated expression of vacuolar proton pump genes and cellular PH in cisplatin resistance. Int J Cancer J Int Cancer 93(6):869–874

Muriel Lopez C, Esteban E, Berros JP, Pardo P, Astudillo A, Izquierdo M, Crespo G, Sanmamed M, Fonseca PJ, Martinez-Camblor P (2012) Prognostic factors in patients with advanced renal cell carcinoma. Clin Genitourin Cancer 10(4):262–270. doi:10.1016/j.clgc.2012.06.005

Murray CM, Hutchinson R, Bantick JR, Belfield GP, Benjamin AD, Brazma D, Bundick RV, Cook ID, Craggs RI, Edwards S, Evans LR, Harrison R, Holness E, Jackson AP, Jackson CG, Kingston LP, Perry MWD, Ross ARJ, Rugman PA, Sidhu SS, Sullivan M, Taylor-Fishwick DA, Walker PC, Whitehead YM, Wilkinson DJ, Wright A, Donald DK (2005) Monocarboxylate transporter MCT1 is a target for immunosuppression. Nat Chem Biol 1(7):371–376. doi:10.1038/Nchembio744

Murri-Plesko MT, Hulikova A, Oosterwijk E, Scott AM, Zortea A, Harris AL, Ritter G, Old L, Bauer S, Swietach P, Renner C (2011) Antibody inhibiting enzymatic activity of tumour-associated carbonic anhydrase isoform IX. Eur J Pharmacol 657(1–3):173–183. doi:10.1016/j.ejphar.2011.01.063

Nakayama Y, Torigoe T, Inoue Y, Minagawa N, Izumi H, Kohno K, Yamaguchi K (2012) Prognostic significance of monocarboxylate transporter 4 expression in patients with colorectal cancer. Exp Ther Med 3(1):25–30. doi:10.3892/etm.2011.361

Neri D, Supuran CT (2011) Interfering with pH regulation in tumours as a therapeutic strategy. Nat Rev Drug Discov 10(10):767–777. doi:10.1038/nrd3554

Nishi T, Forgac M (2002) The vacuolar (H+)-ATPases—nature's most versatile proton pumps. Nat Rev Mol Cell Biol 3(2):94–103.

Nishisho T, Hata K, Nakanishi M, Morita Y, Sun-Wada GH, Wada Y, Yasui N, Yoneda T (2011) The a3 isoform vacuolar type H(+)-ATPase promotes distant metastasis in the mouse B16 melanoma cells. Mol Cancer Res 9(7):845–855. doi:10.1158/1541-7786.MCR-10-0449

Ohta T, Numata M, Yagishita H, Futagami F, Tsukioka Y, Kitagawa H, Kayahara M, Nagakawa T, Miyazaki I, Yamamoto M, Iseki S, Ohkuma S (1996) Expression of 16 kDa proteolipid of vacuolar-type H(+)-ATPase in human pancreatic cancer. Br J Cancer 73(12):1511–1517

Onishi I, Lin PJ, Numata Y, Austin P, Cipollone J, Roberge M, Roskelley CD, Numata M (2012) Organellar (Na+, K+)/H+ exchanger NHE7 regulates cell adhesion, invasion and anchorage-independent growth of breast cancer MDA-MB-231 cells. Oncol Rep 27(2):311–317. doi:10.3892/or.2011.1542

Opavsky R, Pastorekova S, Zelnik V, Gibadulinova A, Stanbridge EJ, Zavada J, Kettmann R, Pastorek J (1996) Human MN/CA9 gene, a novel member of the carbonic anhydrase family: structure and exon to protein domain relationships. Genomics 33(3):480–487

Orlowski J, Grinstein S (2004) Diversity of the mammalian sodium/proton exchanger SLC9 gene family. Pflugers Arch 447(5):549–565. doi:10.1007/s00424-003-1110-3

Ovens MJ, Davies AJ, Wilson MC, Murray CM, Halestrap AP (2010) AR-C155858 is a potent inhibitor of monocarboxylate transporters MCT1 and MCT2 that binds to an intracellular site involving transmembrane helices 7-10. Biochem J 425(3):523–530. doi:10.1042/BJ20091515

Paradiso A, Cardone RA, Bellizzi A, Bagorda A, Guerra L, Tommasino M, Casavola V, Reshkin SJ (2004) The Na+-H+ exchanger-1 induces cytoskeletal changes involving reciprocal RhoA and Rac1 signaling, resulting in motility and invasion in MDA-MB-435 cells. Breast Cancer Res BCR 6(6):R616–R628. doi:10.1186/bcr922

Park HJ, Lyons JC, Ohtsubo T, Song CW (2000) Cell cycle progression and apoptosis after irradiation in an acidic environment. Cell Death Differ 7(8):729–738. doi:10.1038/sj.cdd.4400702

Parker MD, Bouyer P, Daly CM, Boron WF (2008a) Cloning and characterization of novel human SLC4A8 gene products encoding Na+ −driven Cl−/HCO3(−) exchanger variants NDCBE-A, -C, and -D. Physiol Genomics 34(3):265–276. doi:10.1152/physiolgenomics.90259.2008

Parker MD, Musa-Aziz R, Rojas JD, Choi I, Daly CM, Boron WF (2008b) Characterization of human SLC4A10 as an electroneutral Na/HCO3 cotransporter (NBCn2) with Cl- self-exchange activity. J Biol Chem 283(19):12777–12788. doi:10.1074/jbc.M707829200

Parks SK, Pouyssegur J (2015) The Na(+)/HCO3(-) co-transporter SLC4A4 plays a role in growth and migration of colon and breast cancer cells. J Cell Physiol 230(8):1954–1963. doi:10.1002/jcp.24930

Parks SK, Chiche J, Pouyssegur J (2011) pH control mechanisms of tumor survival and growth. J Cell Physiol 226(2):299–308. doi:10.1002/jcp.22400

Parks SK, Chiche J, Pouyssegur J (2013) Disrupting proton dynamics and energy metabolism for cancer therapy. Nat Rev Cancer 13(9):611–623. doi:10.1038/nrc3579

Pastorek J, Pastorekova S, Callebaut I, Mornon JP, Zelnik V, Opavsky R, Zat'ovicova M, Liao S, Portetelle D, Stanbridge EJ et al (1994) Cloning and characterization of MN, a human tumor-associated protein with a domain homologous to carbonic anhydrase and a putative helix-loop-helix DNA binding segment. Oncogene 9(10):2877–2888

Pastorekova S, Casini A, Scozzafava A, Vullo D, Pastorek J, Supuran CT (2004) Carbonic anhydrase inhibitors: the first selective, membrane-impermeant inhibitors targeting the tumor-associated isozyme IX. Bioorg Med Chem Lett 14(4):869–873. doi:10.1016/j.bmcl.2003.12.029

Pena-Llopis S, Vega-Rubin-de-Celis S, Schwartz JC, Wolff NC, Tran TA, Zou L, Xie XJ, Corey DR, Brugarolas J (2011) Regulation of TFEB and V-ATPases by mTORC1. EMBO J 30 (16):3242–3258. doi:10.1038/emboj.2011.257

Perez-Sayans M, Garcia-Garcia A, Reboiras-Lopez MD, Gandara-Vila P (2009) Role of V-ATPases in solid tumors: importance of the subunit C (Review). Int J Oncol 34(6):1513–1520

Pertega-Gomes N, Vizcaino JR, Miranda-Goncalves V, Pinheiro C, Silva J, Pereira H, Monteiro P, Henrique RM, Reis RM, Lopes C, Baltazar F (2011) Monocarboxylate transporter 4 (MCT4) and CD147 overexpression is associated with poor prognosis in prostate cancer. BMC Cancer 11:312. doi:10.1186/1471-2407-11-312

Petersen OH, Tepikin AV (2008) Polarized calcium signaling in exocrine gland cells. Annu Rev Physiol 70:273–299. doi:10.1146/annurev.physiol.70.113006.100618

Phuoc NB, Ehara H, Gotoh T, Nakano M, Kamei S, Deguchi T, Hirose Y (2008) Prognostic value of the co-expression of carbonic anhydrase IX and vascular endothelial growth factor in patients with clear cell renal cell carcinoma. Oncol Rep 20(3):525–530

Pinheiro C, Longatto-Filho A, Scapulatempo C, Ferreira L, Martins S, Pellerin L, Rodrigues M, Alves VA, Schmitt F, Baltazar F (2008) Increased expression of monocarboxylate transporters

1, 2, and 4 in colorectal carcinomas. Virchows Arch 452(2):139–146. doi:10.1007/s00428-007-0558-5

Pinheiro C, Longatto-Filho A, Pereira SM, Etlinger D, Moreira MA, Jube LF, Queiroz GS, Schmitt F, Baltazar F (2009) Monocarboxylate transporters 1 and 4 are associated with CD147 in cervical carcinoma. Dis Markers 26(3):97–103. doi:10.3233/DMA-2009-0596

Pinheiro C, Reis RM, Ricardo S, Longatto-Filho A, Schmitt F, Baltazar F (2010a) Expression of monocarboxylate transporters 1, 2, and 4 in human tumours and their association with CD147 and CD44. J Biomed Biotechnol 2010:427694. doi:10.1155/2010/427694

Pinheiro C, Albergaria A, Paredes J, Sousa B, Dufloth R, Vieira D, Schmitt F, Baltazar F (2010b) Monocarboxylate transporter 1 is up-regulated in basal-like breast carcinoma. Histopathology 56(7):860–867. doi:10.1111/j.1365-2559.2010.03560.x

Pouyssegur J, Franchi A, L'Allemain G, Paris S (1985) Cytoplasmic pH, a key determinant of growth factor-induced DNA synthesis in quiescent fibroblasts. FEBS Lett 190(1):115–119

Putney LK, Barber DL (2003) Na-H exchange-dependent increase in intracellular pH times G2/M entry and transition. J Biol Chem 278(45):44645–44649. doi:10.1074/jbc.M308099200

Rademakers SE, Lok J, van der Kogel AJ, Bussink J, Kaanders JH (2011) Metabolic markers in relation to hypoxia; staining patterns and colocalization of pimonidazole, HIF-1alpha, CAIX, LDH-5, GLUT-1, MCT1 and MCT4. BMC Cancer 11:167. doi:10.1186/1471-2407-11-167

Raghunand N, He X, van Sluis R, Mahoney B, Baggett B, Taylor CW, Paine-Murrieta G, Roe D, Bhujwalla ZM, Gillies RJ (1999) Enhancement of chemotherapy by manipulation of tumour pH. Br J Cancer 80(7):1005–1011. doi:10.1038/sj.bjc.6690455

Rahman T (2012) Dynamic clustering of IP3 receptors by IP3. Biochem Soc Trans 40(2):325–330. doi:10.1042/BST20110772

Reshkin SJ, Bellizzi A, Albarani V, Guerra L, Tommasino M, Paradiso A, Casavola V (2000) Phosphoinositide 3-kinase is involved in the tumor-specific activation of human breast cancer cell Na(+)/H(+) exchange, motility, and invasion induced by serum deprivation. J Biol Chem 275(8):5361–5369

Riganti C, Gazzano E, Polimeni M, Aldieri E, Ghigo D (2012) The pentose phosphate pathway: an antioxidant defense and a crossroad in tumor cell fate. Free Radic Biol Med 53(3):421–436. doi:10.1016/j.freeradbiomed.2012.05.006

Rios EJ, Fallon M, Wang J, Shimoda LA (2005) Chronic hypoxia elevates intracellular pH and activates Na +/H + exchange in pulmonary arterial smooth muscle cells. Am J Physiol Lung Cell Mol Physiol 289(5):L867–L874. doi:10.1152/ajplung.00455.2004

Robey IF, Baggett BK, Kirkpatrick ND, Roe DJ, Dosescu J, Sloane BF, Hashim AI, Morse DL, Raghunand N, Gatenby RA, Gillies RJ (2009) Bicarbonate increases tumor pH and inhibits spontaneous metastases. Cancer Res 69(6):2260–2268. doi:10.1158/0008-5472.CAN-07-5575

Rofstad EK, Mathiesen B, Kindem K, Galappathi K (2006) Acidic extracellular pH promotes experimental metastasis of human melanoma cells in athymic nude mice. Cancer Res 66 (13):6699–6707. doi:10.1158/0008-5472.CAN-06-0983

Romero MF, Hediger MA, Boulpaep EL, Boron WF (1997) Expression cloning and characterization of a renal electrogenic Na+/HCO3− cotransporter. Nature 387(6631):409–413. doi:10.1038/387409a0

Romero MF, Fulton CM, Boron WF (2004) The SLC4 family of HCO 3—transporters. Pflugers Arch 447(5):495–509. doi:10.1007/s00424-003-1180-2

Rooj AK, McNicholas CM, Bartoszewski R, Bebok Z, Benos DJ, Fuller CM (2012) Glioma-specific cation conductance regulates migration and cell cycle progression. J Biol Chem 287(6):4053–4065. doi:10.1074/jbc.M111.311688

Salway JG (2000) Metabolism at a glance. Blackwell Sciences Ltd, Oxford

Sander T, Toliat MR, Heils A, Leschik G, Becker C, Ruschendorf F, Rohde K, Mundlos S, Nurnberg P (2002) Association of the 867Asp variant of the human anion exchanger 3 gene with common subtypes of idiopathic generalized epilepsy. Epilepsy Res 51(3):249–255

Schneiderhan W, Scheler M, Holzmann KH, Marx M, Gschwend JE, Bucholz M, Gress TM, Seufferlein T, Adler G, Oswald F (2009) CD147 silencing inhibits lactate transport and reduces malignant potential of pancreatic cancer cells in in vivo and in vitro models. Gut 58(10):1391–1398. doi:10.1136/gut.2009.181412

Schulze A, Harris AL (2012) How cancer metabolism is tuned for proliferation and vulnerable to disruption. Nature 491(7424):364–373. doi:10.1038/Nature11706

Scott DA, Wang R, Kreman TM, Sheffield VC, Karniski LP (1999) The Pendred syndrome gene encodes a chloride-iodide transport protein. Nat Genet 21(4):440–443. doi:10.1038/7783

Sennoune SR, Luo D, Martinez-Zaguilan R (2004a) Plasmalemmal vacuolar-type H+-ATPase in cancer biology. Cell Biochem Biophys 40(2):185–206. doi:10.1385/CBB:40:2:185

Sennoune SR, Bakunts K, Martinez GM, Chua-Tuan JL, Kebir Y, Attaya MN, Martinez-Zaguilan R (2004b) Vacuolar H+ -ATPase in human breast cancer cells with distinct metastatic potential: distribution and functional activity. Am J Physiol Cell Physiol 286(6):C1443–C1452. doi:10.1152/ajpcell.00407.2003

Sheth VR, Li Y, Chen LQ, Howison CM, Flask CA, Pagel MD (2012) Measuring in vivo tumor pHe with CEST-FISP MRI. Mag Reson Med Official J Soc Mag Reson Med Soc Mag Reson Med 67(3):760–768. doi:10.1002/mrm.23038

Siebels M, Rohrmann K, Oberneder R, Stahler M, Haseke N, Beck J, Hofmann R, Kindler M, Kloepfer P, Stief C (2011) A clinical phase I/II trial with the monoclonal antibody cG250 (RENCAREX(R)) and interferon-alpha-2a in metastatic renal cell carcinoma patients. World J Urol 29(1):121–126. doi:10.1007/s00345-010-0570-2

Slepkov ER, Rainey JK, Sykes BD, Fliegel L (2007) Structural and functional analysis of the Na+/H+ exchanger. Biochem J 401(3):623–633. doi:10.1042/BJ20061062

Sly WS, Hu PY (1995) Human carbonic anhydrases and carbonic anhydrase deficiencies. Annu Rev Biochem 64:375–401. doi:10.1146/annurev.bi.64.070195.002111

Sonveaux P, Vegran F, Schroeder T, Wergin MC, Verrax J, Rabbani ZN, De Saedeleer CJ, Kennedy KM, Diepart C, Jordan BF, Kelley MJ, Gallez B, Wahl ML, Feron O, Dewhirst MW (2008) Targeting lactate-fueled respiration selectively kills hypoxic tumor cells in mice. J Clin Investig 118(12):3930–3942. doi:10.1172/JCI36843

Southgate CD, Chishti AH, Mitchell B, Yi SJ, Palek J (1996) Targeted disruption of the murine erythroid band 3 gene results in spherocytosis and severe haemolytic anaemia despite a normal membrane skeleton. Nat Genet 14(2):227–230. doi:10.1038/ng1096-227

Spugnini EP, Citro G, Fais S (2010) Proton pump inhibitors as anti vacuolar-ATPases drugs: a novel anticancer strategy. J Exp Clin Cancer Res 29:44. doi:10.1186/1756-9966-29-44

Steffan JJ, Cardelli JA (2010) Thiazolidinediones induce Rab7-RILP-MAPK-dependent juxtanuclear lysosome aggregation and reduce tumor cell invasion. Traffic 11(2):274–286. doi:10.1111/j.1600-0854.2009.01012.x

Steffan JJ, Snider JL, Skalli O, Welbourne T, Cardelli JA (2009) Na+/H+ exchangers and RhoA regulate acidic extracellular pH-induced lysosome trafficking in prostate cancer cells. Traffic 10(6):737–753. doi:10.1111/j.1600-0854.2009.00904.x

Sterling D, Casey JR (1999) Transport activity of AE3 chloride/bicarbonate anion-exchange proteins and their regulation by intracellular pH. Biochem J 344(Pt 1):221–229

Stevens TH, Forgac M (1997) Structure, function and regulation of the vacuolar (H+)-ATPase. Annu Rev Cell Dev Biol 13:779–808. doi:10.1146/annurev.cellbio.13.1.779

Stock C, Mueller M, Kraehling H, Mally S, Noel J, Eder C, Schwab A (2007) pH nanoenvironment at the surface of single melanoma cells. Cell Physiol Biochem 20(5):679–686. doi:10.1159/000107550

Stubbs M, Rodrigues L, Howe FA, Wang J, Jeong KS, Veech RL, Griffiths JR (1994) Metabolic consequences of a reversed pH gradient in rat tumors. Cancer Res 54(15):4011–4016

Suo WH, Zhang N, Wu PP, Zhao L, Song LJ, Shen WW, Zheng L, Tao J, Long XD, Fu GH (2012) Anti-tumour effects of small interfering RNA targeting anion exchanger 1 in experimental gastric cancer. Br J Pharmacol 165(1):135–147. doi:10.1111/j.1476-5381.2011.01521.x

Supino R, Scovassi AI, Croce AC, Dal Bo L, Favini E, Corbelli A, Farina C, Misiano P, Zunino F (2009) Biological effects of a new vacuolar-H,-ATPase inhibitor in colon carcinoma cell lines. Ann N Y Acad Sci 1171:606–616. doi:10.1111/j.1749-6632.2009.04705.x

Supuran CT (2008) Carbonic anhydrases: novel therapeutic applications for inhibitors and activators. Nat Rev Drug Discov 7(2):168–181. doi:10.1038/nrd2467

Supuran CT (2010) Carbonic anhydrase inhibitors. Bioorg Med Chem Lett 20(12):3467–3474. doi:10.1016/j.bmcl.2010.05.009

Surfus JE, Hank JA, Oosterwijk E, Welt S, Lindstrom MJ, Albertini MR, Schiller JH, Sondel PM (1996) Anti-renal-cell carcinoma chimeric antibody G250 facilitates antibody-dependent cellular cytotoxicity with in vitro and in vivo interleukin-2-activated effectors. J Immunother Emphasis Tumor Immunol 19(3):184–191

Svastova E, Witarski W, Csaderova L, Kosik I, Skvarkova L, Hulikova A, Zatovicova M, Barathova M, Kopacek J, Pastorek J, Pastorekova S (2012) Carbonic anhydrase IX interacts with bicarbonate transporters in lamellipodia and increases cell migration via its catalytic domain. J Biol Chem 287(5):3392–3402. doi:10.1074/jbc.M111.286062

Swietach P, Wigfield S, Supuran CT, Harris AL, Vaughan-Jones RD (2008) Cancer-associated, hypoxia-inducible carbonic anhydrase IX facilitates CO_2 diffusion. BJU Int 101(Suppl 4):22–24. doi:10.1111/j.1464-410X.2008.07644.x

Swietach P, Patiar S, Supuran CT, Harris AL, Vaughan-Jones RD (2009) The role of carbonic anhydrase 9 in regulating extracellular and intracellular ph in three-dimensional tumor cell growths. J Biol Chem 284(30):20299–20310. doi:10.1074/jbc.M109.006478

Toei M, Saum R, Forgac M (2010) Regulation and isoform function of the V-ATPases. Biochemistry 49(23):4715–4723. doi:10.1021/bi100397s

Tomashek JJ, Brusilow WS (2000) Stoichiometry of energy coupling by proton-translocating ATPases: a history of variability. J Bioenerg Biomembr 32(5):493–500

Tureci O, Sahin U, Vollmar E, Siemer S, Gottert E, Seitz G, Parkkila AK, Shah GN, Grubb JH, Pfreundschuh M, Sly WS (1998) Human carbonic anhydrase XII: cDNA cloning, expression, and chromosomal localization of a carbonic anhydrase gene that is overexpressed in some renal cell cancers. Proc Natl Acad Sci USA 95(13):7608–7613

Ullah MS, Davies AJ, Halestrap AP (2006) The plasma membrane lactate transporter MCT4, but not MCT1, is up-regulated by hypoxia through a HIF-1alpha-dependent mechanism. J Biol Chem 281(14):9030–9037. doi:10.1074/jbc.M511397200

Vaupel P, Okunieff P, Neuringer LJ (1989) Blood flow, tissue oxygenation, pH distribution, and energy metabolism of murine mammary adenocarcinomas during growth. Adv Exp Med Biol 248:835–845

Virkki LV, Wilson DA, Vaughan-Jones RD, Boron WF (2002) Functional characterization of human NBC4 as an electrogenic Na+ −HCO cotransporter (NBCe2). Am J Physiol Cell Physiol 282(6):C1278–C1289. doi:10.1152/ajpcell.00589.2001

Vullo D, Innocenti A, Nishimori I, Pastorek J, Scozzafava A, Pastorekova S, Supuran CT (2005) Carbonic anhydrase inhibitors. Inhibition of the transmembrane isozyme XII with sulfonamides-a new target for the design of antitumor and antiglaucoma drugs? Bioorg Med Chem Lett 15(4):963–969. doi:10.1016/j.bmcl.2004.12.053

Waldegger S, Moschen I, Ramirez A, Smith RJ, Ayadi H, Lang F, Kubisch C (2001) Cloning and characterization of SLC26A6, a novel member of the solute carrier 26 gene family. Genomics 72(1):43–50. doi:10.1006/geno.2000.6445

Walker NM, Simpson JE, Brazill JM, Gill RK, Dudeja PK, Schweinfest CW, Clarke LL (2009) Role of down-regulated in adenoma anion exchanger in HCO3− secretion across murine duodenum. Gastroenterology 136(3):893–901. doi:10.1053/j.gastro.2008.11.016

Wang JQ, Kon J, Mogi C, Tobo M, Damirin A, Sato K, Komachi M, Malchinkhuu E, Murata N, Kimura T, Kuwabara A, Wakamatsu K, Koizumi H, Uede T, Tsujimoto G, Kurose H, Sato T, Harada A, Misawa N, Tomura H, Okajima F (2004) TDAG8 is a proton-sensing and psychosine-sensitive G-protein-coupled receptor. J Biol Chem 279(44):45626–45633. doi:10.1074/jbc.M406966200

Warburg O (1956) On the origin of cancer cells. Science 123(3191):309–314

Watson PH, Chia SK, Wykoff CC, Han C, Leek RD, Sly WS, Gatter KC, Ratcliffe P, Harris AL (2003) Carbonic anhydrase XII is a marker of good prognosis in invasive breast carcinoma. Br J Cancer 88(7):1065–1070. doi:10.1038/sj.bjc.6600796

Wong P, Kleemann HW, Tannock IF (2002) Cytostatic potential of novel agents that inhibit the regulation of intracellular pH. Br J Cancer 87(2):238–245. doi:10.1038/sj.bjc.6600424

Wykoff CC, Beasley NJ, Watson PH, Turner KJ, Pastorek J, Sibtain A, Wilson GD, Turley H, Talks KL, Maxwell PH, Pugh CW, Ratcliffe PJ, Harris AL (2000) Hypoxia-inducible expression of tumor-associated carbonic anhydrases. Cancer Res 60(24):7075–7083

Wykoff CC, Beasley N, Watson PH, Campo L, Chia SK, English R, Pastorek J, Sly WS, Ratcliffe P, Harris AL (2001) Expression of the hypoxia-inducible and tumor-associated carbonic anhydrases in ductal carcinoma in situ of the breast. Am J Pathol 158(3):1011–1019. doi:10.1016/S0002-9440(10)64048-5

Yang YA, Han WF, Morin PJ, Chrest FJ, Pizer ES (2002) Activation of fatty acid synthesis during neoplastic transformation: role of mitogen-activated protein kinase and phosphatidylinositol 3-kinase. Exp Cell Res 279(1):80–90

Yang X, Wang D, Dong W, Song Z, Dou K (2010) Inhibition of Na(+)/H(+) exchanger 1 by 5-(N-ethyl-N-isopropyl) amiloride reduces hypoxia-induced hepatocellular carcinoma invasion and motility. Cancer Lett 295(2):198–204. doi:10.1016/j.canlet.2010.03.001

You H, Jin J, Shu H, Yu B, De Milito A, Lozupone F, Deng Y, Tang N, Yao G, Fais S, Gu J, Qin W (2009) Small interfering RNA targeting the subunit ATP6L of proton pump V-ATPase overcomes chemoresistance of breast cancer cells. Cancer Lett 280(1):110–119. doi:10.1016/j.canlet.2009.02.023

Zhang X, Lin Y, Gillies RJ (2010) Tumor pH and its measurement. J Nucl Med 51(8):1167–1170. doi:10.2967/jnumed.109.068981

Metabolic Features of Cancer Treatment Resistance

Andrea Viale and Giulio F. Draetta

Abstract

A major barrier to achieving durable remission and a definitive cure in oncology patients is the emergence of tumor resistance, a common outcome of different disease types, and independent from the therapeutic approach undertaken. In recent years, subpopulations of slow-cycling cells endowed with enhanced tumorigenic potential and multidrug resistance have been isolated in different tumors, and mounting experimental evidence suggests these resistant cells are responsible for tumor relapse. An in-depth metabolic characterization of resistant tumor stem cells revealed that they rely more on mitochondrial respiration and less on glycolysis than other tumor cells, a finding that challenges the assumption that tumors have a primarily glycolytic metabolism and defective mitochondria. The demonstration of a metabolic program in resistant tumorigenic cells that may be present in the majority of tumors has important therapeutic implications and is a critical consideration as we address the challenge of identifying new vulnerabilities that might be exploited therapeutically.

Keywords

Tumor resistance · Relapse · Cancer stem cell · Quiescence · Metabolism · Mitochondria · Oxidative phosphorylation

A. Viale (✉) · G.F. Draetta
Department of Genomic Medicine, The University of Texas MD Anderson Cancer Center, Houston, TX 77030, USA
e-mail: aviale@mdanderson.org

G.F. Draetta
Department of Molecular and Cellular Oncology, The University of Texas MD Anderson Cancer Center, Houston, TX 77030, USA

1 Overview of Tumor Resistance: Old Concepts and New Foes

One of the major challenges in clinical oncology is disease progression due to the resistance tumor cells eventually develop in response to pharmacological treatments. Indeed, it is well established in clinical practice that even the most impactful cancer therapies, with the notable exception of emerging immune checkpoint therapies, are doomed to fail after a transitory response. Ironically, it was this same problem of acquired drug resistance that inspired ground-breaking work some 50 years ago leading to the development of modern chemotherapy.

In the 1940s, heroic pioneers in the field of clinical oncology, including Louis Goodman, Alfred Gilman, and Sidney Farber, observed extraordinary therapeutic results in children affected by Hodgkin's lymphoma and acute lymphoblastic leukemia upon treatment with nitrogen mustard and aminopterin derivatives (Farber and Diamond 1948; Goodman et al. 1946). Sadly, their excitement was short-lived, and it soon became evident that their greater challenge was not achieving remission, but maintaining it. It would be nearly twenty years before Emil Freireich and Emil Frei would find an answer to overcome drug resistance by applying the principles learned by physicians using combined anti-bacterial therapy, to the treatment of leukemia. By combining two effective cancer drugs having different mechanisms of action, these pioneers made history by curing their young patients, and the modern chemotherapy was born (Frei et al. 1965). Unfortunately, the path to sustained treatment response would not be so clear for all types of cancer, and the effects of chemotherapy on the majority of solid tumors were minimal due to intrinsic or acquired pleiotropic resistance, as will be further explored in this chapter.

While it was acknowledged early on that multiple factors might be responsible for treatment failure (Brockman 1963; Wilson et al. 2006; Holohan et al. 2013; Zahreddine and Borden 2013), it took until the 1980s before the first insights into the molecular mechanisms that could confer drug resistance were described. Seminal work by Victor Ling's laboratory demonstrated that DNA transfer from resistant cells to sensitive cells could confer multidrug resistance (Debenham et al. 1982). Furthermore, they linked this phenomenon to a single gene, MDR1, which encodes the P-glycoprotein efflux pump (Riordan et al. 1985). It is now known that P-glycoprotein represents just one member of a large, 48-member family of ATP-dependent transporters, referred to as the ATP-binding cassette (ABC) family, which includes several other pumps known to be involved in the efflux of drugs (Gottesman 2002). Because overexpression of these transporters seemed to play a role in drug resistance, at least in vitro, for the majority of cancer types, significant effort was put in attempting to counteracting their activity and resensitizing cancer cells to chemotherapy. Promising preclinical results prompted a number of clinical trials in different indications and using different strategies (Persidis 1999). Despite positive outcomes in some trials, poor study design, a lack of information about target engagement, and lack of evidence of the impact that MDR1 inhibitors would have had on the pharmacokinetic properties of chemotherapeutic drugs made it impossible to tease out the actual contribution of MDR1 inhibitors to clinical response (Fisher et al. 1996; Garraway and Chabner 2002; Binkhathlan and Lavasanifar 2013).

Since the identification of MDR1, additional mechanisms responsible for resistance to specific chemotherapeutic drugs have been identified. For example, some drugs need to be metabolically activated within the cell to exert their function, and mutations modulating activating enzymes can impair drug activity. One classic example is represented by cytarabine (ara-C) a nucleoside analog used to treat acute myelogenous leukemia that becomes active when phosphorylated to ara-C triphosphate within tumor cells (Cros et al. 2004; Huang et al. 1991; Kufe and Spriggs 1985). Conversely, drug metabolism may represent a mechanism of resistance. Ara-C, like other deoxycytidine analogs such as gemcitabine, can be inactivated in cells with increased cytidine deaminase activity (Bardenheuer et al. 2005). Other detoxification pathways exploited by tumor cells include drug conjugation with metallothionein or glutathione by specific transferases (Kelley et al. 1988; Mistry et al. 1991).

Despite their different molecular structure, many cytotoxic drugs used in treatment of cancer, as well as radiotherapy, rely on the same basic mechanism of action: killing tumor cells by inducing extensive DNA damage. As a consequence, a cell DNA repair machinery can play an important role in establishing resistance to chemotherapy (Bouwman and Jonkers 2012). For instance, high activity of nucleotide excision repair (NER), a common repair mechanism induced by DNA-damaging drugs, correlates with poor outcomes in different types of cancer (Kwon et al. 2007; Olaussen et al. 2006). The critical need to limit excessive accumulation of DNA damage is further highlighted by a process known as *genetic reversion*, which is the paradoxical restoration in tumor cells of a previously compromised homologous recombination (HR) repair pathway. This is the primary mechanism by which highly sensitive BRCA-mutated cancers become resistant to platinum-based therapies, wherein they acquire reactivating mutations in BRCA genes that restore protein function (Sakai et al. 2008; Edwards et al. 2008). In the same vein, it is interesting that genes responsible for maintaining genomic integrity, mainly represented by regulators of faithful replication and segregation of chromosomes, are modulated in response to chemotherapy and during relapse and could be used as negative prognostic indicators in several tumors (Carter et al. 2006; Swanton et al. 2009; Zhou et al. 2013).

The strategies exploited by tumor cells to survive chemotherapy are numerous, and they often coexist so that they become resistant to diverse treatments. However, an alternative hypothesis has been proposed to suggest that some cancer cells may be intrinsically more resistant to chemotherapy and less sensitive to DNA damage. The rationale that originally resulted in using chemotherapy as a treatment for cancer is based on the assumption that cancer cells are more sensitive than normal cells because they are proliferating much more rapidly than healthy cells. Even today, the assertion that cancer is a *"proliferative"* disease in which *"cells begin to divide without stopping*" is widely disseminated.[1] However mounting evidence suggests a different scenario as first reported in the late 1960s. Bayard Clarkson was conducting pulse-chase experiments in patients with acute myeloid leukemia (AML) receiving ^{3}H-thymidine to evaluate tumor proliferation. His experiment demonstrated that leukemia cells were almost totally post-mitotic, with only a minority (~ 5 %) of blasts actively dividing (Clarkson et al. 1970; Clarkson 1969; Dick 2008). This dividing subpopulation was comprised primarily of fast-cycling cells (doubling time of ~ 24 h), but a smaller fraction of slow-growing cells, or "*dormant cells*" as named by Clarkson, was present and was characterized by quite infrequent divisions (doubling time lasting from weeks to months). Interestingly, because the dormant subpopulation of cells was insensitive to "*anti-proliferative*" chemotherapy drugs and spared by treatment, it was assumed to be responsible for leukemic regrowth and relapse (Clarkson et al. 1970, 1975; Skipper and Perry 1970), a hypothesis that has been supported by numerous studies in more recent years (see below). These were the original studies that confirmed tumors are not homogenous, but harbor functionally heterogeneous subpopulations of cells that respond differently to drug treatment. Unfortunately, the significance of Clarkson's findings remained virtually neglected for over 30 years, as cancer research continued to be dominated by studies of fast-proliferating cells in 2D culture.

The significance of the dormant cell population, now known as tumor-initiating cells or cancer stem cells, has been confirmed in different subtypes of human AML. Indeed, in support of the hypothesis that these cells are responsible for disease relapse, it has been demonstrated that only the quiescent fraction of blasts is endowed with tumorigenic potential, whereas proliferating cells, even when transplanted in high numbers, always fail to engraft (Guan et al. 2003; Ishikawa et al. 2007). Moreover, quiescence is so intimately linked with tumorigenic potential that it can be used as an unbiased method to isolate leukemic stem cells in mouse models (Viale et al. 2009). Subpopulations of tumor cells with features of cancer stem cells have now also been isolated from solid tumors (Al-Hajj et al. 2003; Collins et al. 2005; Kim et al. 2005; Li et al. 2007; O'Brien et al. 2007; Ricci-Vitiani et al. 2007; Singh et al. 2004), and their existence might explain tumor resistance.

Beside quiescence, other functional properties attributed to stem cells may be responsible for the intrinsic drug resistance of tumors. One critical feature is their enhanced drug efflux capacity due to high basal expression of ABC transporters (Hirschmann-Jax et al. 2004; Patrawala et al. 2005; Wang et al. 2007; Liu et al. 2006). Notably, the widespread use of side population analysis, a flow cytometry technique that isolates stem cells based on their ability to induce efflux of cell-permeable dyes, is based on their elevated ABC transporter activity (Zhou et al. 2001). Other stem cell enzymatic markers relevant for their drug resistance are aldehyde dehydrogenases, a group of enzymes having important biological roles, including the detoxification of exogenously and endogenously generated

---

[1]NCI_http://www.cancer.gov What is cancer? http://www.cancer.gov/about-cancer/what-is-cancer.

aldehydes.[2] Thus, high expression of these enzymes in cancer stem cells, especially isoforms of the ALDH1A subfamily, can confer intrinsic resistance to different chemotherapy drugs (Januchowski et al. 2013). Like ABC transporter activity, this is another functional characteristic that can be exploited to isolate stem cells via the Aldefluor fluorescent reagent system that measures aldehyde dehydrogenase activity. A third recognized property of stem cells that can mediate drug resistance is their increased ability to survive DNA damage (Bao et al. 2006; Diehn et al. 2009). Cancer stem cells, like their normal counterparts, seem to activate alternative checkpoints in response to DNA damage that do not induce cell death; rather, this checkpoint response elicits DNA damage tolerance and repair preserving self-renewal (Viale et al. 2009; Beerman et al. 2014; Insinga et al. 2013). Taken together, the existence of multiple biological mechanisms to escape drug toxicities within stem cells—quiescence, drug efflux and conjugation, and tolerance to DNA damage—may explain how this subpopulation of cells can tolerate multiple chemotoxic agents and drive tumor relapse.

Interestingly, there is evidence that cancer stem cells are able to survive most targeted therapies as well. The genomic revolution in clinical oncology has ushered in a new era of personalized medicine where targeted therapies can be chosen based on identification of driver oncogenes and other mutations (Weinstein 2002; Weinstein and Joe 2006). In many cases, these new targeted therapies are remarkably effective and induce tumor remission (Flaherty et al. 2010; Kantarjian et al. 2002). However, despite a robust initial response, all patients relapse indicating that targeting oncogenes is insufficient to completely eradicate tumor cells (Jang and Atkins 2013; Quintas-Cardama et al. 2009; Felsher 2008). Interestingly, cells surviving targeted therapies share at least two features with the cancer stem cells that survive chemotherapy: dormancy and tumorigenic potential. The appearance of secondary acquired mutations of targeted oncogenes (Choi et al. 2010; Gorre et al. 2001; Kobayashi et al. 2005) or activation of collateral signaling effectors (Kapoor et al. 2014; Nazarian et al. 2010; Pettazzoni et al. 2015; Wagle et al. 2011) has also been documented as mechanism of survival.

Thus, mounting evidence points to selection of slow-cycling tumorigenic cells as an outcome common to all treatment modalities (chemotherapy, radiotherapy or targeted therapies), a phenomenon recently coined "*therapy-selected quiescence*" (Wolf 2014). This sets up an intriguing paradox as it seems that treatment resistance, the major challenge of a "*proliferative disease in which cells begin to divide without stopping*," is actually a matter of quiescence.

2 Metabolism of Treatment-Resistant Cells

Like most aspects of tumor biology, the vast majority of studies investigating tumor metabolism have focused on proliferating cells. Only recently, due to an increased interest in tumor heterogeneity, have researchers begun to focus on the

---

[2]Vasiliou_http://www.aldh.org ALDH.ORG. http://www.aldh.org/website/aldh/.

characterization of different subpopulations of tumor cells, such as the quiescent cells that survive treatment. One reason for this delayed interest is likely related to technical issues, as this small, functionally defined population of cells is difficult to isolate and study. However, a second reason is undoubtedly the overwhelming influence that pioneering studies conducted by Otto Warburg had on metabolism research in cancer.

When Warburg performed his first experiment on live tissue slices and discovered that tumors, unlike normal tissues, were able to produce lactate even in the presence of oxygen (Ferreira 2010; Warburg et al. 1927), he was likely unaware of the profound impact his work would have on the entire field of cancer metabolism. Indeed, aerobic glycolysis or *fermentation*, as it was named by Warburg, was confirmed in vivo by Gerty and Carl Cori shortly thereafter and dominated the conversation around cancer cell metabolism for the next several decades (Ferreira 2010; Garber 2006). This idea that fast-proliferating tumor cells are highly glycolytic and depend primarily on lactate production to regenerate NAD^+ to support continued glycolysis for their anabolic metabolism (Vander Heiden et al. 2009) all but eliminated any interest in the role of mitochondria in tumor metabolism. Warburg himself actively contributed to the widely accepted idea that mitochondria might be dysfunctional in cancer cells. Speaking on the origin of cancer before the German Central Committee for Cancer Control in the 1956, Warburg stated, "*When the respiration of body cells has been irreversibly damaged, cancer cells by no means immediately result*" (Warburg 1956), claiming a direct link between suppression of mitochondrial respiration (OXPHOS) and tumor initiation/progression. Quite interestingly, in the same lecture he also considered exceptions asserting, "*when fermentation has become so great that dedifferentiation has commenced, but not so great that the respiratory defect has been fully compensated for energetically by fermentation, we may have cells which indeed look like cancer cells but are still energetically insufficient. Such cells have been referred to as 'sleeping cancer cells'... and will possibly play a role in chemotherapy*" (Warburg 1956). Astonishingly, Warburg predicted the existence of a subpopulation of "*sleeping*" cells relying on mitochondrial respiration in the same speech in which he anointed impaired mitochondrial respiration as the definitive cause of cancer (Warburg 1956). Even if today the full extent of mitochondrial dysfunction, as well as its significance, is not yet completely understood (Wallace 2012), we have isolated Warburg's predicted OXPHOS-competent "*sleeping cancer cells*," and our deep characterization of this subpopulation is painting an increasingly complex picture of tumor metabolism.

Melanoma is an interesting test case for the cancer stem cell hypothesis. Although cancer stem cells have never been formally isolated in melanoma and is still questioned whether these tumors follow the cancer stem cell model (Quintana et al. 2008, 2010), functional heterogeneity among melanoma cells has been recently described based on the expression of the H3K4 demethylase JARID1B (Roesch et al. 2010). Interestingly, cells characterized by high levels of this demethylase have been shown to have a critical role in tumor maintenance; in fact, down-regulation of JARID1B negatively impacts melanoma self-renewal in serial transplantation (Roesch et al. 2010). High expression of JARID1B can be used to identify a subpopulation of slow-cycling melanoma cells characterized by a doubling time of 4 weeks, and cells positive for JARID1B are mutually exclusive for KI67 expression, behaving as Clarkson's *dormant cells.*" Interestingly JARID1B-positive cells are resistant to various drugs, including platinum-based therapy and BRAF inhibitors. Consequently, different pharmacological treatments lead to selection and enrichment of slow-cycling, tumor-maintaining melanoma cells (Roesch et al. 2013). These data represent the first demonstration that a well-defined subpopulation of quiescent cells endowed with multidrug resistance exists even in very aggressive tumors such as melanoma. Unexpectedly, further characterization of JARID1B-positive cells by mass spectrometry revealed de-regulation of their bioenergetics. Indeed, slow-cycling cells were characterized by a sustained up-regulation of proteins involved in the electron transport chain (ETC), the multiprotein enzymatic complexes responsible for mitochondrial respiration, as well as a significant down-regulation of glycolytic enzymes (Roesch et al. 2013). The authors functionally validated these data demonstrating that JARID1B-positive cells had increased oxygen consumption. As their metabolic features suggested, slow-cycling melanoma cells were dependent on mitochondrial respiration for energetics and exposure to OXPHOS inhibitors in combination with BRAF inhibitors or cisplatin extended mouse survival in animal models. These data demonstrated that targeting JARID1B-positive cells by inhibiting mitochondrial respiration is an effective way to overcome intrinsic multidrug resistance in melanoma (Roesch et al. 2013).

In the same year, two other independent studies demonstrated the importance of OXPHOS in melanoma (Vazquez et al. 2013; Haq et al. 2013). A subset of human melanoma characterized by high expression of the mitochondrial master regulator PGC1α (PPARGC1A) was associated with an increased dependency on mitochondrial respiration and a decreased glycolytic profile. Importantly, the tumorigenic potential of PGC1α-expressing cells was strictly connected to the maintenance of the respiratory gene program, and PGC1α expression was proposed as a negative prognostic factor for patient stratification. Interestingly, as the transcription of PGC1α is negatively regulated by continuous activity of oncogenic signaling, melanoma cells treated with BRAF or MAPK inhibitors showed a robust up-regulation of PGC1α and, consequently, induction of mitochondrial biogenesis. These data further supported that pharmacological resistance, specifically to BRAF inhibitors, selects for cells characterized by increased mitochondrial respiration. As was observed in JARID1B-positive cells, OXPHOS inhibitors have been demonstrated to synergize with pharmacological down-regulation of oncogenic signaling in PGC1α-expressing BRAF inhibitor-resistant cells (Haq et al. 2013).

We recently used a mouse model of pancreatic ductal adenocarcinoma (PDAC) in which we could control the expression of mutated Kras, the driver oncogene found in a recent study to be mutated in 98 % of patients (Biankin et al. 2012), in a time- and tissue-specific manner. We confirmed that continuous oncogenic signaling is indispensable for tumor maintenance, as previously reported (Ying et al. 2012; Collins et al. 2012). Indeed, as predicted by the *oncogene addiction* hypothesis (Weinstein 2002), upon Kras inactivation PDA tumors underwent a rapid and apparently complete regression through apoptosis induction. Surprisingly, a small subpopulation of tumor cells organized in ductal-like structures survived extinction of oncogenic signaling, remaining in a dormant state for months until tumors eventually relapsed (Kapoor et al. 2014; Viale et al. 2014). These oncogene extinction-resistant cells (hereafter, "surviving cells" or "resistant cells"), which appeared well-differentiated and polarized upon histological analysis, were positive for stem cell markers and extremely enriched in tumor-initiating cells representing, de facto, a subpopulation of cells with features of cancer stem cells (Viale et al. 2014).

Transcriptome analysis of resistant cells from our inducible mutant Kras mouse model revealed prominent expression of genes involved in mitochondrial function, such as ETC and β-oxidation, as well as in autophagy and lysosome trafficking. Consistently, resistant cells possessed increased mitochondrial mass and hyperpolarized mitochondria and consumed fourfold more oxygen than tumor cells expressing the Kras oncogene. Another important metabolic trait of resistant cells that clearly emerged through metabolomics analysis and Carbon-13 tracing experiments was suppression of glycolytic activity. In fact, not only did surviving cells have low basal levels of glycolysis, but they were unable to up-regulate glycolytic flux even in response to acute inhibition of mitochondrial respiration, a compensation mechanism known as *glycolytic reserve* that is promptly activated by normal or Kras-expressing tumor cells. In the absence of energetic compensation via glycolytic reserve, resistant tumor cells became exquisitely sensitive to inhibition of mitochondrial respiration and succumbed to metabolic stress and underwent apoptosis in response to OXPHOS inhibitors. Consequently, inhibition of mitochondrial respiration effectively eradicated surviving cells and either prevented or delayed tumor relapse in experimental animals (Viale et al. 2014). Notably, it has been demonstrated in mouse models that treatment-naïve PDAC cells expressing stem cell markers have metabolic features similar to those we observed in treatment-surviving cell: increased mitochondrial mass and high mitochondrial transmembrane potential ($\Delta\psi$) (a measure of ETC activity) (Viale et al. 2014). In fact, in unperturbed PDAC tumors mitochondrial activity ($\Delta\psi$) correlates positively with tumorigenic potential and can be used as an unbiased method to isolate cancer stem cells in mouse models.

Importantly, cells with the same metabolic features as resistant cells identified in our inducible Kras mouse model have been isolated from human PDAC tumors after pharmacological inhibition of oncogenic signaling with Mek and PI3K inhibitors (Viale et al. 2014). Moreover, we found a similar subpopulation of quiescent cells positive for stem cell markers and characterized by high mitochondrial content in patient-derived histological samples upon neo-adjuvant chemotherapy and radiation (our unpublished data). We suspect that these tumor cells, spared by and enriched after conventional therapies, might have a role in tumor relapse for patients as well. Like resistant cells from the mouse, resistant human cells had markedly diminished glycolytic activity and, as expected, were extremely sensitive to inhibition of mitochondrial respiration (our unpublished data). This existence of a subpopulation of cells resistant to different treatments characterized by specific metabolic properties has important clinical implications and could explain why both conventional and recently developed targeted therapies are doomed to fail long term despite the astonishing remissions they initially might achieve.

Recent studies in BCR-ABL and FLT3-ITD leukemia led to similar conclusions. Specifically, OXPHOS inhibition in combination with inhibitors of oncogenic signaling such as imatinib or quizartinib exerted synergistic effects in eradicating small molecule inhibitor-resistant leukemic cells (Alvarez-Calderon et al. 2014). Furthermore, in agreement with studies in melanoma and pancreatic tumors, quiescent leukemia-initiating cells were shown to have a lower energetic level compared to proliferating cells, to rely mainly on OXPHOS for energy production, and to have diminished glycolytic reserve. These features make leukemic stem cells highly sensitive to inhibitors targeting mitochondrial respiration (Lagadinou et al. 2013) and are consistent with data demonstrating that genetic or pharmacological down-regulation of molecules involved in mitochondrial protein translation (inhibiting respiration by depleting respiratory complex subunits), selectively kills leukemic stem and progenitor cells (Skrtic et al. 2011). A promising compound with this activity is tigecycline, a glycylcycline antibiotic identified in a chemical screen and currently in phase I clinical studies in patients with acute myeloid leukemia (Skrtic et al. 2011).

One might argue that metabolic features such as increased mitochondrial activity and decreased glycolysis are simply reflective of the particular cell cycle status of resistant cells, namely their quiescence. Indeed, although an extensive review of the literature reveals that glycolysis is not necessarily up-regulated in proliferating tumor cells (Zu and Guppy 2004), it is well established that quiescent, terminally differentiated cells, at least of ectodermal and endodermal origins, rely mainly on mitochondrial respiration for ATP production (Vander Heiden et al. 2009). For this reason, it is critical to account for cell cycle when studying metabolism. To control for the effect of cell cycle on the metabolic analysis of surviving cells resistant to Kras extinction in PDAC, we always compared the properties of quiescent surviving cells with those of quiescent fully formed Kras-expressing spheres. (When grown in 3D culture, PDAC sphere formation is a dynamic and regulated process in which cells expressing Kras exit the cell cycle when tumor spheres are fully formed.) Consequently, and consistently with BrdU uptake and Ki67 staining, analysis of transcriptomic data confirmed that Kras-expressing fully formed spheres did not have elevated expression of pathways related to cell cycle, DNA replication, or cell division relative to surviving cells (Viale et al. 2014), thus allowing us to appropriately attribute metabolic alterations in surviving cells to activation of an autonomous program.

In comparison with unperturbed tumors that show a remarkable capacity for metabolic adaptation in response to environmental changes, shifting energetic metabolism from glycolysis to OXPHOS and vice versa (Viale et al. 2014; Jose et al. 2011), resistant cells are metabolically much less plastic and show specific dependency on mitochondrial respiration, though the reasons why this is the case are not yet clear. However, their diminished metabolic plasticity and increased dependency on mitochondria do not make surviving cells less competent to cope with environmental stresses than other cells in the tumor. Rather, tumor dormancy is a low energetic state characterized by minimal biosynthetic needs relative to highly proliferative cells (Viale et al. 2014; Lagadinou et al. 2013) and, although counterintuitive, OXPHOS dependency in surviving cells might allow them to survive in hypoxic and nutrient-depleted environments that would limit the growth of actively proliferating cells. In fact, studies in intact cells using high-resolution respirometry demonstrated that oxygen concentration is not a limiting factor for mitochondrial respiration until it drops below 1.0 μM, an extremely hypoxic environment ($\sim$0.1 % pO^2; K_M for oxygen at 0.05 $\pm$ 0.01 kPa) (Scandurra and Gnaiger 2010). Taken together, these metabolic properties portend a survival mechanism in resistant cells wherein they take advantage of their ability to completely oxidize carbon skeletons entering the tricarboxylic acid (TCA) cycle to produce sufficient ATP, even in hostile microenvironments where glucose and oxygen are limiting.

That mitochondrial respiration is critical for the maintenance of a fully transformed phenotype is well established, including evidence of the characterization of $\rho 0$ (rho-0) tumor cells, a state in which cells are completely depleted of mitochondrial DNA (mtDNA) and consequently OXPHOS activity. Historically, $\rho 0$ cells were derived by depleting mtDNA upon long-term exposure of cells to low concentrations of ethidium bromide (King and Attardi 1989). Recently, a study using breast cancer and melanoma tumor cells elegantly demonstrated that tumor engraftment upon transplantation in mice as well as lung seeding was dramatically impaired in $\rho 0$ tumor cells, a result consistent with previous studies (Hayashi et al. 1992; Cavalli et al. 1997; Morais et al. 1994). Interestingly, further characterization of circulating cancer cells and metastases derived from $\rho 0$ tumors revealed completely unexpected results, namely a reactivation of mitochondrial respiration through the acquisition of host mtDNA (Tan et al. 2015). These findings demonstrated that recovery of OXPHOS by a subpopulation of cells endowed with tumorigenic and seeding properties is essential to support tumor dissemination and is consistent with another recent work demonstrating that OXPHOS activity and increased mitochondrial biogenesis are critical features of circulating tumor cells during the metastatic process (LeBleu et al. 2014). Combined, these data indicate that mitochondrial respiration in tumors has an essential role in preserving features related to cancer stemness, such as anchorage-independent growth, tumorigenic potential, drug resistance, and invasion.

3 Therapeutic Implications

Our view of tumors has changed substantially in recent years from that of an unorganized mass of highly proliferative cells to our current understanding of tumors as complex tissues comprised of a mosaic of cell types characterized by different functions, varying proliferative potentials, and in diverse states of differentiation. As such, it is no surprise that they are heterogeneous at the metabolic level as well. In light of the abundance of evidence in different cancer models, it is reasonable to assume that a common metabolic program active in a small subset of quiescent/slow-cycling cells may exist in the majority of tumor types. This metabolic program, in which cells rely on mitochondrial respiration instead of glycolysis, would fulfill the low anabolic needs of quiescent cells.

Although our understanding of *"sleeping cancer cells"* remains incomplete, the insight that multidrug resistance in different tumors is associated with a common metabolic phenotype exhibited by quiescent cells represents an important step in identifying new vulnerabilities that might be exploited to overcome drug resistance. As discussed, a wealth of experimental evidence links dormant cells to tumor resistance; thus, up-regulated OXPHOS has become the metabolic signature of multidrug tumor resistance, implying that targeting OXPHOS should be one mechanism to overcome drug resistance. Moreover, the realization that diverse metabolic programs may not only coexist, but become manifest only in response to genetic or pharmacological perturbations, may shed light on rational drug combinations to produce durable drug responses.

As predicted, treatment with OXPHOS inhibitors can eradicate resistant cells (Roesch et al. 2013; Haq et al. 2013; Viale et al. 2014; Alvarez-Calderon et al. 2014). However, the molecular mechanisms responsible for this effect are not yet completely understood. OXPHOS inhibitors induce apoptosis in resistant cells, and some reports suggest that up-regulation of mitochondrial respiration would primarily confer resistance to cancer drugs whose mechanism of action involves apoptosis induced by reactive oxygen species (ROS) (Vazquez et al. 2013). In other words, ROS generation and not ATP depletion would be responsible for the effect of OXPHOS inhibition (Rohlena et al. 2011). Even though it is well established that mitochondria are the main source of cellular ROS, and drugs interfering with ETC activity are potent inducers of ROS generation (Liu and Schubert 2009), we do not believe this is the mechanism responsible for eradication of resistant cells, at least in pancreatic tumors for several reasons. Our data demonstrate that cells resistant to extinction of oncogenic signaling show high basal levels of ROS production and are able to tolerate very well a further increase. Accordingly, resistant cells have higher levels of glutathione than other tumor cells, and their viability is not affected by depletion of glutathione induced by treatment with the γ-glutamylcysteine synthetase inhibitor buthionine sulphoximine. Similarly, effective concentrations of antioxidants, such as vitamin E and tetrakis, are completely unable to attenuate the effects of OXPHOS inhibitors (Viale et al. 2014). As these data indicate that the effect of ROS generation is not responsible for OXPHOS inhibitor-induced apoptosis in resistant cells, a plausible mechanism to explain the effects exerted by OXPHOS inhibitors is induction of metabolic stress. Because cells resistant to the inhibition of the oncogenic pathways have low energetic levels, even minimal inhibition of mitochondrial respiration is able to induce a critical drop in ATP, leading to the activation of AMP-activated protein kinase (AMPKα). In the absence of compensatory mechanisms, such as the up-regulation of glycolysis, this is sufficient to induce apoptosis. Interestingly, inhibition of OXPHOS has no impact on the energetic levels of other cells in the tumor (Viale et al. 2014).

To our knowledge, our work in PDAC represents the first comprehensive evaluation of mechanisms that confer sensitivity to OXPHOS inhibition in quiescent cells. The fact that a subpopulation of cells responsible for tumor maintenance, metastasis, drug resistance, and tumor relapse exists in different tumors and shares a common metabolic program underscores the importance of deeply characterizing the metabolism of these quiescent cells and determining their sensitivity to OXPHOS inhibition to inform on future therapies that may overcome resistance and eradicate cancer.

4 Autophagy, Mitochondrial Respiration, and Tumor Stem Cells: The Ultimate Theory of Tumor Resistance

We have discussed several mechanisms that make cancer stem cells resistant to therapeutic intervention, including quiescence, high efflux capacity, and high intrinsic reducing potential. However, despite the obvious utility of all of these mechanisms to survive drug treatment, none of them—alone or in their totality—can fully explain how cells can endure the intense toxic assault of most chemotherapy regimens, much less how they retain sufficiently healthy subcellular structures to eventually reactivate their anabolic potential to divide and initiate new tumors. Thus, in addition to merely surviving, cells must also clear and promptly repair damaged cellular structures. As we will discuss in this section, cancer stem cells show a marked increase in autophagy in response to stressors, and this appears to be an important characteristic of stem cells that helps them quickly adapt to changes in the environment, including drug treatment. However, autophagy is not the end of the story, and we posit that the essential linchpin of cancer stem cells' endurance is their mitochondrial pool, which does so by integrating the activity of multiple survival pathways.

Macroautophagy, or simply autophagy, is a highly conserved catabolic cellular process by which cytoplasmic components are sequestered in autophagosomes and delivered to lysosomes for enzymatic degradation. Present at basal level in any cell or induced upon stress, autophagy is the major intracellular degradation and recycling pathway responsible for the elimination of damaged or functionally inactive intracellular components and is critical for cell homeostasis and restoration upon perturbations. In recent years, a new function of autophagy behind its organelle "*quality control*" is emerging: Autophagy maintains metabolic homeostasis and represents an alternative source of energetic and anabolic intermediates in response to environmental depletion of substrates (Rabinowitz and White 2010). Accordingly, activation of autophagy has been linked to the survival of tumor cells in response to chemotherapy and radiotherapy and seems to have an important role in tumor resistance (Apel et al. 2008; Hu et al. 2012; Janku et al. 2011; Sui et al. 2013). Tumor cells respond to drug treatment by activating autophagy as a stress response. This is an early event upon treatment, and activation of autophagy represents a common response that has been observed in several different cancer models and in response to treatment by all agents that induce genotoxic and metabolic stress (Viale et al. 2014; Janku et al. 2011; Sui et al. 2013; Lum et al. 2005). Likewise, inhibition of autophagy sensitizes tumor cells to a variety of treatments in preclinical models (Apel et al. 2008; Janku et al. 2011).

It is no surprise that some of the higher-scoring pathways uncovered by the transcriptomic analysis of oncogene extinction-resistant cells in pancreatic tumors involve autophagy and lysosome trafficking. Indeed, autophagic flux was robustly activated in resistant tumorigenic cells and upon analysis by transmission electron microscopy these cells appeared completely loaded with autophagosomes at various stages of development, a condition that was previously observed in other cancer models upon dampening of mitogenic signaling (Viale et al. 2014; Lum et al. 2005). Based on these observations, we hypothesized that autophagy could contribute to the oxidative metabolism of resistant cells, and we subsequently demonstrated that autophagy was responsible for at least 85 % of mitochondrial respiration in resistant cancer stem cells. In fact, inhibition of lysosome acidification, a common pharmacological strategy for impairing autophagy, induced a remarkable drop in oxygen consumption (Viale et al. 2014). It was previously shown that autophagy can contribute to oxidative metabolism in tumor cells (Guo et al. 2011; Yang et al. 2011), but our analysis suggests that the contribution of autophagy to the metabolism of resistant cells may far exceed what was previously documented.

Massive activation of autophagy may also explain some metabolic abnormalities found in carbon-13 (^{13}C) tracing experiments. Following ^{13}C incorporation, we found that resistant cells relied less on glucose and glutamine and more on pyruvate and palmitate to generate tricarboxylic acid (TCA) cycle intermediates and branching metabolites. However, these experiments also illustrated that substantially more carbon in central metabolism was not accounted for by the labeled substrates in resistant cells compared to other tumor cells, suggesting that the TCA cycle in resistant cells was sustained primarily by endogenous, unlabeled substrates derived from autophagic activity versus from the environment (Viale et al. 2014). Consistent with these findings, starvation of resistant cells did not have any negative impact on their oxygen consumption, and respiration remained constant for a long period of time after complete nutrient withdrawal. This behavior is not mirrored by other tumor cells in which starvation rapidly abrogates their mitochondrial respiration (our unpublished data). These findings are compatible with a model in which surviving cells might be more resistant to environmental stresses than other tumor cells, as they can fuel their energetics completely independently from their microenvironment.

To survive treatment with virtually any drug, tumor cells must abruptly shift from an anabolic to a catabolic program, and they are forced to become quiescent and dismantle organelles and integral macromolecular complexes in a short time. The fact that tumor stem cells are quiescent, have more mitochondria, and rely on

OXPHOS for their energetics implies that they can rapidly adopt this phenotype by exploiting an already existing metabolic program. But even if this metabolic adaptation is relatively simple for cancer stem cells, and they can readily shut down their biosynthetic machinery, cells must still cope with the huge amount of substrates produced by protein and lipid degradation upon induction of exceptional autophagic activity. We reason that the most efficient way for a cell to eliminate this abnormal substrate overload would be through complete oxidation to CO_2 in the mitochondrial TCA cycle, arguing that proficient mitochondria might be critical for autophagic processing in resistant cells. However, a continuous regeneration of NAD^+ would be required in order to sustain TCA cycle flux. One possibility is that cancer stem cells may take advantage of futile metabolic cycles in mitochondria to maintain an efficient autophagic activity. Indeed, in resistant tumor stem cells, uncoupling ATP-synthesizing activity of respiratory complex V from the ETC through the expression of uncoupling proteins would ensure continuous regeneration of NAD^+. This hypothesis is consistent with several reports demonstrating that up-regulation of mitochondrial uncoupling proteins (UCPs), inner-membrane mitochondrial anion transporters that regulate mitochondrial transmembrane potential and ETC efficiency through dissipation of the proton gradient without ATP generation, is responsible for drug resistance in different cancer models (Dalla Pozza et al. 2012; Derdak et al. 2008; Mailloux et al. 2010).

Although effects of UCP expression in tumors are commonly linked to decreased oxidative stress, a few interesting reports have described a role for UCPs in regulating lysosome activity. In fact, UCP2 deficiency leads to accumulation of autophagosomes in tumor cells as well as an impaired phagocytosis in macrophages, suggesting a direct link between lysosome degradation activity and mitochondrial ETC (Dando et al. 2013; Park et al. 2011). We do not yet know if tumor stem cells up-regulate UCPs to take advantage of futile metabolic cycles in order to rid themselves of damaged intracellular components upon drug-induced induction of autophagy, but the possibility that they take advantage of their active mitochondrial respiration to survive induction of massive autophagy might explain how they can survive virtually any kind of treatment while other tumor cells succumb (Fig. 1). However, further studies are needed to confirm the validity of this new model.

Thus, data from our laboratory and others have demonstrated that a subpopulation of resistant tumor stem cells can survive a plethora of different treatments to establish relapse tumors. These cells take advantage of a number of apparently preexisting programs that result in quiescence, high drug efflux capacity, and high reducing potential to abrogate the deleterious consequences of ROS. Resistant tumor stem cells also massively induce autophagy, allowing them to shift from an anabolic to a catabolic program and survive in harsh, hypoxic, nutrient-depleted environments by completely oxidizing carbon skeletons entering the TCA cycle through mitochondrial respiration. When one considers the strong evidence that a subpopulation of dormant cells with these characteristics exists in most tumor types, the challenge before us to cure cancer becomes readily apparent. However, based

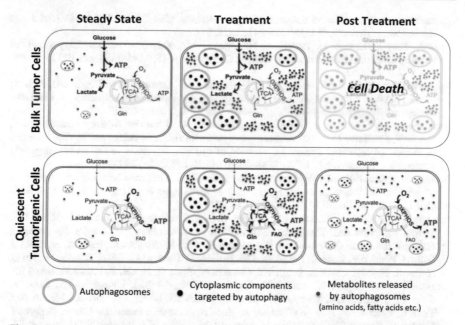

Fig. 1 In tumors, different subpopulations of cells coexist that are characterized by different metabolic programs. Cells representing the bulk of the tumor are mainly glycolytic and minimally dependent on oxidative phosphorylation (OXPHOS); on the other hand, rare quiescent tumor stem cells are more reliant upon mitochondrial respiration. Upon treatment, tumor cells activate an autophagic program in order to eliminate damaged or functionally inactive cellular components. Maintaining an elevated autophagic flux in the event of massive damage, such as in response to pharmacological treatments, is critical for cell survival. Quiescent cells, endowed with high basal OXPHOS, can completely oxidize large amounts of autophagy-derived metabolites entering the TCA cycle, sustaining robust autophagic activity until damaged cytoplasmic components are completely dismantled. Conversely, highly glycolytic proliferating cells devoid of major OXPHOS activity cannot rid themselves of the increased load of autophagy-derived metabolites, become completely burdened with autophagosomes, and undergo apoptosis

on our developing knowledge of cancer stem cell metabolism, we may realize significant clinical impact in the future by drugging the autophagic flux and mitochondria that appear absolutely essential for these cells to survive.

Acknowledgments Authors are thankful to Angela K. Deem for editing the manuscript and apologize for the omission of any primary references.

References

Al-Hajj M, Wicha MS, Benito-Hernandez A, Morrison SJ, Clarke MF (2003) Prospective identification of tumorigenic breast cancer cells. Proc Natl Acad Sci USA 100(7):3983–3988

Alvarez-Calderon F, Gregory MA, Pham-Danis C, DeRyckere D, Stevens BM, Zaberezhnyy V, Hill AA, Gemta L, Kumar A, Kumar V, Wempe MF, Pollyea DA, Jordan CT, Serkova NJ, Graham DK, DeGregori J (2014) Tyrosine kinase inhibition in leukemia induces an altered metabolic state sensitive to mitochondrial perturbations. Clin Cancer Res An Official J Am Assoc Cancer Res. doi:10.1158/1078-0432.CCR-14-2146

Apel A, Herr I, Schwarz H, Rodemann HP, Mayer A (2008) Blocked autophagy sensitizes resistant carcinoma cells to radiation therapy. Cancer Res 68(5):1485–1494. doi:10.1158/0008-5472.CAN-07-0562

Bao S, Wu Q, McLendon RE, Hao Y, Shi Q, Hjelmeland AB, Dewhirst MW, Bigner DD, Rich JN (2006) Glioma stem cells promote radioresistance by preferential activation of the DNA damage response. Nature 444(7120):756–760. doi:10.1038/nature05236

Bardenheuer W, Lehmberg K, Rattmann I, Brueckner A, Schneider A, Sorg UR, Seeber S, Moritz T, Flasshove M (2005) Resistance to cytarabine and gemcitabine and in vitro selection of transduced cells after retroviral expression of cytidine deaminase in human hematopoietic progenitor cells. Leukemia 19(12):2281–2288. doi:10.1038/sj.leu.2403977

Beerman I, Seita J, Inlay MA, Weissman IL, Rossi DJ (2014) Quiescent hematopoietic stem cells accumulate DNA damage during aging that is repaired upon entry into cell cycle. Cell Stem Cell 15(1):37–50. doi:10.1016/j.stem.2014.04.016

Biankin AV, Waddell N, Kassahn KS, Gingras MC, Muthuswamy LB, Johns AL, Miller DK, Wilson PJ, Patch AM, Wu J, Chang DK, Cowley MJ, Gardiner BB, Song S, Harliwong I, Idrisoglu S, Nourse C, Nourbakhsh E, Manning S, Wani S, Gongora M, Pajic M, Scarlett CJ, Gill AJ, Pinho AV, Rooman I, Anderson M, Holmes O, Leonard C, Taylor D, Wood S, Xu Q, Nones K, Fink JL, Christ A, Bruxner T, Cloonan N, Kolle G, Newell F, Pinese M, Mead RS, Humphris JL, Kaplan W, Jones MD, Colvin EK, Nagrial AM, Humphrey ES, Chou A, Chin VT, Chantrill LA, Mawson A, Samra JS, Kench JG, Lovell JA, Daly RJ, Merrett ND, Toon C, Epari K, Nguyen NQ, Barbour A, Zeps N, Australian Pancreatic Cancer Genome I, Kakkar N, Zhao F, Wu YQ, Wang M, Muzny DM, Fisher WE, Brunicardi FC, Hodges SE, Reid JG, Drummond J, Chang K, Han Y, Lewis LR, Dinh H, Buhay CJ, Beck T, Timms L, Sam M, Begley K, Brown A, Pai D, Panchal A, Buchner N, De Borja R, Denroche RE, Yung CK, Serra S, Onetto N, Mukhopadhyay D, Tsao MS, Shaw PA, Petersen GM, Gallinger S, Hruban RH, Maitra A, Iacobuzio-Donahue CA, Schulick RD, Wolfgang CL, Morgan RA, Lawlor RT, Capelli P, Corbo V, Scardoni M, Tortora G, Tempero MA, Mann KM, Jenkins NA, Perez-Mancera PA, Adams DJ, Largaespada DA, Wessels LF, Rust AG, Stein LD, Tuveson DA, Copeland NG, Musgrove EA, Scarpa A, Eshleman JR, Hudson TJ, Sutherland RL, Wheeler DA, Pearson JV, McPherson JD, Gibbs RA, Grimmond SM (2012) Pancreatic cancer genomes reveal aberrations in axon guidance pathway genes. Nature 491(7424):399–405. doi:10.1038/nature11547

Binkhathlan Z, Lavasanifar A (2013) P-glycoprotein inhibition as a therapeutic approach for overcoming multidrug resistance in cancer: current status and future perspectives. Curr Cancer Drug Targets 13(3):326–346

Bouwman P, Jonkers J (2012) The effects of deregulated DNA damage signalling on cancer chemotherapy response and resistance. Nat Rev Cancer 12(9):587–598. doi:10.1038/nrc3342

Brockman RW (1963) Mechanisms of resistance to anticancer agents. Adv Cancer Res 7:129–234

Carter SL, Eklund AC, Kohane IS, Harris LN, Szallasi Z (2006) A signature of chromosomal instability inferred from gene expression profiles predicts clinical outcome in multiple human cancers. Nat Genet 38(9):1043–1048. doi:10.1038/ng1861

Cavalli LR, Varella-Garcia M, Liang BC (1997) Diminished tumorigenic phenotype after depletion of mitochondrial DNA. Cell Growth Differ Mol Biol J Am Assoc Cancer Res 8 (11):1189–1198

Choi YL, Soda M, Yamashita Y, Ueno T, Takashima J, Nakajima T, Yatabe Y, Takeuchi K, Hamada T, Haruta H, Ishikawa Y, Kimura H, Mitsudomi T, Tanio Y, Mano H, Group ALKLCS (2010) EML4-ALK mutations in lung cancer that confer resistance to ALK inhibitors. N Engl J Med 363(18):1734–1739. doi:10.1056/NEJMoa1007478

Clarkson BD (1969) Review of recent studies of cellular proliferation in acute leukemia. Natl Cancer Inst Monogr 30:81–120

Clarkson B, Fried J, Strife A, Sakai Y, Ota K, Okita T (1970) Studies of cellular proliferation in human leukemia. 3. Behavior of leukemic cells in three adults with acute leukemia given continuous infusions of 3H-thymidine for 8 or 10 days. Cancer 25(6):1237–1260

Clarkson BD, Dowling MD, Gee TS, Cunningham IB, Burchenal JH (1975) Treatment of acute leukemia in adults. Cancer 36(2):775–795

Collins AT, Berry PA, Hyde C, Stower MJ, Maitland NJ (2005) Prospective identification of tumorigenic prostate cancer stem cells. Cancer Res 65(23):10946–10951

Collins MA, Bednar F, Zhang Y, Brisset JC, Galban S, Galban CJ, Rakshit S, Flannagan KS, Adsay NV, Pasca di Magliano M (2012) Oncogenic Kras is required for both the initiation and maintenance of pancreatic cancer in mice. J Clin Invest 122(2):639–653. doi:10.1172/JCI59227

Cros E, Jordheim L, Dumontet C, Galmarini CM (2004) Problems related to resistance to cytarabine in acute myeloid leukemia. Leuk Lymphoma 45(6):1123–1132

Dalla Pozza E, Fiorini C, Dando I, Menegazzi M, Sgarbossa A, Costanzo C, Palmieri M, Donadelli M (2012) Role of mitochondrial uncoupling protein 2 in cancer cell resistance to gemcitabine. Biochim Biophys Acta 1823(10):1856–1863. doi:10.1016/j.bbamcr.2012.06.007

Dando I, Fiorini C, Pozza ED, Padroni C, Costanzo C, Palmieri M, Donadelli M (2013) UCP2 inhibition triggers ROS-dependent nuclear translocation of GAPDH and autophagic cell death in pancreatic adenocarcinoma cells. Biochim Biophys Acta 1833(3):672–679. doi:10.1016/j.bbamcr.2012.10.028

Debenham PG, Kartner N, Siminovitch L, Riordan JR, Ling V (1982) DNA-mediated transfer of multiple drug resistance and plasma membrane glycoprotein expression. Mol Cell Biol 2 (8):881–889

Derdak Z, Mark NM, Beldi G, Robson SC, Wands JR, Baffy G (2008) The mitochondrial uncoupling protein-2 promotes chemoresistance in cancer cells. Cancer Res 68(8):2813–2819. doi:10.1158/0008-5472.CAN-08-0053

Dick JE (2008) Stem cell concepts renew cancer research. Blood 112(13):4793–4807. doi:10.1182/blood-2008-08-077941

Diehn M, Cho RW, Lobo NA, Kalisky T, Dorie MJ, Kulp AN, Qian D, Lam JS, Ailles LE, Wong M, Joshua B, Kaplan MJ, Wapnir I, Dirbas FM, Somlo G, Garberoglio C, Paz B, Shen J, Lau SK, Quake SR, Brown JM, Weissman IL, Clarke MF (2009) Association of reactive oxygen species levels and radioresistance in cancer stem cells. Nature 458(7239):780–783. doi:10.1038/nature07733

Edwards SL, Brough R, Lord CJ, Natrajan R, Vatcheva R, Levine DA, Boyd J, Reis-Filho JS, Ashworth A (2008) Resistance to therapy caused by intragenic deletion in BRCA2. Nature 451 (7182):1111–1115. doi:10.1038/nature06548

Farber S, Diamond LK (1948) Temporary remissions in acute leukemia in children produced by folic acid antagonist, 4-aminopteroyl-glutamic acid. N Engl J Med 238(23):787–793. doi:10.1056/NEJM194806032382301

Felsher DW (2008) Tumor dormancy and oncogene addiction. APMIS: Acta Pathologica, Microbiologica, et Immunologica Scandinavica 116(7–8):629–637. doi:10.1111/j.1600-0463.2008.01037.x

Ferreira LM (2010) Cancer metabolism: the Warburg effect today. Exp Mol Pathol 89(3):372–380. doi:10.1016/j.yexmp.2010.08.006

Fisher GA, Lum BL, Hausdorff J, Sikic BI (1996) Pharmacological considerations in the modulation of multidrug resistance. Eur J Cancer 32A(6):1082–1088

Flaherty KT, Puzanov I, Kim KB, Ribas A, McArthur GA, Sosman JA, O'Dwyer PJ, Lee RJ, Grippo JF, Nolop K, Chapman PB (2010) Inhibition of mutated, activated BRAF in metastatic melanoma. N Engl J Med 363(9):809–819. doi:10.1056/NEJMoa1002011

Frei E 3rd, Karon M, Levin RH, Freireich EJ, Taylor RJ, Hananian J, Selawry O, Holland JF, Hoogstraten B, Wolman IJ, Abir E, Sawitsky A, Lee S, Mills SD, Burgert EO Jr, Spurr CL, Patterson RB, Ebaugh FG, James GW 3rd, Moon JH (1965) The effectiveness of combinations of antileukemic agents in inducing and maintaining remission in children with acute leukemia. Blood 26(5):642–656

Garber K (2006) Energy deregulation: licensing tumors to grow. Science 312(5777):1158–1159. doi:10.1126/science.312.5777.1158

Garraway LA, Chabner B (2002) MDR1 inhibition: less resistance or less relevance? Eur J Cancer 38(18):2337–2340

Goodman LS, Wintrobe MM et al (1946) Nitrogen mustard therapy; use of methyl-bis (beta-chloroethyl) amine hydrochloride and tris (beta-chloroethyl) amine hydrochloride for Hodgkin's disease, lymphosarcoma, leukemia and certain allied and miscellaneous disorders. J Am Med Assoc 132:126–132

Gorre ME, Mohammed M, Ellwood K, Hsu N, Paquette R, Rao PN, Sawyers CL (2001) Clinical resistance to STI-571 cancer therapy caused by BCR-ABL gene mutation or amplification. Science 293(5531):876–880. doi:10.1126/science.1062538

Gottesman MM (2002) Mechanisms of cancer drug resistance. Ann Rev Med 53:615–627. doi:10. 1146/annurev.med.53.082901.103929

Guan Y, Gerhard B, Hogge DE (2003) Detection, isolation, and stimulation of quiescent primitive leukemic progenitor cells from patients with acute myeloid leukemia (AML). Blood 101 (8):3142–3149. doi:10.1182/blood-2002-10-3062

Guo JY, Chen HY, Mathew R, Fan J, Strohecker AM, Karsli-Uzunbas G, Kamphorst JJ, Chen G, Lemons JM, Karantza V, Coller HA, Dipaola RS, Gelinas C, Rabinowitz JD, White E (2011) Activated Ras requires autophagy to maintain oxidative metabolism and tumorigenesis. Genes Dev 25(5):460–470. doi:10.1101/gad.2016311

Haq R, Shoag J, Andreu-Perez P, Yokoyama S, Edelman H, Rowe GC, Frederick DT, Hurley AD, Nellore A, Kung AL, Wargo JA, Song JS, Fisher DE, Arany Z, Widlund HR (2013) Oncogenic BRAF regulates oxidative metabolism via PGC1alpha and MITF. Cancer Cell 23 (3):302–315. doi:10.1016/j.ccr.2013.02.003

Hayashi J, Takemitsu M, Nonaka I (1992) Recovery of the missing tumorigenicity in mitochondrial DNA-less HeLa cells by introduction of mitochondrial DNA from normal human cells. Somat Cell Mol Genet 18(2):123–129

Hirschmann-Jax C, Foster AE, Wulf GG, Nuchtern JG, Jax TW, Gobel U, Goodell MA, Brenner MK (2004) A distinct "side population" of cells with high drug efflux capacity in human tumor cells. Proc Natl Acad Sci USA 101(39):14228–14233. doi:10.1073/pnas. 0400067101

Holohan C, Van Schaeybroeck S, Longley DB, Johnston PG (2013) Cancer drug resistance: an evolving paradigm. Nat Rev Cancer 13(10):714–726. doi:10.1038/nrc3599

Hu YL, Jahangiri A, Delay M, Aghi MK (2012) Tumor cell autophagy as an adaptive response mediating resistance to treatments such as antiangiogenic therapy. Cancer Res 72(17): 4294–4299. doi:10.1158/0008-5472.CAN-12-1076

Huang P, Chubb S, Hertel LW, Grindey GB, Plunkett W (1991) Action of 2',2'-difluorodeoxy-cytidine on DNA synthesis. Cancer Res 51(22):6110–6117

Insinga A, Cicalese A, Faretta M, Gallo B, Albano L, Ronzoni S, Furia L, Viale A, Pelicci PG (2013) DNA damage in stem cells activates p21, inhibits p53, and induces symmetric self-renewing divisions. Proc Natl Acad Sci USA 110(10):3931–3936. doi:10.1073/pnas. 1213394110

Ishikawa F, Yoshida S, Saito Y, Hijikata A, Kitamura H, Tanaka S, Nakamura R, Tanaka T, Tomiyama H, Saito N, Fukata M, Miyamoto T, Lyons B, Ohshima K, Uchida N, Taniguchi S, Ohara O, Akashi K, Harada M, Shultz LD (2007) Chemotherapy-resistant human AML stem cells home to and engraft within the bone-marrow endosteal region. Nat Biotechnol 25 (11):1315–1321. doi:10.1038/nbt1350

Jang S, Atkins MB (2013) Which drug, and when, for patients with BRAF-mutant melanoma? Lancet Oncol 14(2):e60–e69. doi:10.1016/S1470-2045(12)70539-9

Janku F, McConkey DJ, Hong DS, Kurzrock R (2011) Autophagy as a target for anticancer therapy. Nat Rev Clin Oncol 8(9):528–539. doi:10.1038/nrclinonc.2011.71

Januchowski R, Wojtowicz K, Zabel M (2013) The role of aldehyde dehydrogenase (ALDH) in cancer drug resistance. Biomed Pharmacother 67(7):669–680. doi:10.1016/j.biopha.2013.04. 005

Jose C, Bellance N, Rossignol R (2011) Choosing between glycolysis and oxidative phosphorylation: a tumor's dilemma? Biochim Biophys Acta 1807(6):552–561. doi:10.1016/j.bbabio. 2010.10.012

Kantarjian H, Sawyers C, Hochhaus A, Guilhot F, Schiffer C, Gambacorti-Passerini C, Niederwieser D, Resta D, Capdeville R, Zoellner U, Talpaz M, Druker B, Goldman J, O'Brien SG, Russell N, Fischer T, Ottmann O, Cony-Makhoul P, Facon T, Stone R, Miller C, Tallman M, Brown R, Schuster M, Loughran T, Gratwohl A, Mandelli F, Saglio G, Lazzarino M, Russo D, Baccarani M, Morra E, International STICMLSG (2002) Hematologic and cytogenetic responses to imatinib mesylate in chronic myelogenous leukemia. N Engl J Med 346(9):645–652. doi:10.1056/NEJMoa011573

Kapoor A, Yao W, Ying H, Hua S, Liewen A, Wang Q, Zhong Y, Wu CJ, Sadanandam A, Hu B, Chang Q, Chu GC, Al-Khalil R, Jiang S, Xia H, Fletcher-Sananikone E, Lim C, Horwitz GI, Viale A, Pettazzoni P, Sanchez N, Wang H, Protopopov A, Zhang J, Heffernan T, Johnson RL, Chin L, Wang YA, Draetta G, DePinho RA (2014) Yap1 activation enables bypass of oncogenic Kras addiction in pancreatic cancer. Cell 158(1):185–197. doi:10.1016/j.cell.2014. 06.003

Kelley SL, Basu A, Teicher BA, Hacker MP, Hamer DH, Lazo JS (1988) Overexpression of metallothionein confers resistance to anticancer drugs. Science 241(4874):1813–1815

Kim CF, Jackson EL, Woolfenden AE, Lawrence S, Babar I, Vogel S, Crowley D, Bronson RT, Jacks T (2005) Identification of bronchioalveolar stem cells in normal lung and lung cancer. Cell 121(6):823–835

King MP, Attardi G (1989) Human cells lacking mtDNA: repopulation with exogenous mitochondria by complementation. Science 246(4929):500–503

Kobayashi S, Boggon TJ, Dayaram T, Janne PA, Kocher O, Meyerson M, Johnson BE, Eck MJ, Tenen DG, Halmos B (2005) EGFR mutation and resistance of non-small-cell lung cancer to gefitinib. N Engl J Med 352(8):786–792. doi:10.1056/NEJMoa044238

Kufe DW, Spriggs DR (1985) Biochemical and cellular pharmacology of cytosine arabinoside. Semin Oncol 12(2 Suppl 3):34–48

Kwon HC, Roh MS, Oh SY, Kim SH, Kim MC, Kim JS, Kim HJ (2007) Prognostic value of expression of ERCC1, thymidylate synthase, and glutathione S-transferase P1 for 5-fluorouracil/oxaliplatin chemotherapy in advanced gastric cancer. Ann Oncol Official J Eur Soc Med Oncol/ESMO 18(3):504–509. doi:10.1093/annonc/mdl430

Lagadinou ED, Sach A, Callahan K, Rossi RM, Neering SJ, Minhajuddin M, Ashton JM, Pei S, Grose V, O'Dwyer KM, Liesveld JL, Brookes PS, Becker MW, Jordan CT (2013) BCL-2 inhibition targets oxidative phosphorylation and selectively eradicates quiescent human leukemia stem cells. Cell Stem Cell 12(3):329–341. doi:10.1016/j.stem.2012.12.013

LeBleu VS, O'Connell JT, Gonzalez Herrera KN, Wikman H, Pantel K, Haigis MC, de Carvalho FM, Damascena A, Domingos Chinen LT, Rocha RM, Asara JM, Kalluri R (2014) PGC-1alpha mediates mitochondrial biogenesis and oxidative phosphorylation in cancer cells to promote metastasis. Nat Cell Biol 16(10):992–1003, 1001–1015. doi:10.1038/ncb3039

Li C, Heidt DG, Dalerba P, Burant CF, Zhang L, Adsay V, Wicha M, Clarke MF, Simeone DM (2007) Identification of pancreatic cancer stem cells. Cancer Res 67(3):1030–1037. doi:10. 1158/0008-5472.CAN-06-2030

Liu Y, Schubert DR (2009) The specificity of neuroprotection by antioxidants. J Biomed Sci 16:98. doi:10.1186/1423-0127-16-98

Liu G, Yuan X, Zeng Z, Tunici P, Ng H, Abdulkadir IR, Lu L, Irvin D, Black KL, Yu JS (2006) Analysis of gene expression and chemoresistance of CD133+ cancer stem cells in glioblastoma. Mol Cancer 5:67. doi:10.1186/1476-4598-5-67

Lum JJ, Bauer DE, Kong M, Harris MH, Li C, Lindsten T, Thompson CB (2005) Growth factor regulation of autophagy and cell survival in the absence of apoptosis. Cell 120(2):237–248. doi:10.1016/j.cell.2004.11.046

Mailloux RJ, Adjeitey CN, Harper ME (2010) Genipin-induced inhibition of uncoupling protein-2 sensitizes drug-resistant cancer cells to cytotoxic agents. PLoS ONE 5(10):e13289. doi:10.1371/journal.pone.0013289

Mistry P, Kelland LR, Abel G, Sidhar S, Harrap KR (1991) The relationships between glutathione, glutathione-S-transferase and cytotoxicity of platinum drugs and melphalan in eight human ovarian carcinoma cell lines. Br J Cancer 64(2):215–220

Morais R, Zinkewich-Peotti K, Parent M, Wang H, Babai F, Zollinger M (1994) Tumor-forming ability in athymic nude mice of human cell lines devoid of mitochondrial DNA. Cancer Res 54 (14):3889–3896

Nazarian R, Shi H, Wang Q, Kong X, Koya RC, Lee H, Chen Z, Lee MK, Attar N, Sazegar H, Chodon T, Nelson SF, McArthur G, Sosman JA, Ribas A, Lo RS (2010) Melanomas acquire resistance to B-RAF(V600E) inhibition by RTK or N-RAS upregulation. Nature 468 (7326):973–977. doi:10.1038/nature09626

O'Brien CA, Pollett A, Gallinger S, Dick JE (2007) A human colon cancer cell capable of initiating tumour growth in immunodeficient mice. Nature 445(7123):106–110

Olaussen KA, Dunant A, Fouret P, Brambilla E, Andre F, Haddad V, Taranchon E, Filipits M, Pirker R, Popper HH, Stahel R, Sabatier L, Pignon JP, Tursz T, Le Chevalier T, Soria JC, Investigators IB (2006) DNA repair by ERCC1 in non-small-cell lung cancer and cisplatin-based adjuvant chemotherapy. N Engl J Med 355(10):983–991. doi:10.1056/NEJMoa060570

Park D, Han CZ, Elliott MR, Kinchen JM, Trampont PC, Das S, Collins S, Lysiak JJ, Hoehn KL, Ravichandran KS (2011) Continued clearance of apoptotic cells critically depends on the phagocyte Ucp2 protein. Nature 477(7363):220–224. doi:10.1038/nature10340

Patrawala L, Calhoun T, Schneider-Broussard R, Zhou J, Claypool K, Tang DG (2005) Side population is enriched in tumorigenic, stem-like cancer cells, whereas ABCG2$^+$ and ABCG2$^-$ cancer cells are similarly tumorigenic. Cancer Res 65(14):6207–6219. doi:10.1158/0008-5472.CAN-05-0592

Persidis A (1999) Cancer multidrug resistance. Nat Biotechnol 17(1):94–95. doi:10.1038/5289

Pettazzoni P, Viale A, Shah P, Carugo A, Ying H, Wang H, Genovese G, Seth S, Minelli R, Green T, Huang-Hobbs E, Corti D, Sanchez N, Nezi L, Marchesini M, Kapoor A, Yao W, Francesco ME, Petrocchi A, Deem AK, Scott K, Colla S, Mills GB, Fleming JB, Heffernan TP, Jones P, Toniatti C, DePinho RA, Draetta GF (2015) Genetic events that limit the efficacy of MEK and RTK inhibitor therapies in a mouse model of KRAS-driven pancreatic cancer. Cancer Res 75(6):1091–1101. doi:10.1158/0008-5472.CAN-14-1854

Quintana E, Shackleton M, Sabel MS, Fullen DR, Johnson TM, Morrison SJ (2008) Efficient tumour formation by single human melanoma cells. Nature 456(7222):593–598. doi:10.1038/nature07567

Quintana E, Shackleton M, Foster HR, Fullen DR, Sabel MS, Johnson TM, Morrison SJ (2010) Phenotypic heterogeneity among tumorigenic melanoma cells from patients that is reversible and not hierarchically organized. Cancer Cell 18(5):510–523. doi:10.1016/j.ccr.2010.10.012

Quintas-Cardama A, Kantarjian H, Cortes J (2009) Imatinib and beyond–exploring the full potential of targeted therapy for CML. Nat Rev Clin Oncol 6(9):535–543. doi:10.1038/nrclinonc.2009.112

Rabinowitz JD, White E (2010) Autophagy and metabolism. Science 330(6009):1344–1348. doi:10.1126/science.1193497

Ricci-Vitiani L, Lombardi DG, Pilozzi E, Biffoni M, Todaro M, Peschle C, De Maria R (2007) Identification and expansion of human colon-cancer-initiating cells. Nature 445(7123):111–115

Riordan JR, Deuchars K, Kartner N, Alon N, Trent J, Ling V (1985) Amplification of P-glycoprotein genes in multidrug-resistant mammalian cell lines. Nature 316(6031):817–819

Roesch A, Fukunaga-Kalabis M, Schmidt EC, Zabierowski SE, Brafford PA, Vultur A, Basu D, Gimotty P, Vogt T, Herlyn M (2010) A temporarily distinct subpopulation of slow-cycling melanoma cells is required for continuous tumor growth. Cell 141(4):583–594. doi:10.1016/j. cell.2010.04.020

Roesch A, Vultur A, Bogeski I, Wang H, Zimmermann KM, Speicher D, Korbel C, Laschke MW, Gimotty PA, Philipp SE, Krause E, Patzold S, Villanueva J, Krepler C, Fukunaga-Kalabis M, Hoth M, Bastian BC, Vogt T, Herlyn M (2013) Overcoming intrinsic multidrug resistance in melanoma by blocking the mitochondrial respiratory chain of slow-cycling JARID1B(high) cells. Cancer Cell 23(6):811–825. doi:10.1016/j.ccr.2013.05.003

Rohlena J, Dong LF, Ralph SJ, Neuzil J (2011) Anticancer drugs targeting the mitochondrial electron transport chain. Antioxid Redox Signal 15(12):2951–2974. doi:10.1089/ars.2011.3990

Sakai W, Swisher EM, Karlan BY, Agarwal MK, Higgins J, Friedman C, Villegas E, Jacquemont C, Farrugia DJ, Couch FJ, Urban N, Taniguchi T (2008) Secondary mutations as a mechanism of cisplatin resistance in BRCA2-mutated cancers. Nature 451(7182):1116–1120. doi:10.1038/nature06633

Scandurra FM, Gnaiger E (2010) Cell respiration under hypoxia: facts and artefacts in mitochondrial oxygen kinetics. Adv Exp Med Biol 662:7–25. doi:10.1007/978-1-4419-1241-1_2

Singh SK, Hawkins C, Clarke ID, Squire JA, Bayani J, Hide T, Henkelman RM, Cusimano MD, Dirks PB (2004) Identification of human brain tumour initiating cells. Nature 432(7015): 396–401

Skipper HE, Perry S (1970) Kinetics of normal and leukemic leukocyte populations and relevance to chemotherapy. Cancer Res 30(6):1883–1897

Skrtic M, Sriskanthadevan S, Jhas B, Gebbia M, Wang X, Wang Z, Hurren R, Jitkova Y, Gronda M, Maclean N, Lai CK, Eberhard Y, Bartoszko J, Spagnuolo P, Rutledge AC, Datti A, Ketela T, Moffat J, Robinson BH, Cameron JH, Wrana J, Eaves CJ, Minden MD, Wang JC, Dick JE, Humphries K, Nislow C, Giaever G, Schimmer AD (2011) Inhibition of mitochondrial translation as a therapeutic strategy for human acute myeloid leukemia. Cancer Cell 20(5):674–688. doi:10.1016/j.ccr.2011.10.015

Sui X, Chen R, Wang Z, Huang Z, Kong N, Zhang M, Han W, Lou F, Yang J, Zhang Q, Wang X, He C, Pan H (2013) Autophagy and chemotherapy resistance: a promising therapeutic target for cancer treatment. Cell Death Dis 4:e838. doi:10.1038/cddis.2013.350

Swanton C, Nicke B, Schuett M, Eklund AC, Ng C, Li Q, Hardcastle T, Lee A, Roy R, East P, Kschischo M, Endesfelder D, Wylie P, Kim SN, Chen JG, Howell M, Ried T, Habermann JK, Auer G, Brenton JD, Szallasi Z, Downward J (2009) Chromosomal instability determines taxane response. Proc Natl Acad Sci USA 106(21):8671–8676. doi:10.1073/pnas.0811835106

Tan AS, Baty JW, Dong LF, Bezawork-Geleta A, Endaya B, Goodwin J, Bajzikova M, Kovarova J, Peterka M, Yan B, Pesdar EA, Sobol M, Filimonenko A, Stuart S, Vondrusova M, Kluckova K, Sachaphibulkij K, Rohlena J, Hozak P, Truksa J, Eccles D, Haupt LM, Griffiths LR, Neuzil J, Berridge MV (2015) Mitochondrial genome acquisition restores respiratory function and tumorigenic potential of cancer cells without mitochondrial DNA. Cell Metab 21(1):81–94. doi:10.1016/j.cmet.2014.12.003

Vander Heiden MG, Cantley LC, Thompson CB (2009) Understanding the Warburg effect: the metabolic requirements of cell proliferation. Science 324(5930):1029–1033. doi:10.1126/science.1160809

Vazquez F, Lim JH, Chim H, Bhalla K, Girnun G, Pierce K, Clish CB, Granter SR, Widlund HR, Spiegelman BM, Puigserver P (2013) PGC1alpha expression defines a subset of human melanoma tumors with increased mitochondrial capacity and resistance to oxidative stress. Cancer Cell 23(3):287–301. doi:10.1016/j.ccr.2012.11.020

Viale A, De Franco F, Orleth A, Cambiaghi V, Giuliani V, Bossi D, Ronchini C, Ronzoni S, Muradore I, Monestiroli S, Gobbi A, Alcalay M, Minucci S, Pelicci PG (2009) Cell-cycle restriction limits DNA damage and maintains self-renewal of leukaemia stem cells. Nature 457 (7225):51–56. doi:10.1038/nature07618

Viale A, Pettazzoni P, Lyssiotis CA, Ying H, Sanchez N, Marchesini M, Carugo A, Green T, Seth S, Giuliani V, Kost-Alimova M, Muller F, Colla S, Nezi L, Genovese G, Deem AK, Kapoor A, Yao W, Brunetto E, Kang Y, Yuan M, Asara JM, Wang YA, Heffernan TP, Kimmelman AC, Wang H, Fleming JB, Cantley LC, DePinho RA, Draetta GF (2014) Oncogene ablation-resistant pancreatic cancer cells depend on mitochondrial function. Nature 514(7524):628–632. doi:10.1038/nature13611

Wagle N, Emery C, Berger MF, Davis MJ, Sawyer A, Pochanard P, Kehoe SM, Johannessen CM, Macconaill LE, Hahn WC, Meyerson M, Garraway LA (2011) Dissecting therapeutic resistance to RAF inhibition in melanoma by tumor genomic profiling. J Clin Oncol Official J Am Soc Clin Oncol 29(22):3085–3096. doi:10.1200/JCO.2010.33.2312

Wallace DC (2012) Mitochondria and cancer. Nat Rev Cancer 12(10):685–698. doi:10.1038/nrc3365

Wang J, Guo LP, Chen LZ, Zeng YX, Lu SH (2007) Identification of cancer stem cell-like side population cells in human nasopharyngeal carcinoma cell line. Cancer Res 67(8):3716–3724. doi:10.1158/0008-5472.CAN-06-4343

Warburg O (1956) On the origin of cancer cells. Science 123(3191):309–314

Warburg O, Wind F, Negelein E (1927) The metabolism of tumors in the body. J Gen Physiol 8 (6):519–530

Weinstein IB (2002) Cancer. Addiction to oncogenes–the Achilles heal of cancer. Science 297 (5578):63–64. doi:10.1126/science.1073096

Weinstein IB, Joe AK (2006) Mechanisms of disease: oncogene addiction–a rationale for molecular targeting in cancer therapy. Nat Clin Pract Oncol 3(8):448–457. doi:10.1038/ncponc0558

Wilson TR, Longley DB, Johnston PG (2006) Chemoresistance in solid tumours. Ann Oncol Official J Eur Soc Med Oncol/ESMO 17(Suppl 10):x315–x324. doi:10.1093/annonc/mdl280

Wolf DA (2014) Is reliance on mitochondrial respiration a "chink in the armor" of therapy-resistant cancer? Cancer Cell 26(6):788–795. doi:10.1016/j.ccell.2014.10.001

Yang S, Wang X, Contino G, Liesa M, Sahin E, Ying H, Bause A, Li Y, Stommel JM, Dell'antonio G, Mautner J, Tonon G, Haigis M, Shirihai OS, Doglioni C, Bardeesy N, Kimmelman AC (2011) Pancreatic cancers require autophagy for tumor growth. Genes Dev 25 (7):717–729. doi:10.1101/gad.2016111

Ying H, Kimmelman AC, Lyssiotis CA, Hua S, Chu GC, Fletcher-Sananikone E, Locasale JW, Son J, Zhang H, Coloff JL, Yan H, Wang W, Chen S, Viale A, Zheng H, Paik JH, Lim C, Guimaraes AR, Martin ES, Chang J, Hezel AF, Perry SR, Hu J, Gan B, Xiao Y, Asara JM, Weissleder R, Wang YA, Chin L, Cantley LC, DePinho RA (2012) Oncogenic Kras maintains pancreatic tumors through regulation of anabolic glucose metabolism. Cell 149(3):656–670. doi:10.1016/j.cell.2012.01.058

Zahreddine H, Borden KL (2013) Mechanisms and insights into drug resistance in cancer. Front Pharmacol 4:28. doi:10.3389/fphar.2013.00028

Zhou S, Schuetz JD, Bunting KD, Colapietro AM, Sampath J, Morris JJ, Lagutina I, Grosveld GC, Osawa M, Nakauchi H, Sorrentino BP (2001) The ABC transporter Bcrp1/ABCG2 is expressed in a wide variety of stem cells and is a molecular determinant of the side-population phenotype. Nat Med 7(9):1028–1034. doi:10.1038/nm0901-1028

Zhou W, Yang Y, Xia J, Wang H, Salama ME, Xiong W, Xu H, Shetty S, Chen T, Zeng Z, Shi L, Zangari M, Miles R, Bearss D, Tricot G, Zhan F (2013) NEK2 induces drug resistance mainly through activation of efflux drug pumps and is associated with poor prognosis in myeloma and other cancers. Cancer Cell 23(1):48–62. doi:10.1016/j.ccr.2012.12.001

Zu XL, Guppy M (2004) Cancer metabolism: facts, fantasy, and fiction. Biochem Biophys Res Commun 313(3):459–465

Tissue-Based Metabolomics to Analyze the Breast Cancer Metabolome

Jan Budczies and Carsten Denkert

Abstract

Mass spectrometry and nuclear magnetic resonance-based metabolomics have been developed into mature technologies that can be utilized to analyze hundreds of biological samples in a high-throughput manner. Over the past few years, both technologies were utilized to analyze large cohorts of fresh frozen breast cancer tissues. Metabolite biomarkers were shown to separate breast cancer tissues from normal breast tissues with high sensitivity and specificity. Furthermore, the metabolome differed between hormone receptor positive (HR+) and hormone receptor negative (HR−) breast cancer, but was unchanged in HER2+ tumors compared to HER2− tumors. New metabolism-related biomarkers were discovered including the 4-aminobutyrate aminotransferase ABAT, where low mRNA expression led to an accumulation of beta-alanine and shortened relapse-free survival. The glutamate-to-glutamine ratio (GGR) represents another new biomarker that was increased in 88 % of HR− tumors and 56 % of HR+ tumors compared to normal breast tissues. The GGR might help to stratify patients for the treatment with specific glutaminase inhibitors that were recently developed and are currently being tested in phase I clinical studies. Surprisingly, 2-hydroxyglutarate (2-HG), initially found to accumulate in isocitrate dehydrogenase (IDH) mutated gliomas and leukemias and described as an oncometabolite, was detected to be drastically increased in several breast carcinomas in the absence of IDH mutations. In summary, metabolomics analysis of breast cancer tissues is a reliable method and has produced many new biological insights that may impact breast cancer diagnostics and treatment over the coming years.

J. Budczies (✉) · C. Denkert
Institute of Pathology, Charité—Universitätsmedizin Berlin, Berlin, Germany
e-mail: jan.budczies@charite.de

C. Denkert
e-mail: carsten.denkert@charite.de

© Springer International Publishing Switzerland 2016
T. Cramer and C.A. Schmitt (eds.), *Metabolism in Cancer*,
Recent Results in Cancer Research 207, DOI 10.1007/978-3-319-42118-6_7

Keywords
Cancer metabolomics · Breast cancer · Beta-alanin · Glutamate-to-glutamine ratio · 2-hydroxyglutarate

Abbreviations and acronyms

2-HG	2-hydroxyglutarate
ABAT	4-aminobutyrate aminotransferase
CMP	Cytidine-monophosphate
Cer	Ceramide
ER	Estrogen receptor
FDG-PET	Fluorodeoxyglucose positron emission tomography
FFPE	Formalin fixation and subsequent paraffin embedding
GC-TOF-MS	Gas chromatography combined with time-of-flight mass spectrometry: a metabolomics platform suitable to investigate up to 200 identified metabolites of the primary metabolism
GGR	Glutamate-to-glutamine ratio
GLS	Glutaminase 1
HR	Hormone receptor
HR-MAS-NMR	High-resolution magic-angle spinning nuclear magnetic resonance: a metabolomics platform for nondestructive tissue analysis
IDH1, IDH2	Isocitrate dehydrogenase 1, isocitrate dehydrogenase 2
KEGG	Kyoto encyclopedia genes and genomes
MS	Mass spectrometry
NMR	Nuclear magnetic resonance
PROFILE	Projection from interaction lattice: a clustering method for the analysis of metabolite changes in the context of the network of enzymatic reactions
PC	Phosphatidylcholine
PE	Phosphatidylethanolamine
PI	Phosphatidylinositol
PgR	Progesterone receptor
SM	Sphingomyelin
TG	Triglyceride
UP-LC-MS	Ultra performance liquid chromatography combined with mass spectrometry: a metabolomics platform for the analysis of up to 300 identified complex lipids
XDH	Xanthine dehydrogenase

1 Introduction

Alterations in cellular energy metabolism are a hallmark of cancer (Hanahan and Weinberg 2011). Metabolic alterations in cancer cells compared to normal cells include the way nutrients such as glucose and glutamine are taken up and processed and the way that important molecular building blocks such as amino acids, lipids, and nucleotides are synthesized (Kroemer and Pouyssegur 2008; Schulze and Harris 2012; Tennant et al. 2009; Vander Heiden et al. 2009). Furthermore, it has been recently discovered that some types of tumor cells produce oncometabolites that are capable of actively driving tumor growth by changing the landscape of DNA and histone methylation via inhibition of methylation modifying proteins (Morin et al. 2014). Some of the metabolic alterations are specific to certain tumor types or more pronounced in some tumor types, while other changes occur commonly in many or all malignant tumors. The aim of cancer metabolomics research was to gain insight into the metabolic alteration in cancer cells and exploit them for better diagnostic and therapeutic strategies. In particular, the aims include: 1. detection of vulnerable structures and targeting of these structures by metabolism-modifying treatments; 2. usage of metabolites as biomarkers for cancer screening or separation of malignant and benign disease; 3. monitoring of treatment in vivo using metabolite imaging methods; and 4. usage of metabolites as biomarkers for tumor classification and selection of patients for targeted therapies.

Breast cancer is the most frequently diagnosed cancer worldwide with an incidence of 8 % in more developed and that of 3 % in less developed areas in woman 74 years and younger (Torre et al. 2015). Furthermore, breast cancer is the leading cause of cancer death in woman aged 20–59 in the US population (Siegel et al. 2016). Seminal work of Perou, Sørlie, and coworkers based on hierarchical clustering of genome-wide expression data has led to the discovery of five molecular subtypes of breast cancer: luminal A, luminal B, HER2-enriched, basal-like, and normal-like breast cancer (Perou et al. 2000; Sorlie et al. 2003; Sørlie et al. 2001). To a large extent, this classification can be reproduced by simpler and less expensive methods including immunohistochemistry for ER, PgR, and HER2 proteins and in situ hybridization for HER2 amplifications that are available in routine pathology laboratories. Today, determination of HR and HER2 status is worldwide standard in breast cancer diagnostics to select patients for endocrine- and HER2-blocking therapies (Goldhirsch et al. 2011).

During the last 20 years, the altered metabolism in breast carcinomas and the resulting alterations in body metabolism have been intensively investigated by metabolomics studies in breast cancer tissues and in body fluids of breast cancer patients (Günther 2015). In this overview, we focus on the most important findings from a translational research perspective and the implication of the metabolomics studies in breast cancer tissues for clinical management.

2   Metabolome and Metabolomics

The metabolome is defined as the totality of all small molecule compounds contained in a biological sample. Metabolomics is the science and analysis technology to investigate the metabolome. During the last decade, there has been a growing interest in cancer metabolomics with more than 450 PubMed-listed publications in 2015. More than 40,000 human metabolites are already annotated in The Human Metabolome Database (www.hmdb.ca) of the Human Metabolome Project that aims to identify and catalog all human metabolites (Wishart et al. 2013).

The metabolism of human tumors can be studied in two ways, either by the analysis of biopsies or surgically removed tumor tissues or indirectly by the analysis of blood, urine, and other body fluids. As the typical tumor burden of early breast cancer compared to the mass body is very low (typically in the range 0.0001–1 %), tumor metabolites are expected to be diluted and to present in the body fluids at low concentrations. Further, metabolite changes in the blood and in other body fluids reflect changes of the body metabolism in response to the disease. Metabolomics analysis of body fluids is certainly an interesting field because they can be obtained noninvasively and easily and possibly can be used for disease screening, disease monitoring, and to detect biomarkers for the response to therapies. However, analysis of cancer tissues represents the most direct way to gain insights into the metabolomics changes in cancer cells.

Table 1 gives an overview on the most important metabolomics studies in human breast cancer tissues. Both nuclear magnetic resonance (NMR)- and mass spectrometry (MS)-based technologies were applied. The latter combined MS with either preceding liquid chromatography (LC) or preceding gas chromatography (GC). An advantage of NMR metabolomics is that it is nondestructive and potentially can be applied in vivo. Advantages of the MS-based methods are their high sensitivity and the identification of a higher number of metabolites compared to the NMR-based methods (Shulaev 2006). Indeed, in the NMR-based studies listed in Table 1, no systematic identification of metabolites and no statistical analysis of metabolite levels were performed.

3   Tissue-Based Metabolomics

Metabolome analysis of cancer tissues is usually performed in fresh-frozen tissues that have been collected during surgery. A critical point of sample collection is the freezing delay time between the cutoff of blood supply during surgery until the sample is frozen for storage. Freezing delay times are difficult to control for intraoperatively acquired human tissues. However, the first MS-based metabolomics studies in large retrospective collections of frozen cancer tissues proved that reliable quantification of metabolites is feasible (Denkert et al. 2006, 2008; Sreekumar et al. 2009). These studies led to the identification of diagnostic and prognostic biomarkers. Additionally, a recent study in patient-derived breast cancer

Table 1 Overview of the most important metabolomics studies in human breast cancer tissues. Only results of large (close to or more than 100 samples) NMR- and MS-based studies were included

Sample type, n (samples)	Method	N (metabolites)	Results	Biomarkers	Reference
82 invasive cancer tissues, 17 DCIS tissues, 106 normal tissues	NMR	–	Separation of invasive cancer and normal tissues with sensitivity 95 % and specificity 96 %	Ratio of N-trimethyl resonance (at 3.25 ppm) to creatine resonance (at 3.05 ppm)	Mackinnon et al. (1997)
160 cancer tissues	HR-MAS-NMR	–	Prediction of ER status, PgR status and nodal status with accuracies 88, 78 and 68 %	PLS analysis without statistical investigation of biomarkers	Giskeødegård et al. (2010)
257 cancer tissues, 10 normal tissues	UPLC-MS	230	Constituents of membrane phospholipids including palmitate-containing phosphatidylcholines are increased in cancer tissues and exhibit the highest concentrations in HR– and in G3 tumors	Increased in HR– tumors: PC (14:0/16:0), PC(16:0/16:0), PC (16:0/20:4), PE(P-16:0/20:4), PE (18:0/20:3), PE(18:0/20:4), PE (P-38:4), PI(18:0/20:4), SM (d18:1/24:0) Increased in HR+ tumors: GlcCer(d18:0/24:0(2-OH)), PE (P-16:0/18:2)	Hilvo et al. (2011)
98 cancer tissues	HR-MAS-NMR	–	Separation of 5-year survivors and nonsurvivors of HR+ breast cancer	Glycerin, lactate	Giskeødegård et al. (2010)
271 cancer tissues, 98 normal tissues	GC-TOF-MS	162	Separation of cancer and normal tissues with sensitivity 95 % and specificity 94 %	Increased in cancer tissues: CMP, AMP, phospho-ethanolamine, taurine, pyrazine 2,5-dihydroxy, creatinine, N-acetylaspartate, hypoxanthine, glycerol-alpha-phosphate, aminoalonate, glutamate, malate, oxoproline Increased in normal tissues: C15:0, C17:0, C24:0, 1-hexadecanol, glycolate, benzoate, hydroxylamine	Budczies et al. (2012)

(continued)

Table 1 (continued)

Sample type, n (samples)	Method	N (metabolites)	Results	Biomarkers	Reference
263 cancer tissues, 65 normal tissues	HR-MAS-NMR	–	Separation of cancer and normal tissues with sensitivity and specificity >90 %	PLS analysis without statistical investigation of biomarkers	Bathen et al. (2013)
204 HR+ cancer tissues and 67 HR− cancer tissues	GC-TOF-MS, whole-genome mRNA expression analysis, immunohisto-chemistry	162	Low ABAT mRNA correlates with beta-alanine accumulation and is a poor prognostic marker in breast cancer, HR + breast cancer and HR− breast cancer	Increased in HR− tumors: ABAT mRNA, beta-alanine, 2-HG, xanthine, trehalose, idonate, uracil, glutamate, glucuronate, fumarate, malate. Increased in HR + tumors: serine, N-methylalanine, citrulline, glycerol, glycerate, 2- aminoadipate, threonate, 3-phosphoglycerate, glutamine	Budczies et al. (2013)
67 cancer tissues, 65 normal tissues (training) and 60 HR− cancer tissues and 29 normal tissues (validation)	UHPLC-MS/MS^2, whole-genome mRNA expression analysis, DNA-methylation analysis	352	MYC-driven accumulation of 2-hydroxyglutarate is associated with poor breast cancer prognosis	2-hydroxyglutarate (2-HG), MYC	Terunuma et al. (2014)
270 cancer tissues, 97 normal tissues	GC-TOF-MS	162	Glutamate-to-glutamine ratio (GGR) is increased in 56 % of HR+ tumor and 88 % of HR− tumors	GRR = ratio of glutamate to glutamine	Budczies et al. (2015)

xenograft models showed that NMR metabolic profiles of tumor samples were reproducible and robust to variation in freezing delay time up to 30 min (Haukaas et al. 2016).

A bottleneck of tissue metabolomics is the dependence on frozen tissues samples, because the standard clinical workflow for tissues sample preparation includes formalin fixation (FF) and subsequent paraffin embedding (PE). Huge collections of FFPE biopsies and surgical tissue samples are archived at pathology institutes, while the number of available frozen tissues is limited. However, methods for metabolite extraction from FFPE tissues have been optimized during the last number of years and first studies demonstrating the feasibility of metabolomics analysis in FFPE tissues have been published (Kelly et al. 2011; Wojakowska et al. 2015).

4 Bioinformatical Strategies for Metabolomics

The analysis workflow of metabolomics data can be divided into two steps: In the first step (data processing), the recorded spectra are deconvoluted to identify and quantify metabolites (Kind et al. 2009; Pluskal et al. 2010). In the second step (data analysis), the metabolite data are analyzed to identify biomarkers and gain information on the biomedical system under study. The workflow and methods for data processing are reviewed in (Smolinska et al. 2012) for NMR data and in (Katajamaa and Oresic 2007) for MS data. In the following, we discuss some general aspects of data analysis, while functional analysis and interpretation in the context of the metabolic pathways is discussed in the next section.

Data analysis usually starts from a data matrix including measurements of some hundred metabolites in the rows, and the investigated samples in the columns. The optimal method for sample normalization in metabolomics is still under discussion and depends on the metabolomics platform. Different strategies have been applied in breast cancer metabolomics including normalization with respect to the sum of all metabolite intensities (Budczies et al. 2012; Denkert et al. 2006, 2008) and normalization with respect to the total protein content (Hilvo et al. 2011). Quantification of metabolites (in terms of concentration) can be achieved by spiking in isotope-labeled internal or external standards.

Methods for the analysis of metabolomics data include metabolite-wise methods and methods of pattern recognition that take into account the correlations between methods. The latter can be divided into methods of unsupervised and supervised learning. In unsupervised learning, patterns are recognized based only on the metabolomics data, while supervised methods include the information of target variables such as diagnostic or prognostic information in the analysis. Many of these methods can be adopted from the analysis of other kinds of –omics data and have been extensively applied and evaluated in whole-genome gene expression analysis (microarray data).

When analyzing metabolite by metabolite to detect differences between two disease states, false positive detection associated with multiple hypotheses testing needs to be taken into account. However, corrections of p values, for example, using the Bonferroni methods for family-wise error control or the Benjamini-Hochberg method for false discovery rate control will be much weaker than for microarray data, because the number compounds in metabolomics data is typically a few hundred compared to thousands of the genes that are investigated in a genome-wide expression analysis. In our recent METAcancer project (Denkert et al. 2012), GC-TOF-MS analysis of 369 breast cancer tissues was performed in a predefined training set (2/3) and validation set (1/3). Metabolite changes were considered detected when found significant in the training set ($p < 0.05$, two-sided t test) and validated when confirmed in the validation set (change in same direction, $p < 0.05$, one-sided t test). Out of a total number of 468 metabolite peaks, 427 changes were detected between cancer and normal tissues and 65 metabolite changes were detected between HR− and HR+ tumors. Out of these, 368 (86 %) and 40 (62 %) could be confirmed in the validation set.

Heatmaps including hierarchical clustering of genes and samples represent the most common method of unsupervised analysis of gene expression data. Heatmaps were also successfully applied to the analysis and visualization of metabolomics data. In the METAcancer study, breast cancer tissues could be well separated from normal breast tissues using hierarchical clustering (Denkert et al. 2012). Furthermore, a new breast cancer subtype, characterized by increased glutamate and decreased glutamine, could be discovered by clustering of breast cancer tissue profiles (Budczies et al. 2015). This finding was obtained in a training data set and could be confirmed in an independent validation data set. Finally, in (Terunuma et al. 2014), a heatmap analysis showed distinct metabolite pattern in HR− breast cancer of patients of West African ancestry compared to patients of European ancestry.

Comprehensive comparison of the performance of supervised learning methods in molecular high-throughput is difficult because of the multitude of methods available and the dependence of some of the methods on external parameters. However, the experiences from the analysis of gene expression data taught that there is a tendency of simple methods (such as nearest centroid classification) to perform comparable to or even outperform more complicated methods (such as support vector machines or neuronal networks). A similar tendency was observed when investigating the separation of malignant tissues of ovarian cancer, breast cancer, and colon cancer from benign tumors or normal tissues (Budczies et al. 2012; Denkert et al. 2006, 2008). Important challenges of supervised pattern recognition in molecular high-throughput data include the following: (1) to avoid overfitting and (2) to secure that the reported classification rates are valid. These challenges should be addressed by a suitable validation protocol. The multiple random cross-validation protocol introduced by Michiels et al. represents the most comprehensive method for validation (Michiels et al. 2005). Using the implementation as R package cancerclass (Budczies et al. 2014), the protocol was used for the analysis of ovarian cancer, colon cancer, and breast cancer metabolomics data (Budczies et al. 2012; Denkert et al. 2006, 2008).

5 Metabolic Pathways and Functional Analysis

Decoding of the metabolic pathways represents one of the greatest achievements of biochemistry during the last 100 years (German et al. 2005). After the sequencing of the human genome was finished in the early 2000s, the richness of metabolic pathway knowledge was integrated with genomic information and is available from databases such as KEGG, HumanCyc, Reactome, and EHMN (Caspi et al. 2014; Croft et al. 2014; Kanehisa et al. 2014; Ma et al. 2007). The atlas of pathways represents a blueprint of the metabolic processes of the cell and includes the maps of how cells produce energy as well as how the molecular building blocks of the cellular structures are synthesized. This information represents a valuable source for the improvement of worldwide health, particularly for cancer research.

Metabolomics and proteomics represent the systems biology tools to study changes in the metabolic pathways acquired by the cancer cell. However, enrichment analysis that is often applied for functional analysis of gene expression data is less helpful in the context of metabolomics data for the following two reasons: 1. Using today's metabolomics platforms, the coverage of the metabolome as well as most of the metabolic pathways is incomplete. This limitation reduces the sensitivity of the enrichment methods. 2. Important changes in a metabolic pathway can involve the change of only a single or a few metabolites. Pathway changes of this kind would be overlooked in an enrichment analysis.

One way to integrate metabolomics data with metabolic pathway knowledge is to visualize the data in a map of enzymatic reactions. MetScape (Karnovsky et al. 2012) is an App for Cytoscape (Cline et al. 2007) providing a framework for visualization and interpretation of metabolomics and gene expression data in the context of the metabolic pathways. Inclusion of a metabolite in a MetScape analysis requires that the chemical structure of the metabolite is known and annotated in KEGG or NHMN. This limitation can be overcome using the bioinformatical tool MetaMapp that includes metabolites not contained in the pathway databases and unknown metabolites into metabolite networks based on chemical similarity and on mass spectral similarity, respectively (Barupal et al. 2012).

METAtarget is another method to integrate metabolomics data with pathway knowledge that focuses on substrate-product pairs of enzymatic reactions (Budczies et al. 2010a). In breast cancer, the equilibrium of glutamine to glutamate and that of glucose to glucose-6-phosphate were shifted to the product in HR− tumors compared to HR+ tumors. A subsequent analysis of gene expression data identified differential expressed enzymes that interconvert between these substrates and products.

PROFILE clustering is a method to visualize and interpret metabolite changes in the context of the network of enzymatic reactions (Denkert et al. 2008). PROFILE clustering uses pathway information from the KEGG database to calculate the distances between all pairs of metabolites included in a metabolomics data set (metabolite distance). To this end, the minimal number of reactions to convert the source metabolite in the target metabolite is calculated for each of these pairs. Then, metabolites are hierarchically clustered with respect to the metabolite distance and the metabolite changes are visualized as a barplot above the clustering. In this way, the complexity of the multidimensional metabolic network is projected to an order along a one-dimensional axis. A PROFILE clustering of breast cancer GC-TOF-MS data is shown in Fig. 1.

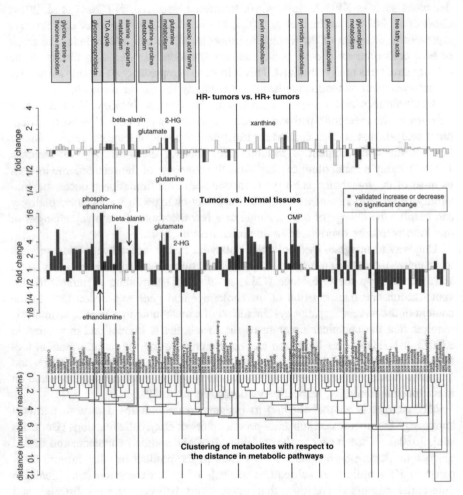

Fig. 1 PROFILE clustering of 271 breast cancer tissues and 98 normal tissues with respect to 129 metabolites. Metabolites were identified and analyzed using GC-TOF-MS-based metabolomics to cover the primary metabolism (Budczies et al. 2012, 2013). Metabolite increases that were detected in the training set (2/3 of data) and could be validated in the validation set (1/3 of data) are marked in *blue*. *Top panel* Metabolic pathways. *Upper panel* Fold changes between HR− and HR+ breast cancer. *Lower panel* Fold changes between breast cancer and normal breast tissues. *Bottom panel* Clustering of metabolites with respect to the distance (number of reactions needed to convert the substrate into the product) in the network of metabolic pathways

6 Comparison of Tumor and Normal Tissues

Metabolomics studies in large tissue cohorts have shown that breast cancer tissues can be separated from normal tissues with high accuracy using both NMR-based and MS-based technologies (Bathen et al. 2013; Budczies et al. 2012; Mackinnon et al. 1997). Typically, sensitivities and specificities around 95 % could be obtained. In general, the –omics approach for tissue analysis is challenged by tissue inhomogeneity causing varying amounts of tumor, glandular, connective, and fat tissues in the samples. However, it turned out that even those samples with tumor content as low as 5 % could be separated from normal tissues using NMR (Bathen et al. 2013). Furthermore, in a GC-TOF-MS study, it was shown that metabolite markers were capable of distinguishing between breast cancer and normal breast tissues as well as between breast cancer and adipose-rich tissues (Budczies et al. 2012). While the NMR studies did not include metabolite identification in a systematic way, GC-TOF-MS identified a list of 20 top primary metabolites (13 tumor tissue markers and 7 normal tissue markers) that separated best between cancer and normal tissues (Table 1). In the same study, it was shown that 79 % of the detected metabolites were significant changes between tumor and normal tissues ($p < 0.05$ in training and validation set). Furthermore, in a lipidomics study using UPLC-MS 70 % of the detected lipids were significantly different in cancer compared to normal tissues even at a very stringent p value cutoff (Hilvo et al. 2011). In detail, mainly membrane phospholipids including PCs, PEs, PIs, SMs, and Cers were altered, while most of the TGs were unaltered. It should be noted that it is unlikely that metabolomics will be used as a primary cancer diagnosis tool, because other less complicated approaches such as classical histopathology are well established for this purpose. Nevertheless, the excellent results for separation of tumor and normal tissue provide a strong validation of the method's approach. In particular, this shows that despite all changes related to tissue handling during and after surgery, metabolic differences are stable enough in the tissue samples to be detectable.

In addition to this method validation, the comparison of normal tissue and tumor tissue provides first insights into the abnormal metabolism of cancer. We have developed PROFILE clustering as a tool for pathway-based analysis of the changes between breast cancer and normal breast tissues in primary metabolism (Fig. 1). The metabolic alterations included upregulation of many amino acids, changes in TCA cycle, changes in glycerophospholipid metabolism, downregulation of the benzoic acid family, upregulation of most of the nucleotides and their phosphates, downregulation of the sugar cluster including sucrose, fructose, and glucose, and downregulation of many free fatty acids. At first sight, the downregulation of fatty acids seems to contradict the upregulation of fatty acid synthase (FASN) and increased de novo fatty acid synthesis that is found in many cancers (Menendez and Lupu 2007). However, this paradox would be resolved by a situation that de novo fatty acid synthesis in breast cancer is increased, but free fatty acids are rapidly metabolized to synthesize membrane phospholipids. Such a situation would be in line with the results of lipidomics study that detected upregulation of various membrane lipids in cancer compared to normal tissues (Hilvo et al. 2011). Furthermore, the GC-TOF-MS data show that equilibrium between ethanolamine and phospho-ethanolamine was shifted toward phospho-ethanolamine in the cancer tissues. The latter reaction is part of the Kennedy pathway, a way for de novo synthesis of PEs and PCs from diacylglycerol as well as ethanolamine and choline, respectively. The shift of the equilibrium from ethanolamine to phospho-ethanolamine is in line with activation of the Kennedy pathway and rapid membrane lipid synthesis from free fatty acids.

7 Metabolic Characterization of the Molecular Subtypes of Breast Cancer

Multiple studies revealed a strong dependence of the breast cancer metabolome on HR status, but not on HER2 status (Budczies et al. 2010b; Budczies et al. 2013; Giskeødegård et al. 2010; Terunuma et al. 2014). Prediction of HR status was feasible using either NMR data or GC-TOF-MS data at sensitivity and specificity >80 %. An overview of important metabolite alterations between HR− and HR+ tumors is given in Fig. 1. The alterations include increase of beta-alanine and xanthine in HR− breast cancer as well as changes of three metabolites in glutamine metabolism. The role of glutamine metabolism and the oncometabolite 2-HG in breast cancer is discussed in the next two sections.

In (Budczies et al. 2013), a correlation analysis of metabolomics data and genome-wide gene expression data was carried out: Interestingly, beta-alanine correlated strongly negative with the mRNA of ABAT, the enzyme that catalyzes the degradation of beta-alanine to malonate semialdehyde ($R_{Spearman} = -0.62$). Furthermore, xanthine correlated strongly positive with the mRNA of XHD, the enzyme that catalyzes the degradation of hypoxanthine to xanthine ($R_{Spearman} = 0.53$). These observations support the notion that distinct transcriptional regulation in HR− and in HR + breast cancer contributes to different metabolomes in these molecular subtypes. Finally, analysis of a large external gene expression data set showed that low ABAT mRNA expression shortened relapse-free survival in breast cancer (hazard ratio = 1.7, p = 3.2E−15). Low ABAT expression remained a negative prognostic marker when analyzed separately in HR+ breast cancer (hazard ratio = 1.5, p = 8.7E−08) and in HR− breast cancer (hazard ratio = 1.3, p = 0.028).

A lipidomics analysis of breast cancer tissues revealed that many phospholipids (PCs, PEs and PIs) were increased in HR− tumors compared to HR+ tumors, while TGs were not different (Table 1). Furthermore, even after normalization for the total phospholipid content, certain phospholipids including PC (14:0/16:0) were increased in HR− tumors (Hilvo et al. 2011). The authors concluded that products of de novo fatty acid synthesis incorporated in membrane lipids were increased in tumor tissues compared to normal tissues and highest in HR− and G3 tumors.

8 Alteration of Glutamine Metabolism

PROFILE clustering of GC-TOF-MS data revealed pronounced changes in glutamine metabolism in breast cancer tissues (Fig. 1): Glutamate and 2-hydroxyglutarate were increased in cancer compared to normal tissues, while alpha-ketoglutarate was decreased. Comparing the molecular subtypes of breast cancer, glutamate and 2-hydroxyglutarate were increased in HR− tumors compared to HR+ tumors, while glutamine was decreased. Furthermore, positive correlation (R = 0.56) of glutamine and glutamate in normal breast tissues changed to a negative correlation (R = −0.42) in breast cancer tissues (Budczies et al. 2015). To quantify the dysregulation of the glutamine metabolism, the glutamate-to-glutamine ratio (GGR) was introduced as a new biomarker. A cutoff point was introduced based on the value of GGR in normal tissues and tissues above the cutoff were considered as glutamate-enriched, while tissues below the cutoff were considered as glutamate normal. Using this definition, 88 % of HR− tumors, 56 % of HR+ tumors, and only 2 % of the normal tissues were glutamate-enriched. GRR may represent a valuable biomarker in the context of a therapeutic inhibition of glutaminase, an enzyme that catalyzes the conversion of glutamine to glutamate. In recent years, specific inhibitors of glutaminase 1 (GLS) including BPTES, 968 and CB-839 have been developed (Gross et al. 2014; Robinson et al. 2007; Wang et al. 2010). CB-839 is currently tested in phase I clinical trials of solid and hematological malignancies. The results on GGR in breast cancer support the notion that glutaminase inhibitors should evaluated in HR− tumors and selected HR+ tumors.

---

9 2-HG as Oncometabolite in Multiple Cancer Types

Three independent studies reported increased level of 2-hydroxyglutarate (2-HG) in breast cancer tissues compared to normal breast tissues (Budczies et al. 2012; Tang et al. 2014; Terunuma et al. 2014). Levels of 2-HG were higher in HR− tumors compared to HR+ tumors and reached the highest levels in triple-negative breast cancer (Budczies et al. 2013; Terunuma et al. 2014). On the other hand, 2-HG is known as oncometabolite in gliomas and leukemias that is produced by gain-of-function mutations in isocitrate dehydrogenase IDH1 or IDH2 (Morin et al. 2014). However, mutations in the IDHs in breast cancer are extremely rare: Analyzing all breast carcinomas in the COSMIC database, only 9/2179 (0.4 %) tumors had IDH1 mutations and only 4/1927 (0.2 %) tumors had IDH2 mutations. So far, only a single metastatic breast cancer case with IDH1 p.R132L was reported and it was shown that 2-HG accumulated in the blood and the urine of this patient (Fathi et al. 2014). However, in (Budczies et al. 2012) and in (Terunuma et al. 2014) about half of the breast carcinomas had elevated 2-HG levels compared to the normal tissues and none of tumors in the latter study had hot spot mutations in the IDH genes. Thus, 2-HG accumulates in many breast carcinomas in an IDH-independent way.

2-HG accumulation is known to provoke a DNA-hypermethylation phenotype and suppress histone-demethylation via the inhibition of alpha-ketoglutarate-dependent dioxygenases (Morin et al. 2014). As shown before in AML cells (Figueroa et al. 2010), DNA-hypermethylation was associated with 2-HG accumulation in breast cancer tissues (Terunuma et al. 2014). Furthermore, in the same study using a MYC activation, it was shown that activated MYC correlated strongly with the accumulation of 2-HG (odds ratio = 11.3). Thus, the authors developed the hypothesis of a MYC-driven accumulation of 2-HG in breast cancer that could be validated in proof-of-principle in vitro experiments. Furthermore, some evidence was obtained that this regulation could be mediated by a mechanism of GLS1 activation by MYC that is mediated via suppression of miR-23a/b that was described before (Gao et al. 2009).

2-HG is a chiral molecule with both enantiomers D-2-HG and L-2-HG occurring in normal tissues at low concentrations (Struys et al. 2004). D-2-HG and not L-2-HG was shown to be produced by mutated IDH1 and IDH2 in gliomas and leukemias (Dang et al. 2009; Ward et al. 2010). Furthermore, it has been shown that D-2-HG accumulates in a subgroup of breast carcinomas (Tang et al. 2014), while L-2-HG has been identified as a putative oncometabolite in renal cancer (Shim et al. 2014). A pilot experiment in a single cell line showed that neither D-2-HG nor L-2-HG was metabolized to any metabolite product in colorectal cancer cells (Gelman et al. 2015). Both 2-HG enantiomers are known to be potent inhibitors of alpha-ketoglutarate-dependent enzymes (Chowdhury et al. 2011; Xu et al. 2011). Thus, although there are cancer-type-specific differences in the action of the D- and the L-enantiomer (Losman et al. 2013), both enantiomers are believed to be capable of acting as oncometabolites through the modification of DNA and histone methylation patterns.

10 Discussion

During the last decade, the research field of cancer metabolomics enjoyed a growing popularity that is reflected in more than 450 PubMed-indexed publications on the subject in 2015. Both MS-based and by NMR-based metabolomics have been established as reproducible technologies that can be used to analyze hundreds of samples in a high-throughput manner (Larive et al. 2015; Liesenfeld et al. 2013). While NMR-based metabolomics has the advantage of leaving samples undestroyed and can by potentially applied in vivo, MS-based methods are more sensitive and allow the identification of a higher number of metabolites. In breast cancer tissues, 162 primary metabolites were identified using GC-MS, 230 complex lipids were identified using UPLC-MS, and a total of 352 metabolites were identified using two separate LC-MS injections (Table 1). On the other hand, the number of identified metabolites in NMR studies on breast cancer tissues was low (10–20) and none of the NMR studies included a systematic identification and statistical analysis of metabolite intensities.

Tissue-based metabolome analysis of breast cancer has revealed new insight into the regulation of metabolism in the molecular subtypes of breast cancer and led to the discovery of new diagnostic and prognostic biomarkers. Metabolomics and lipidomics studies in large cohorts showed that the breast cancer metabolome strongly depends on HR status and on tumor grade, but to a much lesser extend on HER2 status (Budczies et al. 2013; Hilvo et al. 2011). Low ABAT expression was associated with beta-alanine accumulation and shortened relapse-free survival across all subtypes of breast cancer as well as when analyzed separately in the HR+ subtype and in the HR− subtype. Several studies have reported alterations in glutamine metabolism in breast cancer. The glutamate-to-glutamine ratio (GGR) has been introduced as a new biomarker and has been used to stratify breast tissues in either glutamate-enriched or glutamate normal. The GGR could possibly be used as a biomarker for stratification of patients for treatment with one of the new specific glutaminase inhibitors that has recently entered the first clinical studies. A highly interesting finding was the accumulation of 2-HG in some breast carcinoma that has been reported in several studies (Budczies et al. 2013; Tang et al. 2014; Terunuma et al. 2014). 2-HG has been described as oncometabolite in gliomas and leukemias that acts through modification of DNA and histone methylation landscapes. Interestingly, while not associated with IDH mutations in breast cancer, 2-HG accumulation correlated with DNA-hypermethylation and poor outcome in breast cancer (Terunuma et al. 2014). These findings may open new therapeutic options including glutaminase inhibition or treatment with demethylating agents.

A fascinating feature of metabolome analyses is the multitude of three-dimensional spatial structures as well as the multitude of biological functions of the metabolites. However, these properties complicate a comprehensive analysis of the metabolome in comparison with analysis of one-dimensional molecules such as DNA, mRNA, and proteins. In fact, today's metabolomics studies are limited to detection of a few hundred metabolites in the human metabolome that is comprised of several 10,000s of metabolites (Wishart et al. 2013). Several ways to improve the coverage of the metabolome are expected to be implemented during the next years: 1. Extension of the spectral databases by including more metabolite standards. 2. Combination of serval metabolomics platforms. 3. Development of specialized metabolomics platforms for the profiling of specific molecule classes.

In summary, we reported on recent trends of tissue-based metabolomics of breast cancer. Over the past few years, there were several discoveries that were to influence clinical practice of breast cancer diagnosis and treatment in the future: For example, ABAT mRNA expression and beta-alanine accumulation represent prognostic markers, while the glutamate-to-glutamine ratio and/or 2-HG accumulation could be possible predictive markers for treatment with glutaminase inhibitors. Numerous metabolites were found to differ between breast cancer and normal breast tissues. This could potentially be exploited by new methods of screening, by the investigation of body fluids and of imaging using magnetic resonance spectrometry.

A future challenge is to gain a better understanding of the causes of the metabolism deregulation in cancer. Results reported before have shown the power of a combined transcriptomics and metabolomics approach. It should be stressed that the information content in a multiomics data set is much higher than in separate single-omics data sets because of the vast amounts of information contained in the cross-correlations between different molecule types. Therefore, combining metabolomics data with other kinds of –omics data, especially with comprehensive proteomics data ideally obtained from the same tissues, appears a promising approach for a deeper insight into metabolism deregulation and the uncovering of new therapeutic options.

References

Barupal DK, Haldiya PK, Wohlgemuth G et al (2012) Metamapp: mapping and visualizing metabolomic data by integrating information from biochemical pathways and chemical and mass spectral similarity. BMC Bioinform 13:99

Bathen TF, Geurts B, Sitter B et al (2013) Feasibility of MR metabolomics for immediate analysis of resection margins during breast cancer surgery. PLoS ONE 8:e61578

Budczies J, Brockmöller SF, Müller BM et al (2013) Comparative metabolomics of estrogen receptor positive and estrogen receptor negative breast cancer: alterations in glutamine and beta-alanine metabolism. J Proteomics 94:279–288

Budczies J, Denkert C, Müller BM et al. (2010a) Metatarget—extracting key enzymes of metabolic regulation from high-throughput metabolomics data using KEGG reaction information. In: Proceedings of the German Conference on Bioinformatics, GI edn. p. 173

Budczies J, Denkert C, Müller BM et al. (2010b) GC-TOF mass spectrometry reveals strong dependence of breast cancer metabolome on estrogene receptor, but not on HER2 status. In: Proceedings: AACR 101st annual meeting 2010, Cancer Research, vol. 70, p. 5573

Budczies J, Denkert C, Müller BM et al (2012) Remodeling of central metabolism in invasive breast cancer compared to normal breast tissue—a GC-TOFMS based metabolomics study. BMC Genom 13:334

Budczies J, Kosztyla D, von Törne C et al (2014) Cancerclass: an R package for development and validation of diagnostic tests from high-dimensional molecular data. J Stat Softw 59(1):1–19

Budczies J, Pfitzner BM, Györffy B et al (2015) Glutamate enrichment as new diagnostic opportunity in breast cancer. Int J Cancer 136:1619–1628

Caspi R, Altman T, Billington R et al (2014) The MetaCyc database of metabolic pathways and enzymes and the biocyc collection of pathway/genome databases. Nucleic Acids Res 42: D459–D471

Chowdhury R, Yeoh KK, Tian Y et al (2011) The oncometabolite 2-hydroxyglutarate inhibits histone lysine demethylases. EMBO Rep 12:463–469

Cline MS, Smoot M, Cerami E et al (2007) Integration of biological networks and gene expression data using Cytoscape. Nat Protoc 2:2366–2382

Croft D, Mundo AF, Haw R et al (2014) The Reactome pathway knowledgebase. Nucleic Acids Res 42:D472–D477

Dang L, White DW, Gross S et al (2009) Cancer-associated IDH1 mutations produce 2-hydroxyglutarate. Nature 462:739–744

Denkert C, Bucher E, Hilvo M et al (2012) Metabolomics of human breast cancer: new approaches for tumor typing and biomarker discovery. Genome Med 4:37

Denkert C, Budczies J, Kind T et al (2006) Mass spectrometry-based metabolic profiling reveals different metabolite patterns in invasive ovarian carcinomas and ovarian borderline tumors. Cancer Res 66:10795–10804

Denkert C, Budczies J, Weichert W et al (2008) Metabolite profiling of human colon carcinoma–deregulation of TCA cycle and amino acid turnover. Mol Cancer 7:72

Fathi AT, Sadrzadeh H, Comander AH et al (2014) Isocitrate dehydrogenase 1 (IDH1) mutation in breast adenocarcinoma is associated with elevated levels of serum and urine 2-hydroxyglutarate. Oncologist 19:602–607

Figueroa ME, Abdel-Wahab O, Lu C et al (2010) Leukemic IDH1 and IDH2 mutations result in a hypermethylation phenotype, disrupt tet2 function, and impair hematopoietic differentiation. Cancer Cell 18:553–567

Gao P, Tchernyshyov I, Chang T et al (2009) c-Myc suppression of miR-23a/b enhances mitochondrial glutaminase expression and glutamine metabolism. Nature 458:762–765

Gelman SJ, Mahieu NG, Cho K et al (2015) Evidence that 2-hydroxyglutarate is not readily metabolized in colorectal carcinoma cells. Cancer Metab 3:13

German JB, Hammock BD, Watkins SM (2005) Metabolomics: building on a century of biochemistry to guide human health. Metabolomics 1:3–9

Giskeødegård GF, Grinde MT, Sitter B et al (2010) Multivariate modeling and prediction of breast cancer prognostic factors using MR metabolomics. J Proteome Res 9:972–979

Goldhirsch A, Wood WC, Coates AS et al (2011) Strategies for subtypes–dealing with the diversity of breast cancer: highlights of the St. Gallen international expert consensus on the primary therapy of early breast cancer 2011. Ann Oncol 22:1736–1747

Gross MI, Demo SD, Dennison JB et al (2014) Antitumor activity of the glutaminase inhibitor CB-839 in triple-negative breast cancer. Mol Cancer Ther 13:890–901

Günther UL (2015) Metabolomics biomarkers for breast cancer. Pathobiology 82:153–165

Hanahan D, Weinberg RA (2011) Hallmarks of cancer: the next generation. Cell 144:646–674

Haukaas TH, Moestue SA, Vettukattil R et al (2016) Impact of freezing delay time on tissue samples for metabolomic studies. Front Oncol 6:17

Hilvo M, Denkert C, Lehtinen L et al (2011) Novel theranostic opportunities offered by characterization of altered membrane lipid metabolism in breast cancer progression. Cancer Res 71:3236–3245

Kanehisa M, Goto S, Sato Y et al (2014) Data, information, knowledge and principle: back to metabolism in KEGG. Nucleic Acids Res 42:D199–D205

Karnovsky A, Weymouth T, Hull T et al (2012) Metscape 2 bioinformatics tool for the analysis and visualization of metabolomics and gene expression data. Bioinformatics 28:373–380

Katajamaa M, Oresic M (2007) Data processing for mass spectrometry-based metabolomics. J Chromatogr A 1158:318–328

Kelly AD, Breitkopf SB, Yuan M et al (2011) Metabolomic profiling from formalin-fixed, paraffin-embedded tumor tissue using targeted LC/MS/MS: application in sarcoma. PLoS ONE 6:e25357

Kind T, Wohlgemuth G, Lee DY et al (2009) Fiehnlib: mass spectral and retention index libraries for metabolomics based on quadrupole and time-of-flight gas chromatography/mass spectrometry. Anal Chem 81:10038–10048

Kroemer G, Pouyssegur J (2008) Tumor cell metabolism: cancer's Achilles' heel. Cancer Cell 13:472–482

Larive CK, Barding GAJ, Dinges MM (2015) NMR spectroscopy for metabolomics and metabolic profiling. Anal Chem 87:133–146

Liesenfeld DB, Habermann N, Owen RW et al (2013) Review of mass spectrometry-based metabolomics in cancer research. Cancer Epidemiol Biomarkers Prev 22:2182–2201

Losman J, Looper RE, Koivunen P et al. (2013) (R)-2-hydroxyglutarate is sufficient to promote leukemogenesis and its effects are reversible. Science 339:1621–1625

Ma H, Sorokin A, Mazein A et al (2007) The Edinburgh human metabolic network reconstruction and its functional analysis. Mol Syst Biol 3:135

Mackinnon WB, Barry PA, Malycha PL et al (1997) Fine-needle biopsy specimens of benign breast lesions distinguished from invasive cancer ex vivo with proton MR spectroscopy. Radiology 204:661–666

Menendez JA, Lupu R (2007) Fatty acid synthase and the lipogenic phenotype in cancer pathogenesis. Nat Rev Cancer 7:763–777

Michiels S, Koscielny S, Hill C (2005) Prediction of cancer outcome with microarrays: a multiple random validation strategy. Lancet 365:488–492

Morin A, Letouzé E, Gimenez-Roqueplo A et al (2014) Oncometabolites-driven tumorigenesis: from genetics to targeted therapy. Int J Cancer 135:2237–2248

Perou CM, Sørlie T, Eisen MB et al (2000) Molecular portraits of human breast tumours. Nature 406:747–752

Pluskal T, Castillo S, Villar-Briones A et al (2010) MZmine 2: modular framework for processing, visualizing, and analyzing mass spectrometry-based molecular profile data. BMC Bioinformatics 11:395

Robinson MM, McBryant SJ, Tsukamoto T et al (2007) Novel mechanism of inhibition of rat kidney-type glutaminase by bis-2-(5-phenylacetamido-1,2,4-thiadiazol-2-yl) ethyl sulfide (BPTES). Biochem J 406:407–414

Schulze A, Harris AL (2012) How cancer metabolism is tuned for proliferation and vulnerable to disruption. Nature 491:364–373

Shim E, Livi CB, Rakheja D et al (2014) L-2-hydroxyglutarate: an epigenetic modifier and putative oncometabolite in renal cancer. Cancer Discov 4:1290–1298

Shulaev V (2006) Metabolomics technology and bioinformatics. Brief Bioinform 7:128–139

Siegel RL, Miller KD, Jemal A (2016) Cancer statistics, 2016. CA Cancer J Clin 66:7–30

Smolinska A, Blanchet L, Buydens LMC et al (2012) NMR and pattern recognition methods in metabolomics: from data acquisition to biomarker discovery: a review. Anal Chim Acta 750:82–97

Sorlie T, Tibshirani R, Parker J et al (2003) Repeated observation of breast tumor subtypes in independent gene expression data sets. Proc Natl Acad Sci USA 100:8418–8423

Sreekumar A, Poisson LM, Rajendiran TM et al (2009) Metabolomic profiles delineate potential role for sarcosine in prostate cancer progression. Nature 457:910–914

Struys EA, Jansen EEW, Verhoeven NM et al (2004) Measurement of urinary D- and L-2-hydroxyglutarate enantiomers by stable-isotope-dilution liquid chromatography-tandem mass spectrometry after derivatization with diacetyl-L-tartaric anhydride. Clin Chem 50:1391–1395

Sørlie T, Perou CM, Tibshirani R et al (2001) Gene expression patterns of breast carcinomas distinguish tumor subclasses with clinical implications. Proc Natl Acad Sci U. S. A. 98:10869–10874

Tang X, Lin C, Spasojevic I et al (2014) A joint analysis of metabolomics and genetics of breast cancer. Breast Cancer Res 16:415

Tennant DA, Durán RV, Boulahbel H et al (2009) Metabolic transformation in cancer. Carcinogenesis 30:1269–1280

Terunuma A, Putluri N, Mishra P et al (2014) MYC-driven accumulation of 2-hydroxyglutarate is associated with breast cancer prognosis. J Clin Invest 124:398–412

Torre LA, Bray F, Siegel RL et al (2015) Global cancer statistics, 2012. CA Cancer J Clin 65:87–108

Vander Heiden MG, Cantley LC, Thompson CB (2009) Understanding the Warburg effect: the metabolic requirements of cell proliferation. Science 324:1029–1033

Wang J, Erickson JW, Fuji R et al (2010) Targeting mitochondrial glutaminase activity inhibits oncogenic transformation. Cancer Cell 18:207–219

Ward PS, Patel J, Wise DR et al (2010) The common feature of leukemia-associated IDH1 and IDH2 mutations is a neomorphic enzyme activity converting alpha-ketoglutarate to 2-hydroxyglutarate. Cancer Cell 17:225–234

Wishart DS, Jewison T, Guo AC et al (2013) HMDB 3.0–the human metabolome database in
    2013. Nucleic Acids Res 41:D801–D807
Wojakowska A, Marczak Ł, Jelonek K et al (2015) An optimized method of metabolite extraction
    from formalin-fixed paraffin-embedded tissue for GC/MS analysis. PLoS ONE 10:e0136902
Xu W, Yang H, Liu Y et al (2011) Oncometabolite 2-hydroxyglutarate is a competitive inhibitor of
    α-ketoglutarate-dependent dioxygenases. Cancer Cell 19:17–30

Imaging of Tumor Metabolism Using Positron Emission Tomography (PET)

Ivayla Apostolova, Florian Wedel and Winfried Brenner

Abstract

Molecular imaging employing PET/CT enables in vivo visualization, characterization, and measurement of biologic processes in tumors at a molecular and cellular level. Using specific metabolic tracers, information about the integrated function of multiple transporters and enzymes involved in tumor metabolic pathways can be depicted, and the tracers can be directly applied as biomarkers of tumor biology. In this review, we discuss the role of F-18-fluorodeoxyglucose (FDG) as an in vivo glycolytic marker which reflects alterations of glucose metabolism in cancer cells. This functional molecular imaging technique offers a complementary approach to anatomic imaging such as computed tomography (CT) and magnetic resonance imaging (MRI) and has found widespread application as a diagnostic modality in oncology to monitor tumor biology, optimize the therapeutic management, and guide patient care. Moreover, emerging methods for PET imaging of further biologic processes relevant to cancer are reviewed, with a focus on tumor hypoxia and aberrant tumor perfusion. Hypoxic tumors are associated with poor disease control and increased resistance to cytotoxic and radiation treatment. In vivo imaging of hypoxia, perfusion, and mismatch of metabolism and perfusion has the potential to identify specific features of tumor microenvironment associated with poor treatment outcome and, thus, contribute to personalized treatment approaches.

Keywords

Molecular imaging · PET/CT · FDG-PET · Tumor hypoxia

I. Apostolova
Department of Radiology and Nuclear Medicine, Medical School,
Otto-von-Guericke University, Magdeburg A.ö.R., Magdeburg, Germany

F. Wedel · W. Brenner (✉)
Department of Nuclear Medicine, University Medicine Charité, Berlin, Germany
e-mail: Winfried.Brenner@charite.de

© Springer International Publishing Switzerland 2016
T. Cramer and C.A. Schmitt (eds.), *Metabolism in Cancer*,
Recent Results in Cancer Research 207, DOI 10.1007/978-3-319-42118-6_8

1   Introduction

The methods for noninvasive molecular imaging based on nuclear medicine technology, such as positron emission tomography (PET), allow highly sensitive in vivo
visualization of general metabolic processes within cancer cells (Jacobson and Chen
2013). In vivo imaging of tumor cell metabolism with PET has been remarkably
successful in recent years (Plathow and Weber 2008). PET is a modality based on
nuclear medicine technology, with lower spatial resolution than the morphological
imaging methods such as computed tomography (CT) and magnetic resonance
imaging (MRI), however, with excellent sensitivity (Mawlawi and Townsend 2009)
and the potential to provide quantitative information on biochemical processes
within tissues or on specific target macromolecules within the human body
(Jacobson and Chen 2013). Metabolic imaging, in comparison with other molecular
imaging probes, such as receptor ligands, does not target the expression of one
molecule but provides information about the integrated function of multiple
transporters and enzymes involved in a metabolic process (Plathow and Weber
2008). Being able to highlight the areas of abnormal metabolism of cancer cells,
PET in combination with computed tomography (PET/CT) or magnetic resonance
imaging (PET/MRI) has become an important component of cancer imaging and an
integral part of the management of cancer patients.

   Several molecular pathways are responsible for the formation, development, and
aggressiveness of cancer. Oncogenic signaling, loss of suppressor genes, or adaptation to the tumor microenvironment have been shown to result in quantitative and
qualitative alterations of tumor metabolism (Bui and Thompson 2006; Kim and
Dang 2006). The best-known alteration of energy metabolism in cancer cells is
increased glycolysis. F-18-fluorodeoxyglucose (FDG) PET is by far the most
commonly used PET imaging technique enabling the visualization of glucose
metabolism of cancer cells in vivo. FDG PET/CT is widely used for cancer
detection, mostly for staging the extent of spread of newly diagnosed or recurrent
cancers (Kelloff et al. 2005). Numerous clinical trials have also shown that metabolic imaging can significantly impact patient management by improving treatment
planning, especially of image-guided techniques such as radiation therapy and
stereotactic surgery, and monitoring of tumor response to therapy (Plathow and
Weber 2008).

   There is a wide range of FDG uptake in different tumors. Most of the malignant
tumors, such as lung cancer, melanoma, breast cancer, and aggressive lymphoma,
show robustly elevated FDG uptake in the untreated state rendering FDG the most
widely used PET tracer in oncology (Macapinlac 2004; Vansteenkiste 2003; Weber
et al. 2003). Other tumor entities, however, such as early prostate cancer and
well-differentiated hepatocellular carcinoma, differentiated thyroid carcinoma, and
low-grade gliomas have consistently low uptake, and FDG PET has not proven
useful for cancer detection and staging (Khan et al. 2000). The limited applicability
of FDG in these less aggressive or less glycolytic tumors, as well as the limited
specificity, e.g., accelerated energy consumption in inflammation and tissue repair

in response to damage, brings the necessity for imaging probes targeting other pathways and biologic processes. Therefore, a multitude of different tracers have been applied for use in specific tumors, although only very few of these tracers have gained widespread clinical use so far such as F-18-fluoroethyl-tyrosine (FET) in gliomas, Ga-68-labeled somatostatin receptor analogues in neuroendocrine tumors, or choline derivatives and most recently Ga-68-PSMA in prostate cancer.

Besides PET, single-photon emission computed tomography (SPECT) is another modality for imaging cancer that uses radioactive imaging probes. However, SPECT isotopes emit only one photon upon decay and require physical collimation; thus, the SPECT technique is less sensitive than PET and does not allow absolute quantitative measurements of tracer accumulation (Rahmim and Zaidi 2008). In this review, the discussion on metabolic imaging, therefore, is restricted to PET technology. We summarize the use of PET to depict glucose metabolism, hypoxia, and perfusion in tumors and describe the techniques for noninvasive imaging of these processes.

1.1 PET Technique, Radiopharmaceuticals

PET imaging is based on the selection of a suitable carrier molecule, which is specific for the function to be examined. In order to trace the kinetics and spatial distribution of the carrier molecule in the body after intravenous injection and to derive the desired functional information, the carrier molecule is radiolabeled with a positron emitter artificially created, e.g. in a cyclotron. For the purpose of metabolic PET imaging, these are mostly defined biologic molecules, naturally present in the body (e.g., sugar, protein, or DNA elements). The radioactively labeled molecule is than referred to as "tracer." The emitted positron, arising from the radioactive label, travels a short distance (typically 1–2 mm or less) in human tissue until it interacts with target electrons. Both particles are then annihilated, resulting in two opposing 511 keV photons. These γ-ray photons arrive simultaneously at the PET camera detectors, arranged in a ring-shaped array, surrounding the patient. PET enables four-dimensional (three-dimensional spatial and temporal), quantitative determination of the distribution of radioactivity within the human body. The raw PET dataset uses only simultaneously detected, coincident, photon pairs at different projections through the body, from which a 3D image of the radiopharmaceutical concentration can be reconstructed mathematically (Yu and Mankoff 2007). The advantage of the radioactive labeling results from the fact that radioactive decay can be detected with very high sensitivity, so that only very low tracer doses are required for imaging. In PET, typical tracer doses lie in the ng or μg range which is usually far below any pharmacodynamic effects which could interfere with the measured processes. This is an important fact because any relevant interference of the tracer altering the underlying process would distort the measurement itself.

For radiolabeling, various positron emitters are available, including the carbon isotope C-11 and fluorine F-18. These nuclei are commonly found in biologic tissue; thus, the radiopharmaceuticals used in PET technology could be chemically identical to close analogues of naturally occurring molecules (Yu and Mankoff 2007). The advantage of labeling with C-11 is that the chemical properties of the carrier molecule are not altered when a stable C-12 atom is replaced by a radioactive carbon-11 atom. However, the short half-life of C-11 of approximately 20 min requires both an onsite cyclotron production and production of a tracer dose per patient which both hamper routine clinical use. F-18 has a longer half-life (110 min) and, thus, is better applicable in the daily routine even allowing shipment of the tracer within a transportation range of a few hours. However, fluorinated analogues of native biologic molecules may not perfectly match the biochemical properties of the native molecule; therefore, F-18 radiopharmaceuticals often require thorough validation as quantitative markers of the respective biochemistry and molecular biology.

2 F18-FDG PET

In 1924, Otto Warburg observed in thin slices of human and animal tumors ex vivo that tumor cells use glycolysis despite oxygen availability and metabolize more glucose to lactate than normal cells (Warburg et al. 1927). The shift toward lactate production in cancers, even in the presence of adequate oxygen, is termed the Warburg effect or aerobic glycolysis (Bensinger and Christofk 2012; Upadhyay et al. 2013). PET imaging with the glucose analogue FDG verified Warburg's hypothesis of altered glucose metabolism of cancer cells several decades later (Upadhyay et al. 2013). FDG has its origin in the work of Sokoloff et al., who developed a method to measure regional cerebral glucose metabolism in animals using C-14-deoxyglucose autoradiography (Sokoloff 1977). C-14-deoxyglucose is transported into cells in parallel with glucose and is phosphorylated by hexokinase to C-14-deoxyglucose-6-phosphate. Because it lacks a hydroxyl group at the 2 position, it is prevented from being a substrate of enzymes farther down the glycolytic pathway and thus trapped into the cell. This concept was adopted with the development of FDG. FDG, which also lacks a hydroxyl group in the 2 position, enters cancer cells similar to glucose via glucose transporters GLUT1 and GLUT3 and is subsequently phosphorylated by hexokinase to FDG-6-phosphate (Fig. 1a). While glucose-6-phosphate undergoes further isomerization to fructose-6-phosphate in the glycolytic pathway or oxidation to 6-phosphogluconolactone in the pentose phosphate pathway, FDG-6-phosphate cannot be further catabolized due to the lack of an oxygen atom at the C-2 position (Fig. 1b). FDG-6-phosphate is unable to diffuse out of cells due to its negative charge, and dephosphorylation occurs very slowly. Therefore, FDG-6-phosphate becomes trapped within the cell and accumulates at a rate proportional to glucose utilization (Fig. 1) (Phelps et al. 1979; Reivich et al. 1977). Thus, FDG uptake depends on both glucose transporter expression (Szablewski 2013) and hexokinase activity (Mathupala et al. 1997). Unlike glucose, FDG is not reabsorbed after glomerular filtration but excreted with the urine. This contributes to the rapid clearance of FDG from the blood resulting in a low background activity, which is

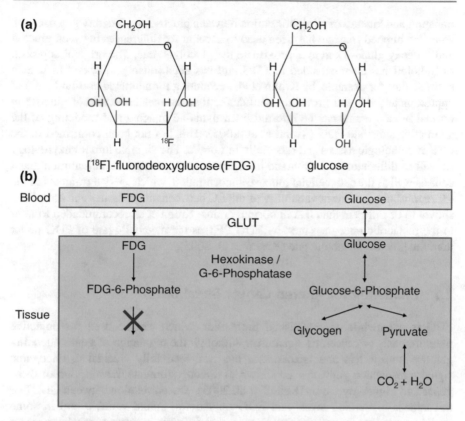

(a)

[^{18}F]-fluorodeoxyglucose (FDG)   glucose

(b)

Fig. 1 **a** The molecules of fluorodeoxyglucose and glucose: a hydroxyl group is exchanged for F-18 at the C-2 position. **b** FDG accumulation: FDG enters the cell via glucose transporters (GLUT) and is subsequently phosphorylated by hexokinase to FDG-6-phosphate. While glucose-6-phosphate undergoes further metabolism, FDG-6-phosphate cannot be further catabolized. FDG-6-phosphate thus becomes trapped in the cell and accumulates at a rate proportional to glucose utilization

important for imaging metabolically active tissues with high contrast (Weber et al. 2003).

It is worth to note that the usage of metabolic tracers is based on the assumption that the tracer follows the same biochemistry, both transport and metabolism, as the compound being traced. This requirement is satisfied when one of the carbon atoms in glucose, for example, is substituted with a C-11 atom (Krohn et al. 2005). However, many C-11-labeled molecules, including C-11 glucose, are rapidly metabolized to C-11-CO_2, which leaves promptly the region being imaged. Different approaches are used as described by Sokoloff et al. to modify the tracer so that the metabolic product is trapped in the tissues. However, the fluorinated derivatives of glucose are chemically distinguishable from the authentic molecules that they are intended to trace. This raised the question how FDG measures the biochemistry of the native substrate glucose. Several studies validated the differences in membrane transport and kinetics of phosphorylation between glucose and 2-deoxy glucose. The so-called lumped constant has been used to account for differences between glucose and 2-deoxy glucose and to approximate the glycolytic rate. The original approach of Sokoloff has been extended to FDG and has been reliably validated in human brain studies for example, by Krohn et al. accounting for a lumped constant (LC) of approximately 0.8 and providing confidence that the metabolic rate of glucose in normal human brains can be inferred from dynamic imaging and modeling of the alternative substrate FDG (Krohn et al. 2005). This has not been confirmed in the case of pathologic tissues and especially in tumors. The variable tumor enzymology as well as differences in membrane transport between malignant and nonmalignant cells as well as the extracellular glucose concentration, which are determinants of the LC, contribute to relevant variability of the LC between tumor tissues and have been shown to be different than that in normal brains. Krohn et al. recommended to refer to the metabolic rates measured by FDG PET as the metabolic rate of FDG rather than the glucose metabolic rate (Krohn et al. 2005).

2.1 Biomarker for Altered Cancer Metabolism

FDG is meanwhile an established biomarker, which can be used for objective measurements of cancerous signatures. Although the regulation of genes encoding glucose transporters and hexokinases has not been fully elucidated, oncogenic signal transduction pathways are shown to directly stimulate transcription of these genes and trigger glycolysis (Kelloff et al. 2005). The correlation between GLUT or hexokinase expression levels and FDG uptake was found variable and in some studies even low, most probably due to the different regulatory mechanisms of transport and phosphorylation (Buerkle and Weber 2008). Also, the correlation between FDG uptake and cellular proliferation has been found to be variable, and in some cases, no significant correlation between tumor FDG uptake and the Ki-67 labeling index was found (Buerkle and Weber 2008). In direct comparison, a much closer correlation between uptake of the thymidine analogue fluorothymidine and Ki-67 was reported (Buck et al. 2003). In several studies, a significant correlation between FDG uptake and grading has been described, but generally the association was weak with considerable overlap between the individual tumor grades. Relatively strong correlations between tumor grading and FDG uptake have been reported in patients with gliomas and sarcomas (Di Chiro et al. 1982; Folpe et al. 2000). Another example is the hepatocellular carcinoma where FDG uptake was associated with the expression of pSTAT3 (signal transducer and activator of transcription 3), HIF-1α, and GLUT1 correlating with poor differentiation and vascular invasion (Mano et al. 2014). In a report differentiating between liver metastases and hepatocellular carcinoma, Izuishi et al. (Izuishi et al. 2014) found variations in the expression of glucose metabolism-related enzymes between HCC and metastases attributed to the origin or degree of differentiation; low FDG uptake in moderately differentiated HCC reflected low GLUT1 and high glucose-6-phosphatase (G6Pase) expression, while high FDG accumulation in poorly differentiated HCC could reflect increased GLUT1 and decreased G6Pase expression. Although cellular proliferation and differentiation influence the FDG uptake, the high metabolic rates of malignant tumors cannot be explained by increased proliferation or loss of cellular differentiation alone (Buerkle and Weber 2008). As recently summarized by Plathow and Weber (Plathow and Weber 2008), there are different models fitting the explanation of the relationship between tumor development and glucose metabolism: (i) upregulation of glycolysis or upregulation of membrane transporters as an adaptation to hypoxia and consequent adaptation through resistance to apoptosis; (ii) increased glycolysis caused by genetic alterations that lead to uncontrolled activity of oncogenes; and (iii) mitochondrial dysfunction. All of these contributing factors are found resulting in the activation of the same effector mechanism, the stabilization of the transcription factor HIF-1 (hypoxia-inducible factor 1) (Plathow and Weber 2008). Hypoxia-inducible factor 1 is a heterodimeric transcription factor consisting of HIF-1α and HIF-1β subunits (Semenza 2003; Semenza 2007). HIF-1α expression and transcriptional activity increase exponentially as cellular O_2 concentration is decreased. Well over 100 target genes that are activated by HIF-1 have been identified which protein products mediate a switch from oxidative to glycolytic metabolism including those genes encoding erythropoietin, glucose transporters, glycolytic enzymes, and vascular endothelial growth factor (Semenza 2013). HIF-1 is also activated in cancer cells by tumor suppressor loss of function (e.g., VHL) and oncogene gain of function (leading to PI3K/AKT/mTOR activity) and mediates metabolic alterations that drive cancer progression and resistance to therapy (Semenza 2013). An increasing number of studies elaborate on the expression patterns of biologic markers related to glucose metabolism. In a xenograft model for colorectal carcinoma, tumor hypoxia and tumor metabolism were linked through the activation of metabolic genes following HIF-1 activation. Chemical induction of functional HIF-1 resulted in increased FDG uptake in vitro and in vivo, colocalized to carbonic anhydrase 9 (CA-9) and GLUT1 expression (Mees et al. 2013). Also in cancer of the tongue, correlations between SUVmax (SUV—standardized uptake value) and expression of HIF-1α, CA-9, and GLUT-1 were observed (Han et al. 2012). Kaira et al. (Kaira et al. 2011) showed that FDG uptake in NSCLC was significantly associated with GLUT1, hexokinase, HIF-1α, vascular endothelial growth factor (VEGF), CD34, p-AKT (protein kinase B), p-mTOR (mechanistic target of rapamycin), and epidermal growth factor receptor (EGFR). PTEN (phosphatidylinositol-3,4,5-trisphosphate 3-phosphatase) expression showed inverse correlation with FDG uptake. In vitro, FDG uptake was markedly decreased by the inhibition of GLUT1 and increased by GLUT1 upregulation by the induction of HIF-1α. NSCLC cells with PTEN loss showed the highest FDG uptake and the least sensitivity to mTOR inhibitors. In a further study, lactate dehydrogenase A (LDHA) was shown to increase FDG accumulation in adenocarcinomas of the lung by the upregulation of GLUT1, but not HK2, expression possibly via the AKT-GLUT1 pathway (Zhou et al. 2014). However, in gastric cancer, SUV was correlated only moderately with HIF-1α, but not with the expression of PCNA (proliferating cell nuclear antigen), HK2, or GLUT1. The authors therefore hypothesized that FDG accumulation represented tissue hypoxia rather than proliferative activity (Takebayashi et al. 2013). In a multitracer study of patients with oral squamous cell carcinoma, an association of HIF-1α expression to uptake of F-18-fluoromisonidazole (FMISO), a marker for cellular hypoxia, was detected, but no significant correlation between FDG and HIF-1α expression was found (Sato et al. 2013). There is increasing evidence that important oncogenes play a key role in regulating cellular glucose metabolism, including Akt1, c-MYC, HER2/neu, Wnt1, or H-Ras (Buerkle and Weber 2008; Alvarez et al. 2014). In NSCLC, SUVmax was shown to be associated with p53 and ERCC1 (excision repair cross-complementing gene 1), p53 expression being the primary predictor for FDG tumor uptake (Duan et al. 2013). It was recently stated that the oncogenic pathway activated within a tumor is a primary determinant of its FDG uptake, mediated by key glycolytic enzymes (Alvarez et al. 2014). The same authors showed that FDG uptake correlated positively with the expression of HK2 and HIF-1α, where the correlation between HK2 and FDG uptake was independent of all variables tested, including the initiating oncogenes. The authors, therefore, suggested that HK2 is an independent predictor of FDG uptake (Alvarez et al. 2014). In breast cancer subtypes, based on estrogen receptor (ER), progesterone receptor (PR), and HER2 expression, quantitative FDG PET parameters were significantly higher in the ER-negative group and the PR-negative group (Yoon et al. 2014). Association between elevated SUVmax and absence of ER expression was also found by other groups, as well as associated with positive HER2 phenotype, pleomorphism, mitosis count, lymphatic invasion, necrosis, and triple receptor negativity, accounting for the variable FGD uptake in breast cancer subtypes (Ekmekcioglu et al. 2013; Garcia Garcia-Esquinas et al. 2014).

2.2 Quantitative Imaging

The uptake of FDG into tissue reflects both transport and phosphorylation of glucose by the cells. Once images are obtained, there are several ways to analyze and interpret the data. The simplest method is qualitative imaging, which, however, fails to take full advantage of the inherently quantitative nature of PET data (Yu and Mankoff 2007). Quantification of tracer uptake or kinetics can be performed on several levels of complexity: semiquantitatively or quantitatively (from dynamic FDG PET acquisitions) using pharmacokinetic analysis. Most commonly, uptake is quantified using the standardized uptake value (SUV), i.e., the FDG tissue activity concentration at a single time point, normalized to the administered activity and a measure for distribution volume such as body weight or lean body mass. However, the FDG concentration measured in a tissue by PET is the summed up signal of three components of radioactive FDG: phosphorylated intracellular FDG, FDG in equilibrium between interstitial and intracellular space, and nonphosphorylated intravascular FDG (Buerkle and Weber 2008). Only the first component, the amount of phosphorylated intracellular FDG, is directly related to the metabolic activity of tumor cells. Static measurements of FDG uptake, which cannot differentiate these three components of the PET signal, therefore are not necessarily correlated with glucose metabolic rates (Buerkle and Weber 2008). The basic assumption underlying the use of the SUV is that at sufficiently late time points after injection, the concentrations of intravascular FDG and nonphosphorylated intracellular FDG become so small that they can be neglected when compared to the concentration of phosphorylated FDG in the tumor tissue (Buerkle and Weber 2008). Although SUV can be easily obtained from routine whole-body PET scans and is widely used in the clinical routine, it can be influenced by a variety of factors other than glucose metabolism, e.g., the time between injection and imaging, the size of the lesion, image reconstruction parameters, and the spatial resolution of the PET scanner but also by patient factors such as kidney function and volume of tumor vasculature. In order to measure metabolic rates of FDG, it is necessary to image the time course of FDG uptake by the tissue and its clearance from the blood. This more sophisticated approach uses dynamic PET imaging and kinetic analysis where the blood clearance curve is obtained in addition to the dynamic PET tumor images and serves as an input to compartment models from which relevant kinetic parameters can be estimated. Dynamic scanning for at least 1 h after injection, as well as an input function based on arterial blood samples, are required (Fig. 2d). In some cases, e.g., thoracic studies, the latter can be derived from large enough vascular structures within the field of view, i.e., image-derived input function, but arterial catheterization and frequent sampling are otherwise required (Kelloff et al. 2005). A mathematical analysis of these data can be used to provide parametric images of the PET tracer's biochemistry in terms of transport parameters and a metabolic flux. The autoradiographic method of Sokoloff for measuring regional metabolic rates of glucose in the brain of rats using C-14-deoxyglucose (Sokoloff 1977) has been modified for human studies using FDG PET by Phelps et al. (Phelps et al. 1979) and Reivich et al. (Reivich et al. 1979). Mostly, the two-tissue three-compartment model with four rate constants K1, k2, k3, and k4 is used, often simplified by assuming that the dephosphorylation rate of FDG-6-phosphate is small enough that it can be ignored (k4 = 0) (Fig. 2e). The metabolic rate of glucose (MRglu) can be calculated from the equation below, where Cglu is the concentration of glucose in plasma and LC is the lumped constant:

$$MRglu = \frac{Cglu}{LC} \times \frac{K1k3}{(k2+k3)}$$

Only later it was recognized that a simplified graphical method for tracer kinetic modeling assuming k4 = 0, the Patlak–Gjedde analysis, can be used that allows the determination of a rate constant (Ki) for the net influx of FDG in the tumor. In contrast to compartment modeling, Patlak–Gjedde analysis is less sensitive to errors in individual data points of the blood– and tissue time–activity curves, leading to more robust and reproducible results (Minn et al. 1995; Weber et al. 2000). For clinical practice and even for most research questions, estimation of SUV is the most relevant approach, while kinetic modeling is the indispensable step in new tracer development and tracer biokinetic studies.

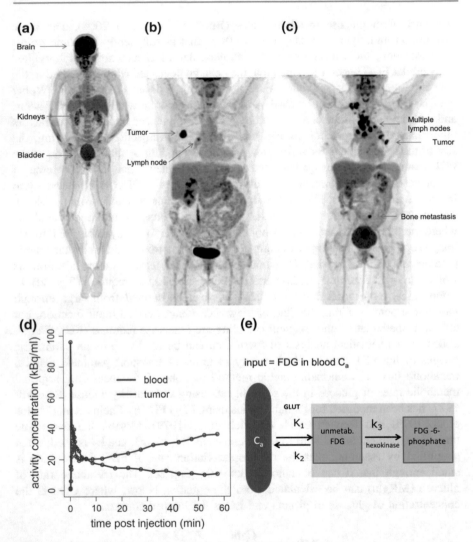

Fig. 2 Normal distribution of FDG (**a**). Increased metabolic activity in lesions of patients with NSCLC stage II B (T2a N1 M0) (**b**) and stage IV (T1b N3 M1) (**c**). **d** Typical time course of FDG uptake by malignant tumors and FDG clearance from the blood. **e** Two-tissue three-compartment model: the kinetic model for the measurement of glucose transport and phosphorylation rate using FDG

There are also studies indicating that kinetic analysis of FDG PET might capture some auxiliary information on tumor biology, especially in regard to the prediction of treatment response. Tseng et al. (Tseng et al. 2004) used a two-tissue compartment model to describe the FDG kinetics in patients with locally advanced breast cancer which obtained FDG PET prior to and after cytotoxic chemotherapy. Chemotherapy induced a decline of tumor growth resulting in a decline of glucose metabolism, above average in comparison with blood flow, identifying the phosphorylation step (k3) as a rate-limiting factor for FDG accumulation. The decline of k3 indicated that the cytotoxic treatment induced a shift in the kinetic pattern of FDG, driven by the rate of phosphorylation (k3) relative to glucose delivery (K1). This implicated that anaerobic metabolism and hypoxia have been diminished with treatment and the tumor-specific aberrant glucose metabolism returned to a state of normal conditions. Tumors that responded to therapy showed a decline in the FDG blood-to-tissue transport (K1) and glucose metabolism (Ki) similar to normal tissue, whereas a persistent imbalance between glucose transport and glucose metabolism was associated with a more aggressive tumor behavior (Tseng et al. 2004).

Besides the quantitative assessment of metabolic parameters, further approaches for advanced tumor characterization are emerging in recent years and summarized under the general term "radiomics" (Lambin et al. 2012). The underlying concept is that human cancers exhibit strong phenotypic differences that can be visualized noninvasively by medical imaging. Radiomics refers to the comprehensive quantification of tumor phenotypes by applying a large number of quantitative image features (Lambin et al. 2012). In the focus of these analyses are the prognostic signatures, capturing intratumor heterogeneity, associated with underlying gene expression patterns (Aerts et al. 2014). Solid cancers have extraordinarily spatial and temporal heterogeneity at different levels: genes, proteins, cells, microenvironment, tissues, and organs (Lambin et al. 2012; Aerts et al. 2014). This limits the use of biopsy-based molecular assays but in contrast gives a huge potential for noninvasive imaging, which has the ability to capture intratumoral heterogeneity in a noninvasive way. Heterogeneity of FDG uptake within the tumor is hypothesized to reflect this biologic variability. Therefore, tumor heterogeneity detected by FDG PET appears promising for the prediction of therapy outcome. Two types of tumor heterogeneities based on FDG PET images are discussed in the literature: (i) heterogeneity of tracer uptake (voxel values) in the tumor lesion and (ii) spatial (shape) irregularity of the lesions (Apostolova et al. 2014). A number of different measures have been proposed for both types, the majority for the characterization of the voxelwise heterogeneity of tracer uptake (Chicklore et al. 2013; Cook et al. 2013; El Naqa et al. 2009; Tixier et al. 2012; Tixier et al. 2014; Tixier et al. 2011). Also, several measures of shape heterogeneity of the FDG uptake in tumors were proposed that have been associated with progression-free survival and overall survival in sarcomas as well as in tumor patients with head and neck cancer and cervical carcinoma (Apostolova et al. 2014; El Naqa et al. 2009; Eary et al. 2008).

2.3   Clinical Applications

FDG PET is mostly used for diagnosis, staging, and restaging of a variety of malignant tumors, including nonsmall-cell lung cancer, melanoma, lymphoma, and esophageal, head and neck, pancreatic, and colorectal cancers. In these and other tumor entities, FDG PET has shown a significant impact on clinical management.

One of the first indications, where FDG PET was evaluated as a noninvasive tool, was the characterization of solitary pulmonary nodules (SPN), demonstrating the use for characterizing the likelihood for malignancy in mass lesions that are not amenable for biopsy. Numerous studies found that PET had relatively high accuracy with the majority of FDG-avid lesions being malignant and nonavid lesions being benign, defining further strategies as active treatment or conservative management (Dewan et al. 1995; Gould et al. 2001; Gupta et al. 1992; Keith et al. 2002).

The role of FDG PET in cancer staging is meanwhile confirmed in many prospective cohorts of patients with different cancer types. In NSCLC, FDG PET/CT is known to lead to stage migration in 30 % of the cases and to modification of treatment in 20–30 %, including avoidance of a futile thoracotomy (Taus et al. 2014). PET/CT has shown better accuracy for mediastinal lymph node staging than conventional imaging with a sensitivity of 85 % and a specificity of 90 %, compared to 61 and 79 % in case of CT, respectively (Gould et al. 2003), as in Fig. 2.

FDG PET/CT is also playing an increasing role in radiotherapy treatment planning. Due to the contrast afforded by differential FDG uptake in cancer cells, while still providing anatomic landmarks, PET/CT offers the potential for improved differentiation of malignant from benign tissues and, therefore, more accurate definition of the gross tumor volume. The advantage of radiotherapy planning based on FDG PET/CT has been demonstrated in esophageal cancer (Leong et al. 2006), rectal cancer (Withofs et al. 2014), and NSCLC (Bradley et al. 2004; Vanuytsel et al. 2000; Mah et al. 2002).

FDG PET/CT is also increasingly used for monitoring tumor response to therapy. One of the major theoretical advantages of PET compared with structural imaging techniques is that there is usually a more rapid decline in tumor metabolism than in tumor size after tumor-specific therapies. Numerous studies have demonstrated that reduction in FDG uptake correlates with subsequent clinical and radiological response, and treatment-induced changes in tumor FDG uptake correlate with patient survival (Allen-Auerbach and Weber 2009). This has led to recommendations for wider use of FDG PET/CT in therapeutic response assessment. For chemotherapy and radiotherapy, there is considerable evidence that reduction of FDG uptake is caused by a loss of viable tumor cells (Weber and Wieder 2006). However, the close relationship between various oncogenic signaling pathways and tumor glucose metabolisms suggests that drugs targeting these signal transduction pathways may have a more direct effect on cellular glucose metabolism (Plathow and Weber 2008). One of the most dramatic examples of these direct effects is the rapid reduction of FDG uptake in gastrointestinal stromal tumors (GISTs) following treatment with an agent that blocks the c-kit oncogene product. Metabolic response can be apparent within 24 h of commencing treatment (Fig. 3) (Treglia et al. 2012; Holdsworth et al. 2007). Moreover, the decrease in FDG uptake after the onset of treatment seems to be related to a positive treatment result and a prolonged progression-free survival. In this setting, the typical morphological response evaluation criteria in solid tumors (RECIST) have been shown to perform poorly, as tumors may become more hypodens on CT and even increase in size after treatment (Holdsworth et al. 2007). Another example of successful application of FDG PET is the management of

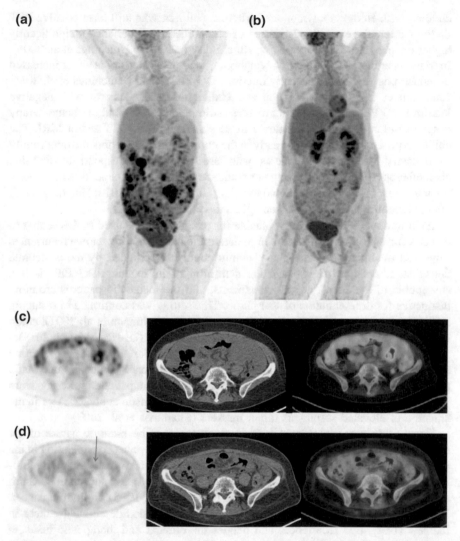

Fig. 3 Patient with a disseminated abdominal GIST tumor: **a** baseline PET, **b** restaging PET 14 days after initiation of imatinib treatment showing metabolic response in the majority of abdominal lesions. Transversal planes of FDG PET, attenuation CT and fusion PET/CT prior to therapy (**c**), and 14 days after initiation of imatinib (**d**)

lymphomas (Cheson et al. 2014). Conventional methods have low specificity for distinguishing viable lymphoma or fibrotic scar tissue after treatment. Since changes in tissue function predate volume changes, it is possible to assess response using functional imaging. Thus, FDG PET is being increasingly used for restaging after treatment of Hodgkin's lymphoma and aggressive non-Hodgkin's lymphomas (Specht 2007). Studies have demonstrated that normalization of FDG PET findings early in the course of chemotherapy is associated with an excellent prognosis in children with Hodgkin's lymphoma, whereas children who still have positive PET findings after, for example, 2 courses of chemotherapy experience a significantly higher relapse rate (Furth et al. 2009; Hutchings et al. 2014; Hutchings et al. 2006). In trials on adults with Hodgkin's lymphoma, treatment intensity is either increased or reduced depending on the early interim FDG PET result (Hutchings et al. 2014; Gallamini et al. 2014; Kostakoglu and Gallamini 2013). Patients with negative interim FDG PET findings do not receive radiotherapy at the end of chemotherapy despite residual morphological tumor masses as depicted by CT and/or MRI. The ability to predict tumor response early in the course of therapy offers the opportunity to intensify treatment in patients who are unlikely to respond to first-line chemotherapy. Conversely, treatment could be shortened in patients who show a favorable response in order to avoid toxicity (Hutchings et al. 2014; Hutchings et al. 2006; Gallamini et al. 2014; Kostakoglu and Gallamini 2013).

After treatment, FDG PET is valuable for restaging and is used in this setting to detect recurrent or residual disease or to determine the extent of a known recurrence. Improved treatment selection and planning could be facilitated by more accurate detection of residual disease and the definition of its extent. FDG PET is less susceptible to the effects of prior treatments, and the metabolic patterns are more distinctive for differentiation of scar tissue. This setting was confirmed in restaging studies of lymphoma (Cheson et al. 2014), NSCLC (Eschmann et al. 2007), colon carcinoma (Kalff et al. 2000), head and neck cancer, etc. (Ul-Hassan et al. 2013).

Another application where FDG PET is of particular utility is in detecting suspected malignancy, for example depicting the primary tumor in metastatic patients with unknown primary (CUP syndrome), or in patients suffering from paraneoplastic syndrome without obvious primary (Wilkinson et al. 2003), or in the situation of elevated suspicious tumor markers (Shammas et al. 2007).

Finally, the potential role of PET in characterizing the biologic nature of the tumor and serving as a prognostic biomarker increasingly gains consideration in the clinical routine. High-grade gliomas had higher FDG uptake than low-grade gliomas, and FDG uptake was shown to be associated with longer survival (Di Chiro et al. 1982; Di Chiro and Bairamian 1988; Patronas et al. 1985). In thyroid cancer, FDG PET signifies the conversion of differentiated to dedifferentiated cancer in iodine-131-negative recurrences and defines the clinical and therapeutic management as well as the prognosis (Caetano et al. 2014). In neuroendocrine tumors, FDG uptake was also shown to be correlated with grading and to be predictive for survival in GEPNET patients (Bahri et al. 2014; Simsek et al. 2014). In various sarcoma entities, FDG PET SUV has been shown a reliable and independent marker for risk assessment and patient survival (Brenner et al. 2004, 2006a, b).

3 Hypoxia and Tumor Perfusion

The most fundamental process of any living cell in all aerobic life forms is the capability to use oxygen in the production of energy. Transported via oxyhemoglobin, oxygen is used in a chain of mitochondrial redox reactions to produce

ATP in oxidative phosphorylation. This principal method is the common ground for both normal healthy cells and tumor cells, because it is the most efficient way to nurture cell metabolism and proliferation in the presence of oxygen. Differences begin to appear when these two entities of cells suffer from deprivation of oxygen. Otherwise, normal cells react to shortage of oxygen by shutting down cell functions and going into apoptosis or necrosis, depending on the circumstances like intensity and length of hypoxia. Tumors with high proliferation rates often outgrow their blood supply and evolve different reactions to hypoxic environments (Krause et al. 2006). The key role in the adaptation process plays the hypoxia-inducible factor 1 (HIF-1) or rather its subunit HIF-1α (Ferrer Albiach et al. 2010). This transcription factor is constantly degraded in the presence of oxygen, but can become stable to promote the transcription of many genes, for instance genes that turn up glycolysis and slow down growth. Mutations concerning the enhanced stability of HIF-1 can be found in many cancer types, especially types with high chances of metastatic growth (Krause et al. 2006; Ferrer Albiach et al. 2010).

The importance and clinical relevance of easy and reliable detection of hypoxia in tumor cells becomes evident when it comes to radiation therapy. The direct correlation between radiation sensitivity of any tissue and an adequate oxidation is a long-known and well-proved principal. The presence of hypoxic areas within a tumor has shown to be a risk factor for poor outcome with high prognostic relevance in a variety of cancer entities (Vaupel and Mayer 2007; Arabi and Piert 2010; Brizel et al. 1996). Modern PET-based imaging, today often merged with CT or even MRI in multimodality imaging, can unveil such areas by using tracers like O-15 water, F-18-fluoromisonidazole (FMISO), and F-18-fluoroazomycin arabinoside (FAZA). In contrast to other direct methods like Eppendorf electrodes and indirect methods like analysis of biomarkers and immunohistochemistry, these PET tracers provide a highly sensitive, easy-to-use, and noninvasive method to delineate perfusion, oxidation, and hypoxia and its spatial distribution within the tumor (Hockel and Vaupel 2001; Wang et al. 2010; Tatum et al. 2006).

3.1　FMISO

FMISO has been widely used to study hypoxic tissue. It originates from misonidazole, a compound of nitroimidazole (Adams and Stratford 1986). Long before it was first used in its F-18-fluorinated form to mark hypoxic cells, misonidazole was successfully applied as a radiosensitizer (Lapi et al. 2009; Jerabek et al. 1986). Until now, there is no gold standard in the synthesis of FMISO, although there are precursor kits commercially available for radiochemical synthesis (Lim and Berridge 1993).

When entering a cell through passive diffusion, FMISO is reduced by nitroreductase enzymes and immediately reoxidized if there is a normal level of oxygen. However, in hypoxic cells, the metabolites of FMISO stay reactive, bind to intracellular proteins covalently, and accumulate (Bourgeois et al. 2011). An accumulation of FMISO or rather its metabolites occurs only in vital cells, while this is not possible in necrotic cells because active enzymes are needed for this process. The overall biochemical properties of FMISO are highly comparable to MISO. Both have similar body clearance and are mainly excreted via the urine. In contrast to MISO, there is less significant metabolization of FMISO in the liver. Furthermore, the LD50 for a 70 kg person differs with 12.5 g for MISO and 6.5 g for FMISO. Although the normal dose administered for imaging purposes can be as high as 1 mg/Kg, there are no known side effects of either MISO or FMISO in this dose range.

Studies show that the selective retention of FMISO in hypoxic cells can be measured starting 1 up to 2.5 h after injection with an optimum tumor/plasma ratio after approximately 2 h (Padhani 2005; Koh et al. 1992). An objective and generally applicable quantification method of tumor/plasma ratio and uptake of FMISO is not yet acknowledged. Depending on variables like cancer entity and its original tissue, a threshold separating hypoxic from normal oxygenated cells can vary. Nevertheless, a ratio of ≥1.4 showed to be a significant indicator of hypoxia in recent studies (Koh et al. 1992). Although FMISO is a reliable and well-tested tracer, this relatively low threshold separating the signal from tumor tissue from a background signal is a drawback for its clinical use (Lapi et al. 2009; Lee et al. 2008). A higher uptake in tumor tissue can be only achieved in tumors with distinct hypoxia, compared to tumors with mild reduction of the oxygen level. Studies showed that 10 mm HG is the lower limit for a reasonable uptake of FMISO, any pO2 above that level prevents a significant tracer accumulation and, thus, a reliable detection of FIMSO uptake.

Besides a mild liver uptake, there is physiological uptake in the bowel. Because of a free diffusion across the blood/brain barrier, there is also a noticeable uptake in the brain. The highest uptake and highest organ dose can be found in the urinary bladder, depending on the applied voiding protocol.

The accumulation of FMISO in hypoxic tumor tissue has been shown by various studies in a wide variety of solid tumors, e.g., brain tumors, head and neck cancer, breast cancer, and NSCLC. To prove that accumulation actually correlates with a low level of oxygen in situ, it has been compared with direct and invasive pO2 measurements using the Eppendorf hypoximeter (Stone et al. 1993). In several studies, this comparison was made and a correlation between a high FMISO uptake and a low pO2 has been reported, even exceeding FDG as a suitable tracer for detecting hypoxia (Lawrentschuk et al. 2005; Gagel et al. 2004; Zimny et al. 2006). Another fundamental requirement for clinical use is the reproducibility of imaging results. A recent study by Okamoto et al. demonstrated a good reproducibility of PET imaging parameters such as SUVmax within a 48-h time frame in a group of 11 patients with untreated head and neck cancer (Okamoto et al. 2013). In addition to such "basic" qualities, FMISO uptake proved to be an independent prognostic factor in pretherapy scans with respect to overall survival in patients with head and neck cancer (Rajendran et al. 2006). Similar results have been shown by studies concerning malignant glioma, breast cancer, NSCLC, and renal carcinoma (Lawrentschuk et al. 2005; Spence et al. 2008; Gagel et al. 2006; Cheng et al. 2013).

3.2 F-18-Fluoroazomycin Arabinoside

FAZA is a close chemical relative of FMISO. It shares a lot of its characteristic features with FMISO. Both tracers are compounds of nitroimidazole. But in contrast to FMISO, FAZA is sugar-coupled. Both follow the same intracellular accumulation pattern. Without sufficient oxygenation, the tracer is reduced to metabolites, which bind covalently to intracellular macromolecules. This reduction requires active enzymes, so a distinction between necrotic and vital but hypo-oxygenated tissue is possible as well.

One of the goals of synthesizing FAZA was to produce a less lipophilic alternative to FMISO. Theoretically, this would lead to a faster uptake in the hypoxic tumor tissue and a better target-to-reference tissue ratio. Since the first reported synthesis by Kumar et al., a number of studies tried to confirm these theoretical advantages in vivo and in vitro (Kumar et al. 1999). The disadvantage of FMISO—a low hypoxic tumor-to-background ratio—is a strength of FAZA: Due to its fast renal elimination, a fast and comparably high contrast can be achieved. An in vivo animal study by Piert et al. proved this theoretical superiority of a less lipophilic tracer by achieving improved tumor-to-blood and tumor-to-muscle ratios (Piert et al. 2005). As an obstacle, many approaches of synthesizing FAZA struggled with a relatively low overall radiochemical yield of only approximately 20 % (Reischl et al. 2005). Newer approaches could improve the yield significantly, and an improved method of a sufficient synthesis following the EU regulations for good clinical practice was published by Ervagi-Vernat et al. (Hayashi et al. 2011; Servagi-Vernat et al. 2014).

However, the quality of FAZA as a tracer for measuring hypoxia is not only determined by its synthesis yield but rather by its ability to reflect the situation in the tumor with reproducible results. Several studies have investigated the biological characteristics of FAZA in vitro and in vivo. Naturally, such studies have to compare a new tracer with more established tracers—in this case with FDG and/or FMISO—and well-established invasive and local methods like an Eppendorf hypoximeter. In one of the first studies on this subject, Busk et al. (2008) examined various carcinoma cell lines with FAZA and found superior uptake characteristics compared to FDG and FMISO. This research group also showed a very good agreement of hypoxia measured by Eppendorf electrodes and uptake of FAZA in an experimental study using xenografted tumors in mice (Busk et al. 2008). Recent studies suggested that FAZA can be of prognostic value concerning response to therapy and survival, although variables like the type of cancer can lessen this prognostic potential (Garcia-Parra et al. 2011). Head and neck cancer comprises a group of tumors which present itself rather inhomogeneous not only varying in tumor location and tumor histology. The tumors often have a high level of heterogeneous structures with varying segments of vital, hypoxic, and necrotic tumor tissue side by side. Mortensen et al. (Mortensen et al. 2012) included 40 patients with this type of cancer and conducted FAZA PET scans prior to and during radiotherapy. After a median follow-up of 19 months, a significantly lower disease-free survival was found in patients with hypoxic tumors. Another recent study by Trinkhaus et al. (Trinkaus et al. 2013), based on a very limited number of patients, investigated FAZA in NSCLC and could not find a prognostic value of FAZA PET scans at all but concluded that more and larger studies are needed to evaluate this matter properly.

The combination of radiotherapy and chemotherapy in a wide variety of protocols—either simultaneously or staggered—is for many kinds of cancers the current gold standard of treatment, e.g., head and neck cancer, anal cancer, and advanced stages of colorectal cancer. The vigorous effect of radiotherapy depends on a sufficient level of oxygenation. Imaging tumor hypoxia, therefore, will remain a major goal for metabolic imaging in the future.

3.3 Tumor Perfusion

As much as the therapeutic effect of radiotherapy depends on oxygenation, chemotherapy depends on a sufficient perfusion of the tumor tissue (Vaupel et al. 1989; Hamilton and Rath 2014). Chemoresistance as a major cause for poor prognosis can be the result of aberrant angiogenesis in many rapidly and heterogeneously growing tumors. In order to depict and quantify this aspect of cancer metabolism, tumor perfusion has to be made visible by functional imaging.

3.4 O-15 Water

The quantification of perfusion with oxygen-15-labeled water and its use as a PET tracer is a well-established approach. But most applications so far are limited to cardiology and neurology (Ter-Pogossian and Herscovitch 1985; Oda and Kondo 1994). Although O-15 water is established as a reliable perfusion tracer in these medical specialities, oncological studies are rare on this topic (Lodge et al. 2008).

A reason for this might be the complicated synthesis and handling of O-15. This radioactive isotope of oxygen is produced by accelerating a deuteron in a cyclotron and let it collide with N-14. The resulting O-15 has a very short half-life of only 122 s, too short to produce it outside the imaging facility or integrate it in a daily clinical routine (Wadsak and Mitterhauser 2010). On the other hand, this extraordinary short half-life opens up the opportunity for very short delays between two imaging sessions. If multiple scans are needed, O-15 water can be reapplied within a few minutes without moving the patient. After intravenous injection, the tracer is freely diffusible, which means the linear relation of signal intensity and concentration of O-15 water in the examined tissue allows the absolute quantification of the tumor perfusion (de Langen et al. 2008).

For every tracer, the reproducibility is a major quality criterion. An established way of measuring reproducibility is the within-subject coefficient of variation (Bland and Altman 1999). In studies on brain and myocardium, O-15 water showed within-subject coefficients of variation in the range of 9–14 % (de Langen et al. 2008). Two often-cited studies of breast cancer patients and intraabdominal tumors by Wilson et al. and Wells et al., respectively, showed that O-15 water PET reaches similar coefficients of variation in tumors as compared to the perfusion of myocardium and brain (de Langen et al. 2008; Wilson et al. 1992; Wells et al. 2003). Wilson et al. calculated a within-subject coefficient of variation of 10 % for breast cancer and Wells et al. 10 % for intra-abdominal tumors.

Whereas in most normal tissues blood flow and metabolism are tightly coupled, in tumors these parameters are often not well matched resulting in an aberrantly high glucose metabolism relative to the blood flow, i.e., a flow–metabolism mismatch (Mankoff et al. 2009). Several studies using combined molecular imaging techniques for measuring regional tumor perfusion and metabolism have recognized the mismatch between blood flow and metabolism as an adverse prognostic marker particularly in terms of prediction of outcome or response to therapy. The combination of FDG and O-15 water PET is a well-studied technique for this task allowing delineation of regional variations in the metabolism–flow ratio on a voxel-by-voxel basis (Apostolova et al. 2014). Due to the short half-life of O-15, both measurements can be performed in a single imaging session without moving the patient Fig. 4. Komar et al. examined 26 patients suspected of suffering from pancreatic cancer by using FDG and O-15 water PET. They found a reduced perfusion in both benign and malignant lesions compared to healthy test subjects and a significantly increased glucose metabolism and decreased blood flow in malignant lesions compared to benign lesions and healthy pancreas. In this study, the combination of both values had a stronger prognostic value than either one parameter (Komar et al. 2009). A combined measurement of glucose metabolism and blood flow was also the concept of a study on 53 patients with advanced breast cancer by Dunnwald et al. (Dunnwald et al. 2008). In this cohort, both a missing decrease and an increase of tumor blood flow between the start of neoadjuvant chemotherapy and the midpoint of therapy was a significant independent risk factor for shorter disease-free survival and overall survival. In patients with advanced cervical carcinoma, a flow–metabolism mismatch as measured by PET with O-15 water and FDG could be detected to a highly variable extent showing a clear association with the histological grading of the tumors (Fig. 4) (Apostolova et al. 2014). So far, the existing data seem to indicate that perfusion imaging with O-15 water in combination with metabolic imaging could be a promising imaging tool for the measurement of flow–metabolism mismatches as a sign of metabolic stress in tumor tissue associated with poor outcome. More studies are needed to evaluate the role of the combination of these two tracers for in vivo selection of patients most likely to benefit from chemotherapy and antivascular therapies and also to measure the effect of various treatment regimens on tumor perfusion and metabolism after treatment.

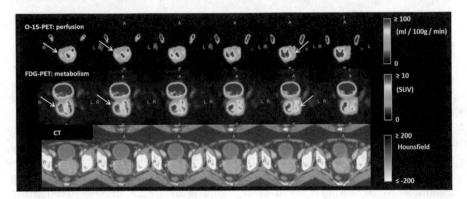

Fig. 4 Regional mismatch (*arrows*) between blood flow measured with O-15 water PET (*upper row*) and glucose metabolism measured with FDG PET (*second row*) in a patient with advanced cervical carcinoma. *Third row* Transversal CT images of the primary tumor

4 Conclusion

Recently, the interest in metabolic imaging has markedly grown, since accumulating evidence supports the clinical value of the methodology for tumor phenotyping as an essential tool for guiding individual patient care. Molecular imaging using metabolic tracers and PET/CT also constitutes a major opportunity in translational research bringing fundamental molecular insights to the clinical area. Correlation between in vivo and in vitro tumor profiling facilitates further understanding of tumor characteristics which plays an essential role in ensuring appropriate monitoring of targeted therapies.

References

Adams GE, Stratford IJ (1986) Hypoxia-mediated nitro-heterocyclic drugs in the radio- and chemotherapy of cancer. An overview. Biochem Pharmacol 35(1):71–76

Aerts HJ, Velazquez ER, Leijenaar RT, Parmar C, Grossmann P, Cavalho S, Bussink J, Monshouwer R, Haibe-Kains B, Rietveld D, Hoebers F, Rietbergen MM, Leemans CR, Dekker A, Quackenbush J, Gillies RJ, Lambin P (2014) Decoding tumour phenotype by noninvasive imaging using a quantitative radiomics approach. Nat Commun 5:4006. doi:10.1038/ncomms5006

Allen-Auerbach M, Weber WA (2009) Measuring response with FDG-PET: methodological aspects. Oncologist 14(4):369–377. doi:10.1634/theoncologist.2008-0119

Alvarez JV, Belka GK, Pan TC, Chen CC, Blankemeyer E, Alavi A, Karp J, Chodosh LA (2014) Oncogene pathway activation in mammary tumors dictates [18F]-FDG-PET uptake. Cancer Res. doi:10.1158/0008-5472.CAN-14-1235

Apostolova I, Steffen IG, Wedel F, Lougovski A, Marnitz S, Derlin T, Amthauer H, Buchert R, Hofheinz F, Brenner W (2014a) Asphericity of pretherapeutic tumour FDG uptake provides independent prognostic value in head-and-neck cancer. Eur Radiol 24(9):2077–2087. doi:10.1007/s00330-014-3269-8

Apostolova I, Hofheinz F, Buchert R, Steffen IG, Michel R, Rosner C, Prasad V, Kohler C, Derlin T, Brenner W, Marnitz S (2014) Combined measurement of tumor perfusion and glucose metabolism for improved tumor characterization in advanced cervical carcinoma. A PET/CT pilot study using [^{15}O]water and [^{18}F]fluorodeoxyglucose. Strahlentherapie und Onkologie: Organ der Deutschen Rontgengesellschaft [et al] 190(6):575–581. doi:10.1007/s00066-014-0611-7

Arabi M, Piert M (2010) Hypoxia PET/CT imaging: implications for radiation oncology. Q J Nucl Med Mol Imaging Off Publ Ital Assoc Nucl Med 54(5):500–509

Bahri H, Laurence L, Edeline J, Leghzali H, Devillers A, Raoul JL, Cuggia M, Mesbah H, Clement B, Boucher E, Garin E (2014) High prognostic value of 18F-FDG PET for metastatic gastroenteropancreatic neuroendocrine tumors: a long-term evaluation. J Nucl Med Off Publ Soc Nucl Med. doi:10.2967/jnumed.114.144386

Bensinger SJ, Christofk HR (2012) New aspects of the Warburg effect in cancer cell biology. Semin Cell Dev Biol 23(4):352–361. doi:10.1016/j.semcdb.2012.02.003

Bland JM, Altman DG (1999) Measuring agreement in method comparison studies. Stat Methods Med Res 8(2):135–160

Bourgeois M, Rajerison H, Guerard F, Mougin-Degraef M, Barbet J, Michel N, Cherel M, Faivre-Chauvet A (2011) Contribution of [64Cu]-ATSM PET in molecular imaging of tumour hypoxia compared to classical [18F]-MISO–a selected review. Nucl Med Rev Central East Eur 14(2):90–95

Bradley J, Thorstad WL, Mutic S, Miller TR, Dehdashti F, Siegel BA, Bosch W, Bertrand RJ (2004) Impact of FDG-PET on radiation therapy volume delineation in non-small-cell lung cancer. Int J Radiat Oncol Biol Phys 59(1):78–86. doi:10.1016/j.ijrobp.2003.10.044

Brenner W, Conrad EU, Eary JF (2004) FDG PET imaging for grading and prediction of outcome in chondrosarcoma patients. Eur J Nucl Med Mol Imaging 31:189–195

Brenner W, Friedrich RE, Gawad KA, Hagel C, von Deimling A., de Wit M, Buchert R, Clausen M, Mautner VF (2006a) Prognostic relevance of FDG PET in patients with neurofibromatosis type-1 and malignant peripheral nerve sheath tumors. Eur J Nucl Med Mol Imaging 33:428–432

Brenner W, Hwang W, Vernon C, Conrad EU, Eary JF (2006b) Risk assessment in liposarcoma patients based on FDG-PET imaging. Eur J Nucl Med Mol Imaging 33:1290–1295

Brizel DM, Scully SP, Harrelson JM, Layfield LJ, Bean JM, Prosnitz LR, Dewhirst MW (1996) Tumor oxygenation predicts for the likelihood of distant metastases in human soft tissue sarcoma. Cancer Res 56(5):941–943

Buck AK, Halter G, Schirrmeister H, Kotzerke J, Wurziger I, Glatting G, Mattfeldt T, Neumaier B, Reske SN, Hetzel M (2003) Imaging proliferation in lung tumors with PET: 18F-FLT versus 18F-FDG. J Nucl Med Off Publ Soc Nucl Med 44(9):1426–1431

Buerkle A, Weber WA (2008) Imaging of tumor glucose utilization with positron emission tomography. Cancer Metastasis Rev 27(4):545–554. doi:10.1007/s10555-008-9151-x

Bui T, Thompson CB (2006) Cancer's sweet tooth. Cancer cell 9(6):419–420. doi:10.1016/j.ccr.2006.05.012

Busk M, Horsman MR, Jakobsen S, Bussink J, van der Kogel A, Overgaard J (2008a) Cellular uptake of PET tracers of glucose metabolism and hypoxia and their linkage. Eur J Nucl Med Mol Imaging 35(12):2294–2303. doi:10.1007/s00259-008-0888-9

Busk M, Horsman MR, Jakobsen S, Keiding S, van der Kogel AJ, Bussink J, Overgaard J (2008b) Imaging hypoxia in xenografted and murine tumors with 18F-fluoroazomycin arabinoside: a comparative study involving microPET, autoradiography, PO2-polarography, and fluorescence microscopy. Int J Radiat Oncol Biol Phys 70(4):1202–1212. doi:10.1016/j.ijrobp.2007.11.034

Caetano R, Bastos CR, de Oliveira IA, da Silva RM, Fortes CP, Pepe VL, Reis LG, Braga JU (2014) Accuracy of PET and PET-CT in the detection of differentiated thyroid cancer recurrence with negative I whole body scan results: a meta-analysis. Head Neck. doi:10.1002/hed.23881

Cheng J, Lei L, Xu J, Sun Y, Zhang Y, Wang X, Pan L, Shao Z, Zhang Y, Liu G (2013) 18F-fluoromisonidazole PET/CT: a potential tool for predicting primary endocrine therapy resistance in breast cancer. J Nucl Med Off Publ Soc Nucl Med 54(3):333–340. doi:10.2967/jnumed.112.111963

Cheson BD, Fisher RI, Barrington SF, Cavalli F, Schwartz LH, Zucca E, Lister TA (2014) Recommendations for initial evaluation, staging, and response assessment of hodgkin and non-hodgkin lymphoma: the lugano classification. J Clin Oncol Off J Am Soc Clin Oncol. doi:10.1200/JCO.2013.54.8800

Chicklore S, Goh V, Siddique M, Roy A, Marsden PK, Cook GJ (2013) Quantifying tumour heterogeneity in 18F-FDG PET/CT imaging by texture analysis. Eur J Nucl Med Mol Imaging 40(1):133–140. doi:10.1007/s00259-012-2247-0

Cook GJ, Yip C, Siddique M, Goh V, Chicklore S, Roy A, Marsden P, Ahmad S, Landau D (2013) Are pretreatment 18F-FDG PET tumor textural features in non-small cell lung cancer associated with response and survival after chemoradiotherapy? J Nucl Med Off Publ Soc Nucl Med 54(1):19–26. doi:10.2967/jnumed.112.107375

de Langen AJ, van den Boogaart VE, Marcus JT, Lubberink M (2008) Use of H2(15)O-PET and DCE-MRI to measure tumor blood flow. Oncologist 13(6):631–644. doi:10.1634/theoncologist.2007-0235

Dewan NA, Reeb SD, Gupta NC, Gobar LS, Scott WJ (1995) PET-FDG imaging and transthoracic needle lung aspiration biopsy in evaluation of pulmonary lesions. A comparative risk-benefit analysis. Chest 108(2):441–446

Di Chiro G, Bairamian D (1988) Brain imaging of glucose utilization in cerebral tumors. Am J Physiol Imaging 3(1):56

Di Chiro G, DeLaPaz RL, Brooks RA, Sokoloff L, Kornblith PL, Smith BH, Patronas NJ, Kufta CV, Kessler RM, Johnston GS, Manning RG, Wolf AP (1982) Glucose utilization of cerebral gliomas measured by [18F] fluorodeoxyglucose and positron emission tomography. Neurology 32(12):1323–1329

Duan XY, Wang W, Wang JS, Shang J, Gao JG, Guo YM (2013) Fluorodeoxyglucose positron emission tomography and chemotherapy-related tumor marker expression in non-small cell lung cancer. BMC Cancer 13:546. doi:10.1186/1471-2407-13-546

Dunnwald LK, Gralow JR, Ellis GK, Livingston RB, Linden HM, Specht JM, Doot RK, Lawton TJ, Barlow WE, Kurland BF, Schubert EK, Mankoff DA (2008) Tumor metabolism and blood flow changes by positron emission tomography: relation to survival in patients treated with neoadjuvant chemotherapy for locally advanced breast cancer. J Clin Oncol Off J Am Soc Clin Oncol 26(27):4449–4457. doi:10.1200/JCO.2007.15.4385

Eary JF, O'Sullivan F, O'Sullivan J, Conrad EU (2008) Spatial heterogeneity in sarcoma 18F-FDG uptake as a predictor of patient outcome. J Nucl Med Off Publ Soc Nucl Med 49(12):1973–1979. doi:10.2967/jnumed.108.053397

Ekmekcioglu O, Aliyev A, Yilmaz S, Arslan E, Kaya R, Kocael P, Erkan ME, Halac M, Sonmezoglu K (2013) Correlation of 18F-fluorodeoxyglucose uptake with histopathological prognostic factors in breast carcinoma. Nucl Med Commun 34(11):1055–1067. doi:10.1097/MNM.0b013e3283658369

El Naqa I, Grigsby P, Apte A, Kidd E, Donnelly E, Khullar D, Chaudhari S, Yang D, Schmitt M, Laforest R, Thorstad W, Deasy JO (2009) Exploring feature-based approaches in PET images for predicting cancer treatment outcomes. Pattern Recogn 42(6):1162–1171. doi:10.1016/j.patcog.2008.08.011

Eschmann SM, Friedel G, Paulsen F, Reimold M, Hehr T, Budach W, Langen HJ, Bares R (2007) 18F-FDG PET for assessment of therapy response and preoperative re-evaluation after neoadjuvant radio-chemotherapy in stage III non-small cell lung cancer. Eur J Nucl Med Mol Imaging 34(4):463–471. doi:10.1007/s00259-006-0273-5

Ferrer Albiach C, Conde Moreno A, Rodriguez Cordon M, Morillo Macias V, Bouche Babiloni A, Beato Tortajada I, Sanchez Iglesias A, Frances Munoz A (2010) Contribution of hypoxia-measuring molecular imaging techniques to radiotherapy planning and treatment.

Clin Transl Oncol Off Publ Fed Span Oncol Soc Nat Cancer Inst Mex 12(1):22–26. doi:10. 1007/s12094-010-0462-3

Folpe AL, Lyles RH, Sprouse JT, Conrad EU 3rd, Eary JF (2000) (F-18) fluorodeoxyglucose positron emission tomography as a predictor of pathologic grade and other prognostic variables in bone and soft tissue sarcoma. Clin Cancer Res Off J Am Assoc Cancer Res 6(4):1279–1287

Furth C, Steffen IG, Amthauer H, Ruf J, Misch D, Schonberger S, Kobe C, Denecke T, Stover B, Hautzel H, Henze G, Hundsdoerfer P (2009) Early and late therapy response assessment with [18F]fluorodeoxyglucose positron emission tomography in pediatric Hodgkin's lymphoma: analysis of a prospective multicenter trial. J Clin Oncol Off J Am Soc Clin Oncol 27(26):4385–4391. doi:10.1200/JCO.2008.19.7814

Gagel B, Reinartz P, Dimartino E, Zimny M, Pinkawa M, Maneschi P, Stanzel S, Hamacher K, Coenen HH, Westhofen M, Bull U, Eble MJ (2004) pO(2) Polarography versus positron emission tomography ([(18)F] fluoromisonidazole, [(18)F]-2-fluoro-2'-deoxyglucose). An appraisal of radiotherapeutically relevant hypoxia. Strahlentherapie und Onkologie: Organ der Deutschen Rontgengesellschaft [et al] 180(10):616–622. doi:10.1007/s00066-004-1229-y

Gagel B, Reinartz P, Demirel C, Kaiser HJ, Zimny M, Piroth M, Pinkawa M, Stanzel S, Asadpour B, Hamacher K, Coenen HH, Buell U, Eble MJ (2006) [18F] fluoromisonidazole and [18F] fluorodeoxyglucose positron emission tomography in response evaluation after chemo-/ radiotherapy of non-small-cell lung cancer: a feasibility study. BMC Cancer 6:51. doi:10.1186/ 1471-2407-6-51

Gallamini A, Barrington SF, Biggi A, Chauvie S, Kostakoglu L, Gregianin M, Meignan M, Mikhaeel GN, Loft A, Zaucha JM, Seymour JF, Hofman MS, Rigacci L, Pulsoni A, Coleman M, Dann EJ, Trentin L, Casasnovas O, Rusconi C, Brice P, Bolis S, Viviani S, Salvi F, Luminari S, Hutchings M (2014) The predictive role of interim positron emission tomography for Hodgkin lymphoma treatment outcome is confirmed using the interpretation criteria of the Deauville five-point scale. Haematologica 99(6):1107–1113. doi:10.3324/ haematol.2013.103218

Garcia Garcia-Esquinas M, Garcia-Saenz JA, Arrazola Garcia J, Enrique Fuentes Ferrer M, Furio V, Rodriguez Rey C, Roman JM, Carreras Delgado JL (2014) 18F-FDG PET-CT imaging in the neoadjuvant setting for stages II–III breast cancer: association of locoregional SUVmax with classical prognostic factors. Q J Nucl Med Mol Imaging Off Publ Ital Assoc Nucl Med 58(1):66–73

Garcia-Parra R, Wood D, Shah RB, Siddiqui J, Hussain H, Park H, Desmond T, Meyer C, Piert M (2011) Investigation on tumor hypoxia in resectable primary prostate cancer as demonstrated by 18F-FAZA PET/CT utilizing multimodality fusion techniques. Eur J Nucl Med Mol Imaging 38(10):1816–1823. doi:10.1007/s00259-011-1876-z

Gould MK, Maclean CC, Kuschner WG, Rydzak CE, Owens DK (2001) Accuracy of positron emission tomography for diagnosis of pulmonary nodules and mass lesions: a meta-analysis. JAMA 285(7):914–924

Gould MK, Sanders GD, Barnett PG, Rydzak CE, Maclean CC, McClellan MB, Owens DK (2003) Cost-effectiveness of alternative management strategies for patients with solitary pulmonary nodules. Ann Intern Med 138(9):724–735

Gupta NC, Frank AR, Dewan NA, Redepenning LS, Rothberg ML, Mailliard JA, Phalen JJ, Sunderland JJ, Frick MP (1992) Solitary pulmonary nodules: detection of malignancy with PET with 2-[F-18]-fluoro-2-deoxy-D-glucose. Radiology 184(2):441–444. doi:10.1148/ radiology.184.2.1620844

Hamilton G, Rath B (2014) A short update on cancer chemoresistance. Wien Med Wochenschr. doi:10.1007/s10354-014-0311-z

Han MW, Lee HJ, Cho KJ, Kim JS, Roh JL, Choi SH, Nam SY, Kim SY (2012) Role of FDG-PET as a biological marker for predicting the hypoxic status of tongue cancer. Head Neck 34(10):1395–1402. doi:10.1002/hed.21945

Hayashi K, Furutsuka K, Takei M, Muto M, Nakao R, Aki H, Suzuki K, Fukumura T (2011) High-yield automated synthesis of [18F]fluoroazomycin arabinoside ([18F]FAZA) for hypoxia-specific tumor imaging. Appl Rad Isot Incl Data Instrum Methods Use Agric Ind Med 69(7):1007–1013. doi:10.1016/j.apradiso.2011.02.025

Hockel M, Vaupel P (2001) Tumor hypoxia: definitions and current clinical, biologic, and molecular aspects. J Natl Cancer Inst 93(4):266–276

Holdsworth CH, Badawi RD, Manola JB, Kijewski MF, Israel DA, Demetri GD, Van den Abbeele AD (2007) CT and PET: early prognostic indicators of response to imatinib mesylate in patients with gastrointestinal stromal tumor. AJR Am J Roentgenol 189(6):W324–W330. doi:10.2214/AJR.07.2496

Hutchings M, Loft A, Hansen M, Pedersen LM, Buhl T, Jurlander J, Buus S, Keiding S, D'Amore F, Boesen AM, Berthelsen AK, Specht L (2006) FDG-PET after two cycles of chemotherapy predicts treatment failure and progression-free survival in Hodgkin lymphoma. Blood 107 (1):52–59. doi:10.1182/blood-2005-06-2252

Hutchings M, Kostakoglu L, Zaucha JM, Malkowski B, Biggi A, Danielewicz I, Loft A, Specht L, Lamonica D, Czuczman MS, Nanni C, Zinzani PL, Diehl L, Stern R, Coleman M (2014) In vivo treatment sensitivity testing with positron emission tomography/computed tomography after one cycle of chemotherapy for Hodgkin lymphoma. J Clin Oncol Off J Am Soc Clin Oncol 32(25):2705–2711. doi:10.1200/JCO.2013.53.2838

Izuishi K, Yamamoto Y, Mori H, Kameyama R, Fujihara S, Masaki T, Suzuki Y (2014) Molecular mechanisms of [18F]fluorodeoxyglucose accumulation in liver cancer. Oncol Rep 31(2):701–706. doi:10.3892/or.2013.2886

Jacobson O, Chen X (2013) Interrogating tumor metabolism and tumor microenvironments using molecular positron emission tomography imaging. Theranostic approaches to improve therapeutics. Pharmacol Rev 65(4):1214–1256. doi:10.1124/pr.113.007625

Jerabek PA, Patrick TB, Kilbourn MR, Dischino DD, Welch MJ (1986) Synthesis and biodistribution of 18F-labeled fluoronitroimidazoles: potential in vivo markers of hypoxic tissue. Int J Radiat Appl Instrum Part A Appl Rad Isot 37(7):599–605

Kaira K, Oriuchi N, Sunaga N, Ishizuka T, Yamamoto N (2011) A systemic review of PET and biology in lung cancer. Am J Transl Res 3(4):383–391

Kalff VV, Hicks R, Ware R, Binns D, McKenzie A (2000) 29. F-18 FDG PET for suspected or confirmed regional recurrence of colon cancer. A prospective study of impact and outcome. Clin Positron Imaging Off J Inst Clin PET 3 4:183

Keith CJ, Miles KA, Griffiths MR, Wong D, Pitman AG, Hicks RJ (2002) Solitary pulmonary nodules: accuracy and cost-effectiveness of sodium iodide FDG-PET using Australian data. Eur J Nucl Med Mol Imaging 29(8):1016–1023. doi:10.1007/s00259-002-0833-2

Kelloff GJ, Hoffman JM, Johnson B, Scher HI, Siegel BA, Cheng EY, Cheson BD, O'Shaughnessy J, Guyton KZ, Mankoff DA, Shankar L, Larson SM, Sigman CC, Schilsky RL, Sullivan DC (2005) Progress and promise of FDG-PET imaging for cancer patient management and oncologic drug development. Clin Cancer Res Off J Am Assoc Cancer Res 11(8):2785–2808. doi:10.1158/1078-0432.CCR-04-2626

Khan MA, Combs CS, Brunt EM, Lowe VJ, Wolverson MK, Solomon H, Collins BT, Di Bisceglie AM (2000) Positron emission tomography scanning in the evaluation of hepatocellular carcinoma. J Hepatol 32(5):792–797

Kim JW, Dang CV (2006) Cancer's molecular sweet tooth and the Warburg effect. Cancer Res 66 (18):8927–8930. doi:10.1158/0008-5472.CAN-06-1501

Koh WJ, Rasey JS, Evans ML, Grierson JR, Lewellen TK, Graham MM, Krohn KA, Griffin TW (1992) Imaging of hypoxia in human tumors with [F-18]fluoromisonidazole. Int J Radiat Oncol Biol Phys 22(1):199–212

Komar G, Kauhanen S, Liukko K, Seppanen M, Kajander S, Ovaska J, Nuutila P, Minn H (2009) Decreased blood flow with increased metabolic activity: a novel sign of pancreatic tumor aggressiveness. Clin Cancer Res Off J Am Assoc Cancer Res 15(17):5511–5517. doi:10.1158/1078-0432.CCR-09-0414

Kostakoglu L, Gallamini A (2013) Interim 18F-FDG PET in Hodgkin lymphoma: would PET-adapted clinical trials lead to a paradigm shift? J Nucl Med Off Publ Soc Nucl Med 54 (7):1082–1093. doi:10.2967/jnumed.113.120451

Krause BJ, Beck R, Souvatzoglou M, Piert M (2006) PET and PET/CT studies of tumor tissue oxygenation. Q J Nucl Med Mol Imaging Off Publ Ital Assoc Nucl Med 50(1):28–43

Krohn KA, Mankoff DA, Muzi M, Link JM, Spence AM (2005) True tracers: comparing FDG with glucose and FLT with thymidine. Nucl Med Biol 32(7):663–671. doi:10.1016/j.nucmedbio.2005.04.004

Kumar P, Stypinski D, Xia H, McEwan AJB, Machulla H-J, Wiebe LI (1999) Fluoroazomycin arabinoside (FAZA): synthesis, 2H and 3H-labelling and preliminary biological evaluation of a novel 2-nitroimidazole marker of tissue hypoxia. J Label Compd Radiopharm 42(1):3–16

Lambin P, Rios-Velazquez E, Leijenaar R, Carvalho S, van Stiphout RG, Granton P, Zegers CM, Gillies R, Boellard R, Dekker A, Aerts HJ (2012) Radiomics: extracting more information from medical images using advanced feature analysis. Eur J Cancer 48(4):441–446. doi:10.1016/j.ejca.2011.11.036

Lapi SE, Voller TF, Welch MJ (2009) Positron emission tomography imaging of hypoxia. PET Clin 4(1):39–47. doi:10.1016/j.cpet.2009.05.009

Lawrentschuk N, Poon AM, Foo SS, Putra LG, Murone C, Davis ID, Bolton DM, Scott AM (2005) Assessing regional hypoxia in human renal tumours using 18F-fluoromisonidazole positron emission tomography. BJU Int 96(4):540–546. doi:10.1111/j.1464-410X.2005.05681.x

Lee NY, Mechalakos JG, Nehmeh S, Lin Z, Squire OD, Cai S, Chan K, Zanzonico PB, Greco C, Ling CC, Humm JL, Schoder H (2008) Fluorine-18-labeled fluoromisonidazole positron emission and computed tomography-guided intensity-modulated radiotherapy for head and neck cancer: a feasibility study. Int J Radiat Oncol Biol Phys 70(1):2–13. doi:10.1016/j.ijrobp.2007.06.039

Leong T, Everitt C, Yuen K, Condron S, Hui A, Ngan SY, Pitman A, Lau EW, MacManus M, Binns D, Ackerly T, Hicks RJ (2006) A prospective study to evaluate the impact of FDG-PET on CT-based radiotherapy treatment planning for oesophageal cancer. Radiother Oncol J Eur Soc Ther Radiol Oncol 78(3):254–261. doi:10.1016/j.radonc.2006.02.014

Lim JL, Berridge MS (1993) An efficient radiosynthesis of [18F]fluoromisonidazole. Appl Radiat Isot Incl Data Instrum Methods Use Agric Ind Med 44(8):1085–1091

Lodge MA, Jacene HA, Pili R, Wahl RL (2008) Reproducibility of tumor blood flow quantification with 15O-water PET. J Nucl Med Off Publ Soc Nucl Med 49(10):1620–1627. doi:10.2967/jnumed.108.052076

Macapinlac HA (2004) FDG PET and PET/CT imaging in lymphoma and melanoma. Cancer J 10 (4):262–270

Mah K, Caldwell CB, Ung YC, Danjoux CE, Balogh JM, Ganguli SN, Ehrlich LE, Tirona R (2002) The impact of (18)FDG-PET on target and critical organs in CT-based treatment planning of patients with poorly defined non-small-cell lung carcinoma: a prospective study. Int J Radiat Oncol Biol Phys 52(2):339–350

Mankoff DA, Dunnwald LK, Partridge SC, Specht JM (2009) Blood flow-metabolism mismatch: good for the tumor, bad for the patient. Clin Cancer Res Off J Am Assoc Cancer Res 15 (17):5294–5296. doi:10.1158/1078-0432.CCR-09-1448

Mano Y, Aishima S, Kubo Y, Tanaka Y, Motomura T, Toshima T, Shirabe K, Baba S, Maehara Y, Oda Y (2014) Correlation between biological marker expression and fluorine-18 fluorodeoxyglucose uptake in hepatocellular carcinoma. Am J Clin Pathol 142(3):391–397. doi:10.1309/AJCPG8AFJ5NRKLLM

Mathupala SP, Rempel A, Pedersen PL (1997) Aberrant glycolytic metabolism of cancer cells: a remarkable coordination of genetic, transcriptional, post-translational, and mutational events that lead to a critical role for type II hexokinase. J Bioenerg Biomembr 29(4):339–343

Mawlawi O, Townsend DW (2009) Multimodality imaging: an update on PET/CT technology. Eur J Nucl Med Mol Imaging 36(Suppl 1):S15–S29. doi:10.1007/s00259-008-1016-6

Mees G, Dierckx R, Vangestel C, Laukens D, Van Damme N, Van de Wiele C (2013) Pharmacologic activation of tumor hypoxia: a means to increase tumor 2-deoxy-2-[18F] fluoro-D-glucose uptake? Mol Imaging 12(1):49–58

Minn H, Zasadny KR, Quint LE, Wahl RL (1995) Lung cancer: reproducibility of quantitative measurements for evaluating 2-[F-18]-fluoro-2-deoxy-D-glucose uptake at PET. Radiology 196(1):167–173. doi:10.1148/radiology.196.1.7784562

Mortensen LS, Johansen J, Kallehauge J, Primdahl H, Busk M, Lassen P, Alsner J, Sorensen BS, Toustrup K, Jakobsen S, Petersen J, Petersen H, Theil J, Nordsmark M, Overgaard J (2012) FAZA PET/CT hypoxia imaging in patients with squamous cell carcinoma of the head and neck treated with radiotherapy: results from the DAHANCA 24 trial. Radiother Oncol J Eur Soc Ther Radiol Oncol 105(1):14–20. doi:10.1016/j.radonc.2012.09.015

Oda Y, Kondo M (1994) Noninvasive quantitative evaluation of regional myocardial blood flow and tissue fraction with O-15-labeled water and carbon dioxide inhalation using dynamic positron emission tomography. Nihon rinsho Japan J Clin Med 52 Suppl (Pt 1):614–621

Okamoto S, Shiga T, Yasuda K, Ito YM, Magota K, Kasai K, Kuge Y, Shirato H, Tamaki N (2013) High reproducibility of tumor hypoxia evaluated by 18F-fluoromisonidazole PET for head and neck cancer. J Nucl Med Off Publ Soc Nucl Med 54(2):201–207. doi:10.2967/jnumed.112.109330

Padhani AR (2005) Where are we with imaging oxygenation in human tumours? Cancer Imaging Off Publ Int Cancer Imaging Soc 5:128–130. doi:10.1102/1470-7330.2005.0103

Patronas NJ, Di Chiro G, Kufta C, Bairamian D, Kornblith PL, Simon R, Larson SM (1985) Prediction of survival in glioma patients by means of positron emission tomography. J Neurosurg 62(6):816–822. doi:10.3171/jns.1985.62.6.0816

Phelps ME, Huang SC, Hoffman EJ, Selin C, Sokoloff L, Kuhl DE (1979) Tomographic measurement of local cerebral glucose metabolic rate in humans with (F-18) 2-fluoro-2-deoxy-D-glucose: validation of method. Ann Neurol 6(5):371–388. doi:10.1002/ana.410060502

Piert M, Machulla HJ, Picchio M, Reischl G, Ziegler S, Kumar P, Wester HJ, Beck R, McEwan AJ, Wiebe LI, Schwaiger M (2005) Hypoxia-specific tumor imaging with 18F-fluoroazomycin arabinoside. J Nucl Med Off Publ Soc Nucl Med 46(1):106–113

Plathow C, Weber WA (2008) Tumor cell metabolism imaging. J Nucl Med Off Publ Soc Nucl Med 49(Suppl 2):43S–63S. doi:10.2967/jnumed.107.045930

Rahmim A, Zaidi H (2008) PET versus SPECT: strengths, limitations and challenges. Nucl Med Commun 29(3):193–207. doi:10.1097/MNM.0b013e3282f3a515

Rajendran JG, Schwartz DL, O'Sullivan J, Peterson LM, Ng P, Scharnhorst J, Grierson JR, Krohn KA (2006) Tumor hypoxia imaging with [F-18] fluoromisonidazole positron emission tomography in head and neck cancer. Clin Cancer Res Off J Am Assoc Cancer Res 12 (18):5435–5441. doi:10.1158/1078-0432.CCR-05-1773

Reischl G, Ehrlichmann W, Bieg C, Solbach C, Kumar P, Wiebe LI, Machulla HJ (2005) Preparation of the hypoxia imaging PET tracer [18F]FAZA: reaction parameters and automation. Appl Radiat Isot Incl Data Instrum Methods Use Agric Ind Med 62(6):897–901. doi:10.1016/j.apradiso.2004.12.004

Reivich M, Kuhl D, Wolf A, Greenberg J, Phelps M, Ido T, Casella V, Fowler J, Gallagher B, Hoffman E, Alavi A, Sokoloff L (1977) Measurement of local cerebral glucose metabolism in man with 18F-2-fluoro-2-deoxy-d-glucose. Acta Neurol Scand Suppl 64:190–191

Reivich M, Kuhl D, Wolf A, Greenberg J, Phelps M, Ido T, Casella V, Fowler J, Hoffman E, Alavi A, Som P, Sokoloff L (1979) The [18F]fluorodeoxyglucose method for the measurement of local cerebral glucose utilization in man. Circ Res 44(1):127–137

Sato J, Kitagawa Y, Yamazaki Y, Hata H, Okamoto S, Shiga T, Shindoh M, Kuge Y, Tamaki N (2013) 18F-fluoromisonidazole PET uptake is correlated with hypoxia-inducible factor-1alpha expression in oral squamous cell carcinoma. J Nucl Med Off Publ Soc Nucl Med 54(7):1060–1065. doi:10.2967/jnumed.112.114355

Semenza GL (2003) Targeting HIF-1 for cancer therapy. Nat Rev Cancer 3(10):721–732. doi:10.1038/nrc1187

Semenza GL (2007) Hypoxia-inducible factor 1 (HIF-1) pathway. Science's STKE: signal transduction knowledge environment 2007 (407):cm8. doi:10.1126/stke.4072007cm8

Semenza GL (2013) HIF-1 mediates metabolic responses to intratumoral hypoxia and oncogenic mutations. J Clin Invest 123(9):3664–3671. doi:10.1172/JCI67230

Servagi-Vernat S, Differding S, Hanin FX, Labar D, Bol A, Lee JA, Gregoire V (2014) A prospective clinical study of (1)(8)F-FAZA PET-CT hypoxia imaging in head and neck squamous cell carcinoma before and during radiation therapy. Eur J Nucl Med Mol Imaging 41(8):1544–1552. doi:10.1007/s00259-014-2730-x

Shammas A, Degirmenci B, Mountz JM, McCook BM, Branstetter B, Bencherif B, Joyce JM, Carty SE, Kuffner HA, Avril N (2007) 18F-FDG PET/CT in patients with suspected recurrent or metastatic well-differentiated thyroid cancer. J Nucl Med Off Publ Soc Nucl Med 48(2):221–226

Simsek DH, Kuyumcu S, Turkmen C, Sanli Y, Aykan F, Unal S, Adalet I (2014) Can complementary 68 Ga-DOTATATE and 18F-FDG PET/CT establish the missing link between histopathology and therapeutic approach in gastroenteropancreatic neuroendocrine tumors? J Nucl Med Off Publ Soc Nucl Med. doi:10.2967/jnumed.114.142224

Sokoloff L (1977) Relation between physiological function and energy metabolism in the central nervous system. J Neurochem 29(1):13–26

Specht L (2007) 2-[18F]fluoro-2-deoxyglucose positron-emission tomography in staging, response evaluation, and treatment planning of lymphomas. Semin Radiat Oncol 17(3):190–197. doi:10.1016/j.semradonc.2007.02.005

Spence AM, Muzi M, Swanson KR, O'Sullivan F, Rockhill JK, Rajendran JG, Adamsen TC, Link JM, Swanson PE, Yagle KJ, Rostomily RC, Silbergeld DL, Krohn KA (2008) Regional hypoxia in glioblastoma multiforme quantified with [18F]fluoromisonidazole positron emission tomography before radiotherapy: correlation with time to progression and survival. Clin Cancer Res Off J Am Assoc Cancer Res 14(9):2623–2630. doi:10.1158/1078-0432.CCR-07-4995

Stone HB, Brown JM, Phillips TL, Sutherland RM (1993) Oxygen in human tumors: correlations between methods of measurement and response to therapy. Summary of a workshop held 19–20 Nov 1992, at the National Cancer Institute, Bethesda, Maryland. Radiation Res 136(3):422–434

Szablewski L (2013) Expression of glucose transporters in cancers. Biochim Biophys Acta 1835 (2):164–169. doi:10.1016/j.bbcan.2012.12.004

Takebayashi R, Izuishi K, Yamamoto Y, Kameyama R, Mori H, Masaki T, Suzuki Y (2013) [18F] Fluorodeoxyglucose accumulation as a biological marker of hypoxic status but not glucose transport ability in gastric cancer. J Exp Clin Cancer Res CR 32:34. doi:10.1186/1756-9966-32-34

Tatum JL, Kelloff GJ, Gillies RJ, Arbeit JM, Brown JM, Chao KS, Chapman JD, Eckelman WC, Fyles AW, Giaccia AJ, Hill RP, Koch CJ, Krishna MC, Krohn KA, Lewis JS, Mason RP, Melillo G, Padhani AR, Powis G, Rajendran JG, Reba R, Robinson SP, Semenza GL, Swartz HM, Vaupel P, Yang D, Croft B, Hoffman J, Liu G, Stone H, Sullivan D (2006) Hypoxia: importance in tumor biology, noninvasive measurement by imaging, and value of its measurement in the management of cancer therapy. Int J Radiat Biol 82(10):699–757. doi:10.1080/09553000601002324

Taus A, Aguilo R, Curull V, Suarez-Pinera M, Rodriguez-Fuster A, Rodriguez de Dios N, Pijuan L, Zuccarino F, Vollmer I, Sanchez-Font A, Belda-Sanchis J, Arriola E (2014) Impact of 18F-FDG PET/CT in the treatment of patients with non-small cell lung cancer. Arch Bronconeumol 50(3):99–104. doi:10.1016/j.arbres.2013.09.017

Ter-Pogossian MM, Herscovitch P (1985) Radioactive oxygen-15 in the study of cerebral blood flow, blood volume, and oxygen metabolism. Semin Nucl Med 15(4):377–394

Tixier F, Le Rest CC, Hatt M, Albarghach N, Pradier O, Metges JP, Corcos L, Visvikis D (2011) Intratumor heterogeneity characterized by textural features on baseline 18F-FDG PET images predicts response to concomitant radiochemotherapy in esophageal cancer. J Nucl Med Off Publ Soc Nucl Med 52(3):369–378. doi:10.2967/jnumed.110.082404

Tixier F, Hatt M, Le Rest CC, Le Pogam A, Corcos L, Visvikis D (2012) Reproducibility of tumor uptake heterogeneity characterization through textural feature analysis in 18F-FDG PET. J Nucl Med Off Publ Soc Nucl Med 53(5):693–700. doi:10.2967/jnumed.111.099127

Tixier F, Hatt M, Valla C, Fleury V, Lamour C, Ezzouhri S, Ingrand P, Perdrisot R, Visvikis D, Cheze Le Rest C (2014) Visual versus quantitative assessment of intratumor 18F-FDG PET uptake heterogeneity: prognostic value in non-small cell lung cancer. J Nucl Med Off Publ Soc Nucl Med. doi:10.2967/jnumed.113.133389

Treglia G, Mirk P, Stefanelli A, Rufini V, Giordano A, Bonomo L (2012) 18F-Fluorodeoxyglucose positron emission tomography in evaluating treatment response to imatinib or other drugs in gastrointestinal stromal tumors: a systematic review. Clin Imaging 36 (3):167–175. doi:10.1016/j.clinimag.2011.08.012

Trinkaus ME, Blum R, Rischin D, Callahan J, Bressel M, Segard T, Roselt P, Eu P, Binns D, MacManus MP, Ball D, Hicks RJ (2013) Imaging of hypoxia with 18F-FAZA PET in patients with locally advanced non-small cell lung cancer treated with definitive chemoradiotherapy. J Med Imaging Radiat Oncol 57(4):475–481. doi:10.1111/1754-9485.12086

Tseng J, Dunnwald LK, Schubert EK, Link JM, Minoshima S, Muzi M, Mankoff DA (2004) 18F-FDG kinetics in locally advanced breast cancer: correlation with tumor blood flow and changes in response to neoadjuvant chemotherapy. J Nucl Med Off Publ Soc Nucl Med 45 (11):1829–1837

Ul-Hassan F, Simo R, Guerrero-Urbano T, Oakley R, Jeannon JP, Cook GJ (2013) Can (18) F-FDG PET/CT reliably assess response to primary treatment of head and neck cancer? Clin Nucl Med 38(4):263–265. doi:10.1097/RLU.0b013e31828165a8

Upadhyay M, Samal J, Kandpal M, Singh OV, Vivekanandan P (2013) The Warburg effect: insights from the past decade. Pharmacol Ther 137(3):318–330. doi:10.1016/j.pharmthera.2012.11.003

Vansteenkiste JF (2003) PET scan in the staging of non-small cell lung cancer. Lung Cancer 42 (Suppl 1):S27–S37

Vanuytsel LJ, Vansteenkiste JF, Stroobants SG, De Leyn PR, De Wever W, Verbeken EK, Gatti GG, Huyskens DP, Kutcher GJ (2000) The impact of (18)F-fluoro-2-deoxy-D-glucose positron emission tomography (FDG-PET) lymph node staging on the radiation treatment volumes in patients with non-small cell lung cancer. Radiother Oncol J Eur Soc Ther Radiol Oncol 55(3):317–324

Vaupel P, Mayer A (2007) Hypoxia in cancer: significance and impact on clinical outcome. Cancer Metastasis Rev 26(2):225–239. doi:10.1007/s10555-007-9055-1

Vaupel P, Kallinowski F, Okunieff P (1989) Blood flow, oxygen and nutrient supply, and metabolic microenvironment of human tumors: a review. Cancer Res 49(23):6449–6465

Wadsak W, Mitterhauser M (2010) Basics and principles of radiopharmaceuticals for PET/CT. Eur J Radiol 73(3):461–469. doi:10.1016/j.ejrad.2009.12.022

Wang W, Lee NY, Georgi JC, Narayanan M, Guillem J, Schoder H, Humm JL (2010) Pharmacokinetic analysis of hypoxia (18)F-fluoromisonidazole dynamic PET in head and neck cancer. J Nucl Med Off Publ Soc Nucl Med 51(1):37–45. doi:10.2967/jnumed.109.067009

Warburg O, Wind F, Negelein E (1927) The metabolism of tumors in the body. J Gen Physiol 8 (6):519–530

Weber WA, Wieder H (2006) Monitoring chemotherapy and radiotherapy of solid tumors. Eur J Nucl Med Mol Imaging 33(Suppl 1):27–37. doi:10.1007/s00259-006-0133-3

Weber WA, Schwaiger M, Avril N (2000) Quantitative assessment of tumor metabolism using FDG-PET imaging. Nucl Med Biol 27(7):683–687

Weber WA, Petersen V, Schmidt B, Tyndale-Hines L, Link T, Peschel C, Schwaiger M (2003) Positron emission tomography in non-small-cell lung cancer: prediction of response to chemotherapy by quantitative assessment of glucose use. J Clin Oncol Off J Am Soc Clin Oncol 21(14):2651–2657. doi:10.1200/JCO.2003.12.004

Wells P, Jones T, Price P (2003) Assessment of inter- and intrapatient variability in C15O2 positron emission tomography measurements of blood flow in patients with intra-abdominal cancers. Clin Cancer Res Off J Am Assoc Cancer Res 9(17):6350–6356

Wilkinson MD, Fulham MJ, Heard RN, McCaughan BC, McCarthy SW (2003) FDG-PET in paraneoplastic neuropathy. Neurology 60(10):1668

Wilson CB, Lammertsma AA, McKenzie CG, Sikora K, Jones T (1992) Measurements of blood flow and exchanging water space in breast tumors using positron emission tomography: a rapid and noninvasive dynamic method. Cancer Res 52(6):1592–1597

Withofs N, Bernard C, Van der Rest C, Martinive P, Hatt M, Jodogne S, Visvikis D, Lee JA, Coucke PA, Hustinx R (2014) FDG PET/CT for rectal carcinoma radiotherapy treatment planning: comparison of functional volume delineation algorithms and clinical challenges. J Appl Clin Med Phys/Am Coll Med Phys 15(5):4696. doi:10.1120/jacmp.v15i5.4696

Yoon HJ, Kang KW, Chun IK, Cho N, Im SA, Jeong S, Lee S, Jung KC, Lee YS, Jeong JM, Lee DS, Chung JK, Moon WK (2014) Correlation of breast cancer subtypes, based on estrogen receptor, progesterone receptor, and HER2, with functional imaging parameters from (6)(8) Ga-RGD PET/CT and (1)(8)F-FDG PET/CT. Eur J Nucl Med Mol Imaging 41(8):1534–1543. doi:10.1007/s00259-014-2744-4

Yu EY, Mankoff DA (2007) Positron emission tomography imaging as a cancer biomarker. Expert Rev Mol Diagn 7(5):659–672. doi:10.1586/14737159.7.5.659

Zhou X, Chen R, Xie W, Ni Y, Liu J, Huang G (2014) Relationship between 18F-FDG accumulation and lactate dehydrogenase a expression in lung adenocarcinomas. J Nucl Med Off Publ Soc Nucl Med. doi:10.2967/jnumed.114.145490

Zimny M, Gagel B, DiMartino E, Hamacher K, Coenen HH, Westhofen M, Eble M, Buell U, Reinartz P (2006) FDG–a marker of tumour hypoxia? A comparison with [18F] fluoromisonidazole and pO2-polarography in metastatic head and neck cancer. Eur J Nucl Med Mol Imaging 33(12):1426–1431. doi:10.1007/s00259-006-0175-6

Quantitative Analysis of Cancer Metabolism: From pSIRM to MFA

Christin Zasada and Stefan Kempa

Abstract

Metabolic reprogramming is a required step during oncogenesis and essential for cellular proliferation. It is triggered by activation of oncogenes and loss of tumor suppressor genes. Beside the combinatorial events leading to cancer, common changes within the central metabolism are reported. Increase of glycolysis and subsequent lactic acid formation has been a focus of cancer metabolism research for almost a century. With the improvements of bioanalytical techniques within the last decades, a more detailed analysis of metabolism is possible and recent studies demonstrate a wide range of metabolic rearrangements in various cancer types. However, a systematic and mechanistic understanding is missing thus far. Therefore, analytical and computational tools have to be developed allowing for a dynamic and quantitative analysis of cancer metabolism. In this chapter, we outline the application of pulsed stable isotope resolved metabolomics (pSIRM) and describe the interface toward computational analysis of metabolism.

Keywords

Cancer metabolism · Metabolomics · Proteomics · Metabolic flux analysis · Quantitative mass spectrometry · Pulsed stable isotope resolved metabolomics (pSIRM)

List of Abbreviations

CCM	Central carbon metabolism
DDA	Data-dependent analysis

C. Zasada · S. Kempa (✉)
Integrative Proteomics and Metabolomics, Max-Delbrueck-Center
of Molecular Medicine in the Helmholtz Association, Robert-Roessle-Str. 10,
13125 Berlin, Germany
e-mail: stefan.kempa@mdc-berlin.de

© Springer International Publishing Switzerland 2016
T. Cramer and C.A. Schmitt (eds.), *Metabolism in Cancer*,
Recent Results in Cancer Research 207, DOI 10.1007/978-3-319-42118-6_9

FBA	Flux balance analysis
FBP	Fructose-1,6-bisphosphate
FGFR1	Fibroblast growth factor receptor 1
GC-MS	Gas-chromatography–mass spectrometry
Glc	Glucose
HIF-1	Hypoxia-inducible factor 1
ICAT	Isotopic-coded affinity tag
ID-MS	Isotope dilution mass spectrometry
INST-MFA	Nonstationary metabolic flux analysis
iTRAQ	Isobaric tags for relative and absolute quantification
Lac	Lactic acid
LC-MS	Liquid chromatography–mass spectrometry
MFA	Metabolic flux analysis
MID	Mass isotopomer distribution
MS	Mass spectrometry
N	Stoichiometric matrix
NMR	Nuclear magnetic resonance
Oct-4	Octamer-binding transcription factor 4
PEP	Phosphoenolpyruvic acid
PK	Pyruvate kinase
PKM1/PKM2	Pyruvate kinase isoform M1/isoform M2
PPP	Pentose phosphate pathway
pSIRM	Pulsed stable isotope resolved metabolomics
Pyr	Pyruvic acid
ROS	Reactive oxygen species
SILAC	Stable isotope labeling of amino acids in cell culture
STAT3	Signal transducers and activators of transcription
TCA cycle	Tricarboxylic acid cycle
TIS	Therapy-induced senescence
TMFA	Thermodynamics-based metabolic flux analysis

1 Introduction

The central carbon metabolism (CCM) is a super-pathway consisting of glycolysis, the pentose phosphate pathway (PPP), the tricarboxylic acid cycle (TCA) cycle, and amino acid metabolism and provides all energy and redox equivalents for cell growth, survival, and function. The requirements for dividing cells are different from differentiated tissues, e.g., muscle or neurons. Generally, the same pathways provide energy to meet the demands but there are clear differences of the usage of central metabolic pathways. Beside energy, dividing cells also need building blocks for cell growth. Therefore, the metabolic system is tuned to generate energy and precursors for biosynthesis in the same pathway at the same time. During the transition from quiescence to proliferation, the metabolic enzymes change their 'identity' as isoenzymes with different biochemical properties are expressed (Cho-Chung and Nesterova 2005; Yeung et al. 2008). Although not much is known about the kinetic parameters of the majority of CCM isoenzymes, some examples demonstrate the impact of altered biochemical properties.

Already more than 100 years ago it became evident that malignant transformation is closely linked to metabolic reprogramming. And since then it is under debate if the metabolic reprogramming during oncogenesis is partially retrograde to the metabolic shift during stem cell differentiation. Stem cells vastly reprogram their metabolism during differentiation (Prigione et al. 2011; Panopoulos et al. 2012). As mitochondria are immature in pluripotent stem cells, mitochondrial ripening starts early in differentiation and a shift toward mitochondrial metabolism was observed in a number of studies (Cho et al. 2006; Prigione et al. 2010). Due to this tight connection between mitochondrial integrity and differentiation processes, it becomes clear that mitochondrial defects are able to cause developmental impairments. Although it is not finally approved that the metabolic program of stem cells is similar to that of cancer cells at the molecular level, the high level of lactate production is shared between both (Ito and Suda 2014). This effect has been described already at the beginning of the twentieth century by Otto Warburg and is now termed Warburg Effect (Warburg et al. 1927). In the recent literature, many links between the Warburg Effect and cellular proliferation can be found (Hsu and Sabatini 2008; DeBerardinis et al. 2008). However, this link is not definitive as Dörr et al. (2013), for example, demonstrated a high degree of lactate production in senescent lymphoma cells.

Within the glycolytic pathway, several enzymes are changing their identity during oncogenic metabolic reprogramming (Fig. 1). Until now there is no clear map of these events, even more it is expected that different tumor entities also vary in their isoenzyme arsenal. In addition, it must be expected that there is heterogeneity within the same tumor subtype. This may or may not directly depend on the alterations of oncogenes and tumor suppressor genes. Finally, the question must be asked how many different ways exist to upregulate glycolysis during metabolic reprogramming.

One of the commonly modulated enzymes in stem and cancer cells is the central glycolytic enzyme pyruvate kinase (PK). It catalyzes the ATP-producing reaction of glycolysis—the conversion of phosphoenolpyruvic acid (PEP) to pyruvic acid. The enzyme belongs to a family of four tissue-dependently expressed homo-tetrameric isoenzymes—L, R, M1, and M2 (Noguchi et al. 1987; Harada et al. 1978). The latter ones are splice variants derived from the same mRNA. Whereas PKM2, expressed in embryonic cells and dividing cells, contains exon 10, the constitutively expressed PKM1 possesses exon 9 (Fig. 1, dotted box) (Noguchi et al. 1987). The exchange of 56 amino acids modifies dramatically the kinetic properties by creating a binding pocket for fructose-1,6-bisphosphate (FBP), an allosteric activator of PKM2 and intermediate of glycolysis (Ashizawa et al. 1991; Mazurek et al. 2005). Diminished level of FBP induces the dissociation of the PKM2 tetramer into the dimeric form

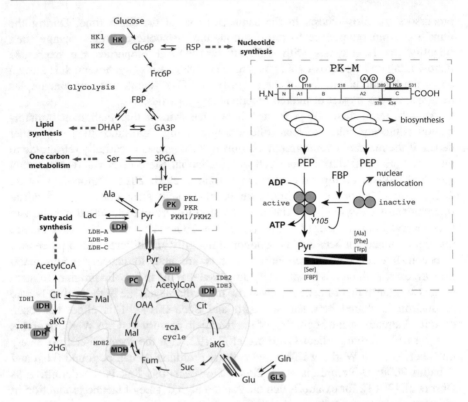

Fig. 1 Activation of oncogenes or inactivation of tumor suppressor genes leads to an adjustment of the CCM. Enzymes such as HK2, LDHA, or PKM are regulated by transcription factors that are known key players of tumorigenesis. The adaption of carbon routing is a central element to meet the cellular requirements for maintenance and biosynthesis. Compartment-specific isoenzymes facilitate the same reaction at different places in the cell with specific secondary functions such as the gain-of-function mutation of IDH1 producing the onco-metabolite 2HG. Level of CCM intermediates itself also regulate protein activity in addition to transcriptional activation, e.g., conformational change of PKM2. On the one hand, the active, tetrameric form of PKM2, induced by high level of FPB, drives the production of ATP, and on the other, the dimeric counterpart owns nuclear function and supports the generation of precursors for biosynthesis by blocking the conversion of PEP to Pyr. *2HG* 2-hydroxyglutarate; *3PGA* 3-phosphoglycerate; *aKG* alpha-ketoglutarate; *FBP* fructose-1,6-bisphosphate; *GLS* glutaminase; *HIF-1* hypoxia-inducible factor 1; *HK* hexokinase; *IDH* isocitrate dehydrogenase; *LDH* lactate dehydrogenase; *MDH* malate dehydrogenase; *PC* pyruvate carboxylase; *PDH* pyruvate dehydrogenase; *PEP* phosphoenolpyruvate; *PK* pyruvate kinase

with low affinity to PEP. Activation of the FGF receptor promotes PKM2 phosphorylation at tyrosine Y105 and induces the release of FBP from the pocket, inactivating the enzyme (Mazurek 2007; Christofk et al. 2008). The presence of an inactive PKM2 variant seems to be counterintuitive to the metabolic phenotype of tumor cells, but this metabolic mode enables a sufficient fueling of biosynthetic pathways or precursors for the reduction of ROS. A similar allosteric regulation of PKM2 was observed for amino acids, e.g., serine. Serine binds to PKM2 and promotes aerobic glycolysis until the serine pool reaches a certain threshold. The binding of serine drops and carbons derived from glucose are redirected from glycolysis into serine synthesis by 'drossling' dimerization of PKM2. Beside its catabolic task, PKM2 interacts with several members of signaling cascades after translocation into the nucleus. Nuclear PKM2 supports cell survival and proliferation by the interaction with transcription factors such as Oct4, signal transducer and activator of transcription (STAT3) and global players beta-catenin and hypoxia-inducible factor 1 (HIF-1) (Lee et al. 2008; Luo et al. 2011; Gao et al. 2012; Yang et al. 2011). Interestingly, also the less active, dimeric PKM2 variant acts as a protein kinase and activates STAT3 using PEP for phosphorylation (Gao et al. 2012).

Increased conversion of glucose into lactic acid is not necessarily linked with enhanced proliferation. Therapy-induced senescence (TIS) of B lymphoma cells (Eµ–myc, Bcl2 positive) showed a hypermetabolic state—enhanced pathway activity of glycolysis, fatty acid synthesis, and oxidative phosphorylation—after treatment with the topoisomerase inhibitor adriamycin (ADR) (Dörr et al. 2013). The transition from PKM2 to PKM1 is in agreement with the block in cell growth. At the same time, Dörr et al. observed an elevated glycolytic flux, demonstrating that PKM2 is not a positive regulator of the glycolytic flux on its own.

2 Quantifying (Cancer) Metabolism

2.1 Quantitative Omics-Methodologies

In order to understand these processes at the molecular level, it is necessary to analyze the metabolic program of cancer cells and stem cells quantitatively and in high resolution. The flexibility of cellular metabolism in terms of turnover and pathway usage requires studying cellular metabolism dynamically and in absolute quantities. In addition, the molecular composition of the biochemical network of enzymes has to be determined by quantitative proteomics approaches. This is challenging from both the analytical and the computational perspectives. Remarkably, already in the 1960s, scientists applied labeled substrates in rat experiments to understand metabolic regulation of the liver (Hoberman and D'Adamo 1960; Bailey et al. 1968; Heath and Threlfall 1968). In the absence of high performance computers typical for today, only by using basic biochemistry methods different research groups reported metabolic flux analyses. The development of mass spectrometry marks a milestone in omics-technologies, especially proteomics and metabolomics. Within ten years, a high number of methods were established for the analysis and quantification of the proteome. Looking at the publication numbers, one could anticipate the same effect in the field of metabolomics only a few steps ahead. Decoding the molecular mechanisms of metabolic regulation in cancer will require the determination of both, proteins (the players) and metabolites (the balls).

2.1.1 Proteomics Methods

The way we look at the proteome, the 'processors' of the cell, changed a lot since Beadle and Tatum proposed the 'one gene, one polypeptide' hypothesis in the 1940s (Beadle and Tatum 1941). Within the last 75 years, scientists discovered post-translational regulatory events, e.g., alternative splicing and post-translational modification of proteins. Even a single modification transforms the mode of action of the protein and therefore its role within the cellular context. Different proteomic approaches provide the identification of high and middle abundant proteins in a complex mixture by liquid chromatography coupled with mass spectrometry (LC-MS). Three approaches are distinguished regarding their technical properties of protein identification and quantification: direct, targeted and shotgun proteomics, the latter two being the most frequently applied methods (Domon and Aebersold 2010). The targeted method requires prior knowledge about the physico-chemical properties of the proteins of interest. The shotgun or discovery approach surveys all present and detectable ions (data-dependent analysis (DDA)) and is widely applied in large-scale studies or generation of protein inventories (Beausoleil et al. 2006; De Godoy et al. 2008; Denny et al. 2008). Identification of proteins is based on the comparison of *in silico* and experimentally derived proteolytic peptides. None of these methods are able to provide a complete measurement of the proteome with all modifications, because the expression and processing of proteins remains dependent on the status of the cell or tissue. Nevertheless, the continuously developing equipment resulted in possible identification of around 3.500 proteins in one shotgun LC-MS run. A huge fraction of those proteins belongs to the GOBP-class 'metabolic processes' and cover CCM-associated proteins.

A broad variety of methods were developed for quantification of the proteome. Commonly used for relative quantification are methods such as stable isotope labeling of amino acids in cell culture (SILAC) and isotope-coded affinity tag (ICAT). Both methods rely on the introduction of isotopic labels via metabolic labeling (Ong et al. 2002; Gygi et al. 1999). The quantification is carried out by the comparison of coeluting light and heavy peptides at the MS/MS-level. The absolute quantification of proteins requires either the application of isobaric tags for relative and absolute quantification (iTRAQ) or internal isotopic-labeled standards (Wiese et al. 2007; Ross 2004). The latter approach results in the highest accuracy, but is limited to a selected number of peptides that are spiked in prior the LC-MS analysis. Least accurate but worthwhile are label-free quantification approaches: 'Spectral counting' and 'XICs' (Ono et al. 2006; Old et al. 2005; Ishihama et al. 2005).

2.1.2 Quantifying the Metabolome

In order to address the beforehand-raised questions, reliable and quantitative methods have to be applied in a panel of *in vitro* and *in vivo* analyses and metabolic settings have to be measured. During the last years, we have developed methods to decode the dynamics of metabolism in both scenarios in a qualitative and quantitative manner. One challenge of metabolomics is the huge variety of the molecules such as hexoses, phosphates, amino acids, and lipids with different physico-chemical properties.

Mass spectrometry (MS) and nuclear magnetic resonance (NMR) spectroscopy display complimentary advantages and disadvantages when applied for metabolomics analyses. MS coupled with chromatographic methods strikes with high sensitivity, and the huge number and variety of detected compounds and well-maintained databases for a reliable identification. The MS disadvantage—less accurate quantification—is the major argument supporting NMR. NMR approaches facilitate the metabolite analysis in liquid state and can be performed even in intact tissue samples (Pan and Raftery 2007). Nevertheless, the NMR approach suffers from a lack of sensitivity in comparison with MS. The chemical modification of the sample prior to the measurement and the optimization of technical parameter lead to a high coverage of CCM intermediates in a single MS measurement (Pietzke et al. 2014).

Indeed, the majority of metabolic studies rely on the relative analysis of peak areas or intensities in the chromatogram of two biological conditions, e.g., wild type versus knockout of a given gene. Dang and colleagues observed a gain-of-function mutation of the cytosolic enzyme isocitrate dehydrogenase IDH1 while comparing tumor with untransformed control samples. In glioma cells, the enzyme facilitates—beside the catalyzation of citrate to alpha-ketoglutarate (αKG)—also the conversion of αKG to the onco-metabolite 2-hydroxyglutarate (2HG) (Dang et al. 2009). The qualitative evaluation of chromatograms may be sufficient in the case of a global shutdown of a specific pathway, but probably does not reliably reflect minor quantitative changes of metabolism.

The application of stable isotopes is a possibility to improve quantification in LC- or GC-MS measurements. Supplementing samples with known amounts of labeled substances, an approach termed isotope dilution mass spectrometry (IDMS), provides direct quantification (Heumann 1992; Fassett and Paul 1989). Quantitative analysis by IDMS is limited to the availability of labeled substances. However, external measured standards in dilution series undergo this limitation and, if well implemented, cover the complete dynamic range within an experiment. A global snapshot of the metabolic state, especially when comparing healthy versus disease, helps to gain insights into metabolic regulation. However, the picture remains rather static. A more dynamic examination is mandatory in regard understanding how oncogene activation and nutrient availability adjust metabolism to meet their requirements (Fan et al. 2013).

2.1.3 Quantifying Metabolic Dynamics in Cancer Cells

Glycolysis, TCA cycle, and pentose phosphate pathway link the uptake of nutrients with neighboring parts of metabolism, resulting in facilitation of lipid, amino acid, and nucleotide synthesis. In the late 1960s, first studies were carried out with radioactive isotopes to track routing within the CCM (Wenzel et al. 1966). Nowadays, application of stable isotope-labeled substrates provides a more robust monitoring of nutrient utilization. Labeled substrates are applied as tracers, either supplemented in cell culture media or directly injected into tissues (Fan et al. 2009).

Dependent on the substrate (e.g., glucose, glutamine, and essential amino acids) and the kind of label (e.g., carbon-13, nitrogen-15), the turnover changes the atomic composition of the intermediates and allows to monitor the routing of single atoms along the pathway within the CCM (Pietzke and Kempa 2014). Beside the robust evaluation of isotopic incorporation by GC-MS, such measurements provide also the quantification of the intermediates within the same analysis. Even structurally similar components (glucose and fructose, hexose phosphates) are distinguishable and quantifiable in a separate manner (Kempa et al. 2009).

Metabolism is the fastest responder to environmental perturbations. Therefore, it is necessary to maintain a constant supplementation of the major nutrients during a metabolomics experiment. We developed a workflow minimizing external stress factors and sampling time for cell culture experiments (Pietzke et al. 2014). After isotope incorporation, the use of a washing buffer containing major carbon-sources and salts for just a few seconds, maintains osmolarity and major nutrient supply, reduces the carryover of extracellular metabolites of the media and supports a robust determination of intracellular metabolites. Adding ice-cold, 50 % methanol quenches subsequently all cellular processes and initiates the extraction process. This procedure allows a fast handling of the cell harvest and short incubation times of isotope-labeled substrates in a range of one minute. This is a prerequisite to cover/identify the distinct dynamics within CCM.

The evaluation of isotope incorporation in specific branching points highlights differences in CCM pathway activity and reveals distinct modes how carbons are incorporated in glycolytic and TCA cycle intermediates (Fig. 2). Whereas all transformed (HEK293 cells) or tumorigenic cell lines (HeLa, HCT116) route glucose-derived carbons into lactic acid (the Warburg Effect), T98G cells only marginally use aerobic glycolysis. Out of this panel of four cell lines, HeLa cells exclusively facilitate ^{13}C-incorporation into fructose metabolism. HEK293 cells prefer to route carbon backbones into alanine and serine and toward glycerol metabolism (Fig. 2).

2.2 Computational Modeling Approaches in Metabolic Flux Analysis

The analysis of metabolic fluxes requires the application of computational tools. Cyclic pathway structures such as the TCA cycle and reversible reactions make a direct quantitative evaluation of the metabolic flux based on stable isotope incorporation impossible. A qualitative analysis at defined branching points in the CCM may be feasible, but a reliable evaluation of cofactor status requires the integration of quantitative data in a computational framework. A number of methods were established during the last decade. Those methods differ widely with regard to their requirements. Genome-scale kinetic models, mirroring the enzyme kinetics, are rare. Parameters describing the catalytic activity of an enzyme are often context dependent, e.g., on temperature and pH. Due to the lack of knowledge of kinetic

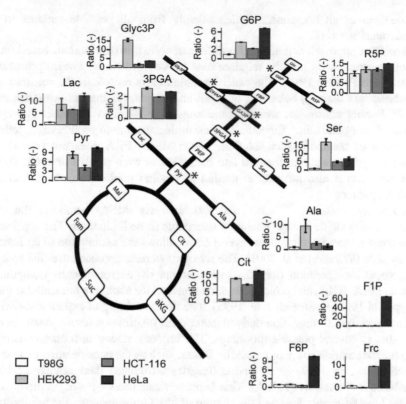

Fig. 2 Quantitative metabolomics analysis in combination with the application of ^{13}C-substrates revealed distinct routing of carbons through the CCM in four different cell lines (data are taken from Pietzke et al. 2014). HeLa cells show an elevated shuttling of ^{13}C-glucose into fructose metabolism, whereas HEK293 cells route carbons toward amino and fatty acid synthesis. Opposing to T98G cells, all other cell lines show a higher conversion of glucose into lactate and citrate. Ratios represent ^{13}C-glucose-labeled quantities per 1×10^6 cells in relation to T98G cell levels

parameters, kinetic models often cover subnetworks of the metabolism (Niedenführ et al. 2015). The development of flux balance (FBA) (Varma and Palsson 1994) and ^{13}C-metabolic flux analysis (MFA) (Wiechert 2001) was mainly driven by the optimization of prokaryotic-based product synthesis. Nowadays, it is the challenge to adapt existing modules to the complex, compartment-separated nature of eukaryotic cells (Buescher et al. 2015). Independent from cell type and complexity or computational approaches, all models are based on translation of the biochemical network in a stoichiometric matrix N (m × r). Educts and products (m) are assigned according to their molar ratios for each reaction (r). The composition of this matrix is well known, due to the broad biochemical knowledge, and invariant regarding time, kinetics and concentration of compounds (Niedenführ et al. 2015). The stoichiometric matrix allows the topological analysis and reflects the structural relationship of all components independently from all possible variants in the biochemical network.

The FBA approach determines feasible metabolic flux distributions based on the stoichiometric matrix and the requirements for cell growth by linear optimization (Varma and Palsson 1994). The stoichiometric matrix represents the constraints of metabolic flux map: (i) balancing of input and output, (ii) inequalities (Orth et al. 2010). Further constraints are exchange rates such as glucose uptake. The application of an optimization function, like assuming maximum proliferation, reduces the space of metabolic flux solution. Nevertheless, FBA does not provide the evaluation of intracellular forward and backward or even parallel or cyclic fluxes. This approach is also not able to predict metabolite pool sizes or to account for regulatory effects.

Therefore, classical ^{13}C-metabolic flux analysis (MFA) combines the FBA approach with stable isotope tracing to overcome those limitations. The application of isotopic tracers for more than several hours allows the assumption of an isotopic steady state (Wiechert et al. 2001). The retrobiosynthetic approach uses the analysis of isotopic incorporation from the measurement of macromolecular compounds, such as DNA, RNA, and proteins, to draw conclusions about the intracellular fluxes (Szyperski 1995; Eisenreich et al. 1993). The iterative fitting of experiment-derived and network-based simulation-derived isotopomer distribution results ideally in one, in reality in several possible flux maps. The implementation of further constraints reduces the solution space of metabolic fluxes, such as the consideration of reaction directionality. In 2007, Henry and colleagues introduced thermodynamics-based metabolic flux analysis (TMFA). This approach calculates—based on the concentration level of metabolites and the change of free Gibbs energy—the feasibility of metabolic flux directions. Reversibility of reactions becomes restricted with regard to their likeliness and reduces the number of possible flux maps (Henry et al. 2007).

However, stationary ^{13}C-MFA has two major disadvantages: the duration of the labeling experiment and—with that—high experimental costs. Recent developments in intracellular metabolite quantification and mathematical approaches pave the way for a nonstationary MFA approach (INST-MFA) (Nöh and Wiechert 2011; Young et al. 2007). Still, this approach relies on the metabolic steady-state assumption of the system. The only time-variable component is the routing of stable isotopes that has to be monitored in a time series study (Nöh et al. 2007). Prerequisites are the determination of intracellular pool sizes and uptake/production rates. The network has to fulfill the carbon balance and atom transitions have to be defined for each reaction of the network, similar to the stationary approach. Benefits of this approach are the determination of backward and forward fluxes and an estimate of pool sizes of intermediates that are not measurable (Wahl et al. 2008). Still, a few challenges remain, e.g., the implementation of subcellular compartments, the complex formula of the media, and (eventually) slow labeling times in the INST-MFA framework (Zamboni 2011).

3 Conclusion

Driven by the improvements of bioanalytical techniques and computational methods to analyze the dynamics of metabolism, detailed analyses of cellular metabolism became possible. Today's challenge is to develop tools that are affordable, simple to apply, and rapid. At the same time these tools have to provide a detailed picture of cancer metabolism to broaden our understanding of oncogenic transformations at a mechanistic level. For example, the combination of pulsed stable isotope resolved metabolomics and mathematical approaches, as instationary metabolic flux analysis may present a possibility to streamline dynamic and quantitative metabolomics analyses of cancer, both *in vitro* and *in vivo*. By combining such complex approaches, we hope to find key-regulated steps in cancer metabolism that are oncogene dependent and predictive for therapy responses.

References

Ashizawa K, Willingham MC, Liang CM, Cheng SY (1991) In vivo regulation of monomer-tetramer conversion of pyruvate kinase subtype M2 by glucose is mediated via fructose 1,6-bisphosphate. J Biol Chem 266:16842–16846

Bailey E, Stirpe F, Taylor CB (1968) Regulation of rat liver pyruvate kinase. The effect of preincubation, pH, copper ions, fructose 1,6-diphosphate and dietary changes on enzyme activity. Biochem J 108:427–436

Beadle GW, Tatum EL (1941) Genetic control of biochemical reactions in neurospora. Proc Natl Acad Sci USA 27:499–506. doi:10.1086/281267

Beausoleil SA, Villén J, Gerber SA et al (2006) A probability-based approach for high-throughput protein phosphorylation analysis and site localization. Nat Biotechnol 24:1285–1292. doi:10.1038/nbt1240

Buescher JM, Antoniewicz MR, Boros LG et al (2015) A roadmap for interpreting ^{13}C metabolite labeling patterns from cells. Curr Opin Biotechnol 34:189–201. doi:10.1016/j.copbio.2015.02.003

Cho YM, Kwon S, Pak YK et al (2006) Dynamic changes in mitochondrial biogenesis and antioxidant enzymes during the spontaneous differentiation of human embryonic stem cells. Biochem Biophys Res Commun 348:1472–1478. doi:10.1016/j.bbrc.2006.08.020

Cho-Chung YS, Nesterova MV (2005) Tumor reversion: protein kinase A isozyme switching. Ann N Y Acad Sci 1058:76–86. doi:10.1196/annals.1359.014

Christofk HR, Vander Heiden MG, Wu N et al (2008) Pyruvate kinase M2 is a phosphotyrosine-binding protein. Nature 452:181–186

Dang L, White DW, Gross S et al (2009) Cancer-associated IDH1 mutations produce 2-hydroxyglutarate. Nature 462:739–744. doi:10.1158/1538-7445.AM10-33

De Godoy LMF, Olsen JV, Cox J et al (2008) Comprehensive mass-spectrometry-based proteome quantification of haploid versus diploid yeast. Nature 455:1251–1254. doi:10.1038/nature07341

DeBerardinis RJ, Lum JJ, Hatzivassiliou G, Thompson CB (2008) The biology of cancer: metabolic reprogramming fuels cell growth and proliferation. Cell Metab 7:11–20. doi:10.1016/j.cmet.2007.10.002

Denny P, Hagen FK, Hardt M et al (2008) The proteomes of human parotid and submandibular/sublingual gland salivas collected as the ductal secretions. J Proteome Res 7:1994–2006. doi:10.1021/pr700764j

Domon B, Aebersold R (2010) Options and considerations when selecting a quantitative proteomics strategy. Nat Biotechnol 28:710–721. doi:10.1038/nbt.1661

Dörr JR, Yu Y, Milanovic M et al (2013) Synthetic lethal metabolic targeting of cellular senescence in cancer therapy. Nature 501:421–425. doi:10.1038/nature12437

Eisenreich W, Strauss G, Werz U et al (1993) Retrobiosynthetic analysis of carbon fixation in the phototrophic eubacterium *Chloroflexus aurantiacus*. Eur J Biochem 215:619–632. doi:10.1111/j.1432-1033.1993.tb18073.x

Fan TWM, Lane AN, Higashi RM et al (2009) Altered regulation of metabolic pathways in human lung cancer discerned by ^{13}C stable isotope-resolved metabolomics (SIRM). Mol Cancer 8:41. doi:10.1186/1476-4598-8-41

Fan TW, Lorkiewicz P, Sellers K et al (2013) Stable isotope-resolved metabolomics and applications for drug development. Pharmacol Ther 133:366–391. doi:10.1016/j.pharmthera.2011.12.007.Stable

Fassett JD, Paul J (1989) Isotope dilution mass spectrometry for accurate elemental analysis. Anal Chem 61:643–649. doi:10.1021/ac00185a715

Gao X, Wang H, Yang JJ et al (2012) Pyruvate kinase M2 regulates gene transcription by acting as a protein kinase. Mol Cell 45:598–609. doi:10.1016/j.molcel.2012.01.001

Gygi SP, Rist B, Gerber S et al (1999) Quantitative analysis of complex protein mixtures using isotope-coded affinity tags. Nat Biotechnol 17:994–999. doi:10.1038/13690

Harada K, Saheki S, Wada K, Tanaka T (1978) Purification of four pyruvate kinase isozymes of rats by affinity elution chromatography. Biochim Biophys Acta 524:327–339. doi:10.1016/0005-2744(78)90169-9

Heath DF, Threlfall CJ (1968) The interaction of glycolysis, gluconeogenesis and the tricarboxylic acid cycle in rat liver in vivo. Biochem J 110:337–362

Henry CS, Broadbelt LJ, Hatzimanikatis V (2007) Thermodynamics-based metabolic flux analysis. Biophys J 92:1792–1805. doi:10.1529/biophysj.106.093138

Heumann KG (1992) Isotope dilution mass spectrometry. Int J Mass Spectrom Ion Process 118–119:575–592. doi:10.1016/0168-1176(92)85076-C

Hoberman HD, D'Adamo AF (1960) Coupling of oxidation of substrates reductive biosyntheses. J Biol Chem 235:2

Hsu PP, Sabatini DM (2008) Cancer cell metabolism: Warburg and beyond. Cell 134:703–707. doi:10.1016/j.cell.2008.08.021

Ishihama Y, Oda Y, Tabata T et al (2005) Exponentially modified protein abundance index (emPAI) for estimation of absolute protein amount in proteomics by the number of sequenced peptides per protein. Mol Cell Proteomics 4:1265–1272. doi:10.1074/mcp.M500061-MCP200

Ito K, Suda T (2014) Metabolic requirements for the maintenance of self-renewing stem cells. Nat Rev Mol Cell Biol 15:243–256. doi:10.1038/nrm3772

Kempa S, Hummel J, Schwemmer T et al (2009) An automated GCxGC-TOF-MS protocol for batch-wise extraction and alignment of mass isotopomer matrixes from differential 13C-labelling experiments: a case study for photoautotrophic-mixotrophic grown Chlamydomonas reinhardtii cells. J Basic Microbiol 49:82–91. doi:10.1002/jobm.200800337

Lee J, Kim HK, Han Y-M, Kim J (2008) Pyruvate kinase isozyme type M2 (PKM2) interacts and cooperates with Oct-4 in regulating transcription. Int J Biochem Cell Biol 40:1043–1054. doi:10.1016/j.biocel.2007.11.009

Luo W, Hu H, Chang R et al (2011) Pyruvate kinase M2 Is a PHD3-stimulated coactivator for hypoxia-inducible factor 1. Cell 145:732–744. doi:10.1016/j.cell.2011.03.054

Mazurek S (2007) Pyruvate kinase type M2: a key regulator within the tumour metabolome and a tool for metabolic profiling of tumours. In: Proceedings of the Ernst Schering Found Symposium, pp 99–124

Mazurek S, Boschek CB, Hugo F, Eigenbrodt E (2005) Pyruvate kinase type M2 and its role in tumor growth and spreading. Semin Cancer Biol 15:300–308. doi:10.1016/j.semcancer.2005.04.009

Niedenführ S, Wiechert W, Nöh K et al (2015) How to measure metabolic fluxes: a taxonomic guide for ^{13}C fluxomics. Curr Opin Biotechnol 34:82–90. doi:10.1016/j.copbio.2014.12.003

Noguchi T, Yamada K, Inoue H et al (1987) The L- and R-type isozymes of rat pyruvate kinase are produced from a single gene by use of different promoters. J Biol Chem 262:14366–14371

Nöh K, Wiechert W (2011) The benefits of being transient: isotope-based metabolic flux analysis at the short time scale. Appl Microbiol Biotechnol 91:1247–1265. doi:10.1007/s00253-011-3390-4

Nöh K, Grönke K, Luo B et al (2007) Metabolic flux analysis at ultra short time scale: isotopically non-stationary ^{13}C labeling experiments. J Biotechnol 129:249–267. doi:10.1016/j.jbiotec.2006.11.015

Old WM, Meyer-Arendt K, Aveline-Wolf L et al (2005) Comparison of label-free methods for quantifying human proteins by shotgun proteomics. Mol Cell Proteomics 4:1487–1502. doi:10.1074/mcp.M500084-MCP200

Ong S-E, Blagoev B, Kratchmarova I et al (2002) Stable isotope labeling by amino acids in cell culture, SILAC, as a simple and accurate approach to expression proteomics. Mol Cell Proteomics 1:376–386. doi:10.1074/mcp.M200025-MCP200

Ono M, Shitashige M, Honda K et al (2006) Label-free quantitative proteomics using large peptide data sets generated by nanoflow liquid chromatography and mass spectrometry. Mol Cell Proteomics 5:1338–1347. doi:10.1074/mcp.T500039-MCP200

Orth JD, Thiele I, Palsson BØ (2010) What is flux balance analysis? Nat Biotechnol 28:245–248. doi:10.1038/nbt.1614

Pan Z, Raftery D (2007) Comparing and combining NMR spectroscopy and mass spectrometry in metabolomics. Anal Bioanal Chem 387:525–527. doi:10.1007/s00216-006-0687-8

Panopoulos AD, Yanes O, Ruiz S et al (2012) The metabolome of induced pluripotent stem cells reveals metabolic changes occurring in somatic cell reprogramming. Cell Res 22:168–177. doi:10.1038/cr.2011.177

Pietzke M, Kempa S (2014) Pulsed stable isotope-resolved metabolomic studies of cancer cells. Methods Enzymol. doi:10.1016/B978-0-12-801329-8.00009-X (1st ed)

Pietzke M, Zasada C, Mudrich S, Kempa S (2014) Decoding the dynamics of cellular metabolism and the action of 3-bromopyruvate and 2-deoxyglucose using pulsed stable isotope-resolved metabolomics. Cancer Metab 2:9. doi:10.1186/2049-3002-2-9

Prigione A, Fauler B, Lurz R et al (2010) The senescence-related mitochondrial/oxidative stress pathway is repressed in human induced pluripotent stem cells. Stem Cells 28:721–733. doi:10.1002/stem.404

Prigione A, Lichtner B, Kuhl H et al (2011) Human induced pluripotent stem cells harbor homoplasmic and heteroplasmic mitochondrial DNA mutations while maintaining human embryonic stem cell-like metabolic reprogramming. Stem Cells 29:1338–1348. doi:10.1002/stem.15

Ross PL (2004) Multiplexed protein quantitation in saccharomyces cerevisiae using amine-reactive isobaric tagging reagents. Mol Cell Proteomics 3:1154–1169. doi:10.1074/mcp.M400129-MCP200

Szyperski T (1995) Biosynthetically directed fractional ^{13}C-labeling of proteinogenic amino acids. An efficient analytical tool to investigate intermediary metabolism. Eur J Biochem 232:433–448. doi:10.1111/j.1432-1033.1995.tb20829.x

Varma A, Palsson BO (1994) Metabolic flux balancing: basic concepts, scientific and practical use. Nat Biotechnol 12:994–998

Wahl SA, Nöh K, Wiechert W (2008) ^{13}C labeling experiments at metabolic nonstationary conditions: an exploratory study. BMC Bioinformatics 9:152. doi:10.1186/1471-2105-9-152

Warburg O, Wind F, Negelein E (1927) The metabolism of tumors in the body. J Gen Physiol 8:519–530

Wenzel M, Joel I, Oelkers W (1966) Application of double isotope labeling to the study of the metabolism of ascites tumor cells under the influence of glyceraldehyde. Adv Tracer Methodol 3:223–231

Wiechert W (2001) ^{13}C metabolic flux analysis. Metab Eng 206:195–206

Wiechert W, Mo M, Petersen S, De Graaf AA (2001) A universal framework for ^{13}C metabolic flux analysis. Metab Eng 283:265–283

Wiese S, Reidegeld K, Meyer HE, Warscheid B (2007) Protein labeling by iTRAQ: a new tool for quantitative mass spectrometry in proteome research. Proteomics 7:340–350. doi:10.1002/pmic.200600422

Yang W, Xia Y, Ji H et al (2011) Nuclear PKM2 regulates β-catenin transactivation upon EGFR activation. Nature 478:118–122. doi:10.1038/nature10598

Yeung SJ, Pan J, Lee MH (2008) Roles of p53, MYC and HIF-1 in regulating glycolysis—the seventh hallmark of cancer. Cell Mol Life Sci 65:3981–3999. doi:10.1007/s00018-008-8224-x

Young JD, Walther JL, Antoniewicz MR et al (2007) An elementary metabolite unit (EMU) based method of isotopically nonstationary flux analysis. Biotechnol Bioeng 99:686–699. doi:10.1002/bit

Zamboni N (2011) ^{13}C metabolic flux analysis in complex systems. Curr Opin Biotechnol 22:103–108. doi:10.1016/j.copbio.2010.08.009

Mathematical Modeling of Cellular Metabolism

Nikolaus Berndt and Hermann-Georg Holzhütter

Abstract

Cellular metabolism basically consists of the conversion of chemical compounds taken up from the extracellular environment into energy (conserved in energy-rich bonds of organic phosphates) and a wide array of organic molecules serving as catalysts (enzymes), information carriers (nucleic acids), and building blocks for cellular structures such as membranes or ribosomes. Metabolic modeling aims at the construction of mathematical representations of the cellular metabolism that can be used to calculate the concentration of cellular molecules and the rates of their mutual chemical interconversion in response to varying external conditions as, for example, hormonal stimuli or supply of essential nutrients. Based on such calculations, it is possible to quantify complex cellular functions as cellular growth, detoxification of drugs and xenobiotic compounds or synthesis of exported molecules. Depending on the specific questions to metabolism addressed, the methodological expertise of the researcher, and available experimental information, different conceptual frameworks have been established, allowing the usage of computational methods to condense experimental information from various layers of organization into (self-)consistent models. Here, we briefly outline the main conceptual frameworks that are currently exploited in metabolism research.

Keywords

Mathematical modeling · Cellular metabolism · Kinetic model · Statistical models · Stoichiometric models

N. Berndt · H.-G. Holzhütter (✉)
Institute of Biochemistry—Computational Systems Biochemistry Group,
Charité–Universitätsmedizin Berlin, 10117 Berlin, Germany
e-mail: hergo@charite.de

© Springer International Publishing Switzerland 2016
T. Cramer and C.A. Schmitt (eds.), *Metabolism in Cancer*,
Recent Results in Cancer Research 207, DOI 10.1007/978-3-319-42118-6_10

221

1 Metabolic Modeling

Cells are highly heterogeneous structures whose functionality depends on the spatial and temporal organization of its constituents. Cellular metabolism describes the chemical conversion of small molecules (metabolites) by intracellular catalysts (enzymes) to provide the ingredients necessary for self-maintenance, growth, and reproduction.

Organization principles of metabolic systems are manifold. In principle, the enzymatic activities depend on local metabolite and enzyme concentrations, but myriads of environmental and structural factors influencing enzymatic activities exist. Spatial segregation of metabolic activity between compartments allows the creation of specific microenvironments through the controlled exchange of metabolites and ions to facilitate and enable special metabolic functions. For example, mitochondria generate an electrical field and a pH gradient across their inner mitochondrial membrane enabling the generation of ATP (Mitchell 1966). Metabolites can also be directly handed over between metabolic enzymes forming metabolons to avoid dilution and direct metabolic fluxes. In Arabidopsis thaliana, it was shown that glycolytic enzymes dynamically associate with mitochondria to directly channel substrate into mitochondrial respiration (Graham et al. 2007). This complex formation can be so advanced that multienzyme complexes can be regarded as single enzymes (like complex I of the respiratory chain, the pyruvate dehydrogenase complex, or the α-ketoglutarate dehydrogenase complex). Also within cellular compartments, concentration gradients of important metabolites exist and are being used for targeted transport within the cell. ATP generated by the mitochondria has to be transferred to the main energy utilization sites in the cell, e.g., the ion pumps at the plasma membrane. ATP transport is accomplished by diffusion along the concentration gradient between the mitochondrial and the plasma membrane. This mechanism is reinforced by the use of creatine phosphate. The creatine kinase, reversibly catalyzing the transfer of the high-energy phosphate from ATP to creatine, is located at the mitochondrial membrane and at the place of ATP utilization. Due to its phosphorylation potential, its higher cytosolic concentration, and its higher diffusion rate, it greatly facilitates energy transduction with the cell (Wallimann et al. 1992). Stochastic variations in different constituents might also influence the systemic behavior, but the extent and importance remain controversial. In general, it is unfeasible to try to take every regulatory property into account, but a main challenge is the identification of the regulatory aspects needed in understanding/modeling the specific system at hand.

The goals of metabolic modeling are to construct simplified representations of the real world that allow to sort existing knowledge, unravel basic principles, and predict cellular responses or to identify metabolic targets for intervention. Depending on available information and the specific question at hand, various mathematical representations have been established, allowing usage of computational methods for condensing experimental information from various layers of molecular organization (mRNA and protein profiles, metabolite patterns, and enzyme-kinetic data) into consistent network models of cellular metabolism.

1.1 Statistical Models

The simplest approach in metabolic modeling consists in the correlation of different trades with the aim to unravel possible interdependencies. The big advantage of this method lies in the fact that it does not rely on prior knowledge and is not limited to any kind of specific information. It is often used by correlating large datasets such as transcriptome, proteome, or metabolome data of tissue or blood with clinical parameters (e.g., cancer versus healthy) through *multivariate statistical analyses* with the goal of biomarker identification (see, e.g., Dhanasekaran et al. 2001; Bartels and Tsongalis 2009). The disadvantage is that correlations do not provide insight into the regulation of metabolic networks leaving it to more sophisticated methods to translate them into causal relationships.

1.2 Stoichiometric Models

Stoichiometric models represent basically whole-cell reaction graphs encompassing all currently known metabolites and biochemical reactions in a given cell type that can be inferred from genomic and biochemical information sources. Constraint-based modeling (CBM)/flux balance analysis (FBA) relies on a description of a metabolic network connecting metabolites M_i with reactions v_j. The metabolic network is represented by a stoichiometric matrix S_{ij} composed of the stoichiometric coefficients with which metabolites (i) are produced or consumed in reactions (j). An important additional assumption is stationarity; i.e., the time-dependent variation of metabolite concentrations is assumed to be zero. Since the stoichiometric matrix is assumed to be constant as well, fluxes are also constant. These model inputs and assumptions can be conveniently phrased as

$$\frac{d\,[M_i]}{dt} = \sum_{i=1}^{n} S_{ij} v_j = 0$$

$$\frac{d\,S_{ij}}{dt} = 0$$

$$\frac{dv_j}{dt} = 0$$

With the additional assumption that fluxes are independent of metabolite concentrations, the flux space completely decouples from the metabolite concentrations which are then often completely neglected. To arrive at nonzero flux distribution, uptake and release fluxes have to be assigned. The most often used assignment is a stoichiometric combination of target fluxes, so-called objective function, as well as possible uptake and waste fluxes. To restrict the possible flux distributions relating input with output fluxes, optimality criteria are applied which can be interpreted as design principles. Often used optimality criteria are the maximization of yield per substrate, the maximization of growth rate or biomass production, or the minimization of fluxes (see, e.g., Edwards and Palsson 1998; Holzhutter 2004). This can be interpreted as an optimal usage of limited resources or a minimization of effort by the cell. To take observed protein abundances/gene expression profiles into account, additional constraints limiting the maximal capacity of a reaction can be assigned

$$-lb_j \leq v_j \leq ub_j$$

A further important constraint arising from basic thermodynamics (but often disregarded in CBM applications) is the condition that the direction of fluxes has to be in concordance with the sign of the change of the Gibbs free energy (i.e., $v_j > 0$ if $\Delta G < 0$) (Hoppe et al. 2007). This is achieved by assigning a range of feasible concentration ranges to each metabolite and a ΔG value to each reaction to exclude all flux distributions where it is impossible to find a set of metabolite concentrations that render all reactions thermodynamically feasible.

In short, FBA modeling needs a stoichiometric network defining all possible reactions, an objective function defining the metabolic output, and optimality criteria defining design principles and predicts flux distributions that can be compared to experimental data.

1.2.1 Applications

The commonly asked question to be answered by FBA and constraint-based modeling can be phrased as follows:

Is a given metabolic network able to fulfill a chosen objective function and what is the resulting flux distribution?

A first application of FBA models was the identification of gaps in our current knowledge by testing whether a given substrate can be converted into a given product (see, e.g., Gille et al. 2010). If the conversion is not possible, one or more reaction steps in the network are missing. This is often used to predict phenotype (= metabolic output) from genotype (= protein profiles/gene expression data) (see, e.g., Price et al. 2004). Secondly, one wishes to identify essential reactions for a given objective function which upon elimination from the network (= synthetic knockout) renders at least one physiologically relevant biochemical reaction impossible. Essential reactions are regarded as convenient targets for the development of drugs impairing the metabolism of pathogens (see, e.g., Bazzani et al. 2012;Huthmacher et al. 2010). Interestingly, it was shown that defects of enzymes catalyzing essential reactions are frequently associated with the occurrence of metabolic disorders (Gille et al. 2010).

Experimental validation for flux distributions comes from labeling experiments. By feeding labeled substrates (such as ^{13}C-glucose or ^{15}N-amino acids) in a metabolic system, the time-resolved label distributed can be used to trace the fluxes within the network if the label fate maps of the underlying reaction networks are known. If a stationary metabolic state can be assumed, FBA models can be used to simulate the time-dependent label distribution under the assumptions that enzymatic activity is independent of the label. When trying to model label distributions in a metabolic network, a major problem arises from incomplete information since not all metabolites are measured simultaneously, as well as the restriction of the metabolic network if it is not genome-scale. Usually, this is handled by the introduction of suitable effluxes from the label that are regarded as free parameters and fitted to the measured label distributions. In general, label experiments are powerful tools for the investigation of metabolic systems, but their interpretation has to be performed with care as incomplete information and non-stationarity spoil the applicability of CBM/FBA models.

1.2.2 Challenges

Network reconstruction: Starting point for CBM/FBA is the reconstruction of the network under consideration which is identical to the stoichiometric matrix. There is no general method for the network reconstruction. Often subnetworks of genome-scale 'master networks,' encompassing all known enzymatically catalyzed reactions inferred from genomic information, are used. Based on gene expression profiles of the cells or the tissue under consideration, active subnetworks are defined. This method is an all or nothing assignment—a reaction/enzyme is present or it is absent. This means that cutoff values for the expression values have to be assigned, which is far from straightforward and unbiased. Another problem arises from the fact that non-enzymatic reactions, as, for example, the proton leak in the respiratory chain for heat production, are not considered. In general, it is not obvious that this method yields a functional network. First, the correctness of the 'master network' is rarely checked so that unconnected metabolites or disconnected parts might exist. Second, so-called housekeeping enzymes, that are enzymes considered to be constitutively expressed and inevitable necessary for proper cellular function, often show very low expression levels and might be omitted using such an approach. Therefore, function tests and manual curation have to be evoked to ensure a proper network topology. A further restriction is that only stoichiometrically fixed reactions can be modeled, and if no fixed stoichiometry exists, it has to be postulated. For example, the P/O quotient connecting NADH utilization with ATP production is usually fixed to a ratio of 1:2.5. However, this is most certainly not true as the efficiency and the relative share of ATP production from protons (not considered explicitly) strongly depend on the metabolic state.

Choosing the appropriate objective function and optimality criteria: As metabolism enables a large number of diverse cellular functions at the same time, the assumption of a single flux objective does not seem realistic. For example, hepatocytes show many equally important functions reported for the human liver (disregarding the uttermost complex self-maintaining functions) that, in principal, have to be included into a reliable CSB model (Gille et al. 2010). Furthermore, metabolic objectives are highly variable and depend on the external conditions. While economically favorable minimization of fluxes is an often employed optimality criteria, other selection criteria such as stability, variability, robustness, and adaptability are neglected. For example, hepatocytes have glucose homeostasis as an objective, which is realized as glucose production or utilization depending on external conditions. Here, the ability to respond to external changes has to be optimized a concept that can hardly be put in an FBA model. Likewise, microorganisms may strive for optimal usage of resources during shortcoming times, but focus on fast growth regardless of efficiency in times of ample substrate supply (Schuetz et al. 2007). In general, the objective function and the optimality criteria used dictate a rational that has to be justified by experimental data. As an illustrative example, the study by Schuetz et al. can be considered (Schuetz et al. 2012). Comparing 44 experimentally determined *in vivo* MFA datasets for E. coli to predicted flux distributions based on a large number of putative flux objectives and linear combinations of these objectives, the best concordance was achieved by a combination of three flux objectives: maximizing biomass production, maximizing ATP generation, and minimizing reaction fluxes across the network. But still the relative proportions among these three objectives varied between different cells putting the predictive value of such multi-objective function into question.

Conclusion: The striking advantage of CBM is its simplicity: It requires only knowledge of the stoichiometry of the network and thus can be immediately applied to whole-cell network reconstructions without caring for the kinetic and regulatory properties of enzymes and transporters. A common view is that considering very large networks enables to study 'systemic' behavior. It is noteworthy that his argument contradicts the basic idea of modularity, stating that a complex system can be broken down to simpler components like pathways or single reactions. Interestingly, often enough only small parts of a metabolic network are used for fulfillment of a specific objective function/metabolic task, which are close or identical to the classical metabolic pathways. Furthermore, mathematical concepts like minimal flux modes reintroduce modularity through the definition of minimal subsets that build a basis for the solution space.

On the downside, metabolite levels, regulatory properties, and time dependencies are completely neglected. The predicted flux distributions are hypothetical and have to be considered with great care, because they are essentially dictated by the subjective choice of the flux objectives and presumed optimality principles. However, through the teleological nature of the objective functions and the optimality principles used, CBM/FBA allows an interpretation in terms of cellular aims and evolutionary development.

2   Kinetic Models

Importantly, regulation of enzymatic activity does not only proceed through changes in enzyme abundance, but by substrate availability, allosteric regulation, or hormonal control as well. Furthermore, not all enzymes in a given pathway are equally important in the regulation of the flux through the pathway; rather, key regulatory enzymes control the metabolic activity and many key regulatory enzymes do not exhibit significant changes in their abundance despite large flux changes (ter Kuile and Westerhoff 2001; Vogt et al. 2002). This strongly contradicts the structure of CNB/FBA models, where all regulatory aspects are neglected and the enzymes differ only through their position in the network. Thus, to understand metabolic regulation in detail, the concept of kinetic modeling needs to be employed.

Kinetic models also start with a metabolic network represented by a stoichiometric matrix S_{ij} connecting metabolites M_i and reactions v_j. The time-dependent changes in metabolite concentrations are given by

$$\frac{d[M_i]}{dt} = \sum_{i=1}^{n} S_{ij} v_j$$

In contrast to CBM/FBA, the reaction rates v_j, representing enzymatically catalyzed reactions, are not constant but depend on external and internal conditions. Regulatory principles might comprise kinetic regulation through substrate availability and allosteric regulation, enzyme modifications (e.g., through interconversion), changes in enzyme abundance, or other regulatory mechanisms such as complex formation, spatial confinement, and activation by ions. In general, local enzyme and metabolite concentrations are replaced by mean concentrations in a given cellular compartment, and the assumption is made that the concentrations are high enough so that stochastic variations can be neglected. Once the rate equations for the v_j are fixed, the only additional information needed is the external medium compositions (e.g., metabolite, ion, or hormone concentrations). Compared to CBM/FBA, no objective function or optimality criteria have to be chosen, as any functional output and internal state are deterministically dictated by the properties of the constituents and the external conditions.

2.1 Applications

Kinetic models can predict the metabolic state of a cell in dependence of external and internal conditions. The metabolic state consists of intracellular metabolite concentrations, the internal flux state, and exchange fluxes in dependence of external parameters (like substrate supply) and internal state (enzyme abundance, phosphorylation state, metabolite concentrations). As they allow time-dependent simulations, interpretation of label experiments with kinetic models does not rely on the assumption of a metabolic steady state vastly expanding the scope of application. A kinetic model should be regarded as an in silico representation of a cell that can be used analogous to cells by experimentalists.

An important function of kinetic models lies in the explanation of experimental data based on the functional properties of the enzymes constituting the model (see, e.g., Berndt et al. 2015). They allow to test whether our understanding of the molecular basis is sufficient to quantitatively and qualitatively explain systemic behavior or whether we are missing important regulatory mechanism.

In addition, kinetic models can be used to extrapolate from the regime in which they were built to predict the behavior of the model system under different conditions (see, e.g., Berndt et al. 2012). Similar to CBM/FBA models, kinetic models can be used to predict phenotype from genotype albeit in a quantitative time-dependent manner without evoking the underlying optimization principles.

2.2 Challenges

Network reconstruction: Metabolic networks underlying kinetic models are constructed from pathways with known input–output relation. Since the enzymes constituting the pathway have to be characterized in great detail, the problem of disconnected pathways does not exist.

Choosing the rate equations: Kinetic rate equations rely on detailed biochemical information of the reaction kinetics and enzyme regulatory properties for the individual reactions and transport processes, while conceptually easy and straightforward kinetic modeling is only applicable for systems that are very well characterized. The main difficulty consists in the selection of the regulatory properties considered. In general, kinetic models focus on a small number of reactions pertinent to the metabolic function of interest embedded in central metabolism by keeping essential cofactors constant. The main (and sole) challenge in kinetic modeling lies in choosing the appropriate rat laws. In principle, the rate equations depend on (local) metabolite and enzyme concentration, but myriads of environmental factors such the ionic surrounding (pH value, calcium, magnesium, phosphate, potassium, sodium, chloride, etc., concentration), protein modifications (such as phosphorylation state, complex formation), or phospholipid environment (for membranous proteins) influence enzymatic activities. It is unfeasible to try to take every regulatory property into account. It is the main task to identify and consider the regulatory aspects needed in understanding/modeling the specific system at hand. Therefor, kinetic models are often tailored to answer specific questions within a limited application range: To investigate the glycolytic capacity of a cell, it is usually unnecessary to consider changes in the ionic strength, while models of the respiratory chain rely on this information. Besides the problem of limiting the regulatory principles considered, another problem arises from either a lack of detailed information or (seemingly) contradictory experimental findings. If one was to set up a kinetic model without the assumption of homogeneity in enzyme and metabolite concentrations, experimental data need to be present. Similarly, only the regulatory properties that are known and characterized (such as allosteric regulator with their Ki-values) can be considered. If it is only known that a specific drug inhibits a certain metabolic function but not which enzyme is inhibited in which way, it is not possible to model the effect (although one might identify possible mechanisms that would result in the observed systemic effect, thereby giving hints what should be validated experimentally).

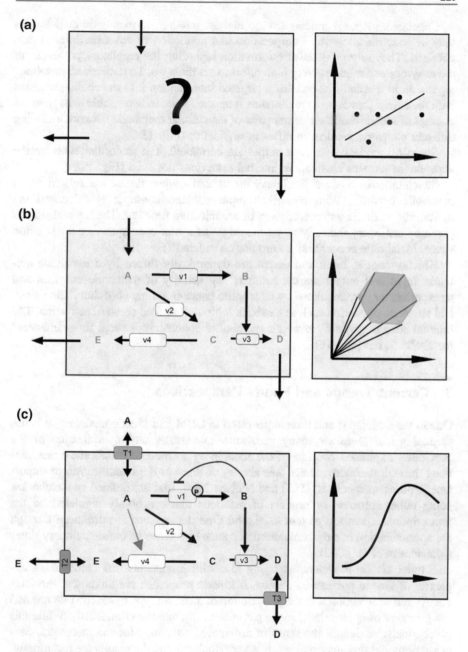

Fig. 1 Graphic representation of different modeling approaches **a** Statistical models; **b** Stoichiometric models; **c** Kinetic models

Conclusion: Kinetic models are the highest standard in metabolic modeling as they in principle allow the incorporation and prediction of any experimental data obtained. They rely on detailed information regarding the regulatory properties of the enzymes and require a very high effort to set them up. To understand metabolic regulation in a detail that enables a targeted intervention into metabolic processes with the aim to prevent or cure diseases, it appears to be indispensable to expand the concept of kinetic models to larger parts of the cellular metabolic network including different pathways resident in different organelles (Fig. 1).

Statistical models: Input and output are correlated. The molecular basis for the dependence remains elusive. An internal state does not exist (Fig. 1a).

Stoichiometric models: Stationary input and output fluxes are related via a metabolic network. With prescribed input, metabolic output is calculated via optimization criteria and prescription of an objective function. The internal state is given by stationary fluxes lying on the edges of a flux cone spanning the solution space. Metabolite concentrations are not considered (Fig. 1b).

Kinetic models: Input and output are dynamically linked by a metabolic network. Input and output are the result of the activity of metabolic enzymes and transporters that are regulated by metabolite concentrations, phosphorylation state, and allosteric regulation such as feedback inhibition or feed-forward activation. The internal state is given by dynamic metabolite concentrations and time-dependent metabolic fluxes (Fig. 1c).

3 Current Trends and Future Perspectives

Due to the difficulties and limitations faced in CBM and kinetic modeling, it is not surprising that there are many variations, extensions, and hybridization of the previously explained concepts. FBA models try to include metabolite concentrations through thermodynamics (see above) or through the identification of important 'reporter metabolites' (Patil and Nielsen 2005) that are defined as metabolites being either substrate or product of reactions that are highly regulated on the transcription/expression/protein level, and time dependence is introduced through the approximation of a dynamic metabolic state by a series of quasi-stationary states (Mahadevan et al. 2002).

Similar kinetic modelers are trying to establish concepts that reduce the complexity of kinetic parameters needed. If kinetic properties are unknown, shortcuts are evoked which consist in the prescription of a certain type of kinetic rate law and the guessing of appropriate kinetic parameters (Smallbone et al. 2010). While this gives kinetic models in the sense of differential equations for the metabolite concentrations and flux distributions, it is very doubtful that the results are meaningful. More promising simplifications arise from the observation that not all enzymes are equally important for the regulation of a metabolic function. If the regulatory enzymes can be identified, it is feasible to model these regulatory enzymes in greater detail, while the rest of the enzymes can be treated via simplified kinetics.

Indeed, it was shown (Bulik et al. 2009) that this gives fairly good approximations compared to a fully kinetic model.

Hybrid models try to combine aspects of both modeling approaches. Structural kinetic modeling tries to extract kinetic parameters from stability considerations (Grimbs et al. 2007), while other approaches try to combine CBM with kinetic models by replacing parts of a large metabolic network by kinetic models, which then define the exchange fluxes used in the larger FBA parts.

Expanding the scope: Through the availability of omics data sets a whole universe of regulatory layers comprising signaling nodes, transcription factors, mRNA and miRNA has been opened. Not surprisingly all these layers are extensively studied. Gene regulatory networks try to unravel the principles underlying the interaction between transcription factors, gene products, and effectors (miRNA), signaling networks try to understand the activation of transcription factors in response to different stimuli, and ontology-based models try to translate changes occurring on the mRNA level into changes in the cellular metabolism. While in metabolic modeling the topology of the metabolic network is worked out, a major challenge for these approaches lies in the construction of the network topology itself.

Closing remarks: Currently, we are witnessing a renewed scientific interest in the study of the cellular metabolism especially with respect to medical application. Two complementary strategies are pursuit to unravel the principles underlying metabolic regulation. CBM/FBA models try to include more and more information through omics datasets, trying to combine gene regulation, signaling, and metabolism neglecting non-topological aspects of the underlying regulation, while kinetic models try to include much more of the regulatory principles but are limited to rather small systems.

An important aspect of future modeling approaches will be the progression from the cellular to the tissue and organ level. Coupling of heterogeneous cells by blood perfusion taking heterogeneous substrate supply and mutual interaction between cells into account will lead to multi-scale models appropriate for the understanding of clinical relevant phenomena such as in fatty liver disease, neurodegenerative diseases, or progressive tissue damage after an acute infarction (heart and brain).

References

Bartels CL, Tsongalis GJ (2009) MicroRNAs: novel biomarkers for human cancer. Clin Chem 55 (4):623–631

Bazzani S, Hoppe A, Holzhutter HG (2012) Network-based assessment of the selectivity of metabolic drug targets in Plasmodium falciparum with respect to human liver metabolism. BMC Syst Biol 6:118

Berndt, N, Kann O, Holzhutter HG (2015) Physiology-based kinetic modeling of neuronal energy metabolism unravels the molecular basis of NAD(P)H fluorescence transients. J Cereb Blood Flow Metab

Berndt N, Bulik S, Holzhutter HG (2012) Kinetic modeling of the mitochondrial energy metabolism of neuronal cells: the impact of reduced alpha-ketoglutarate dehydrogenase activities on atp production and generation of reactive oxygen species. Int J Cell Biol 2012:757594

Bulik S et al (2009) Kinetic hybrid models composed of mechanistic and simplified enzymatic rate laws–a promising method for speeding up the kinetic modelling of complex metabolic networks. FEBS J 276(2):410–424

Dhanasekaran SM et al (2001) Delineation of prognostic biomarkers in prostate cancer. Nature 412 (6849):822–826

Edwards JS, Palsson BO (1998) How will bioinformatics influence metabolic engineering? Biotechnol Bioeng 58(2–3):162–169

Gille C et al (2010) HepatoNet1: a comprehensive metabolic reconstruction of the human hepatocyte for the analysis of liver physiology. Mol Syst Biol 6:411

Graham JW et al (2007) Glycolytic enzymes associate dynamically with mitochondria in response to respiratory demand and support substrate channeling. Plant Cell 19(11):3723–3738

Grimbs S et al (2007) The stability and robustness of metabolic states: identifying stabilizing sites in metabolic networks. Mol Syst Biol 3:146

Holzhutter HG (2004) The principle of flux minimization and its application to estimate stationary fluxes in metabolic networks. Eur J Biochem 271(14):2905–2922

Hoppe A, Hoffmann S, Holzhutter HG (2007) Including metabolite concentrations into flux balance analysis: thermodynamic realizability as a constraint on flux distributions in metabolic networks. BMC Syst Biol 1:23

Huthmacher C et al (2010) Antimalarial drug targets in Plasmodium falciparum predicted by stage-specific metabolic network analysis. BMC Syst Biol 4:120

Mahadevan R, Edwards JS, Doyle FJ 3rd (2002) Dynamic flux balance analysis of diauxic growth in Escherichia coli. Biophys J 83(3):1331–1340

Mitchell P (1966) Chemiosmotic coupling in oxidative and photosynthetic phosphorylation. Biol Rev Camb Philos Soc 41(3):445–502

Patil KR, Nielsen J (2005) Uncovering transcriptional regulation of metabolism by using metabolic network topology. Proc Natl Acad Sci USA 102(8):2685–2689

Price ND, Reed JL, Palsson BO (2004) Genome-scale models of microbial cells: evaluating the consequences of constraints. Nat Rev Microbiol 2(11):886–897

Schuetz R, Kuepfer L, Sauer U (2007) Systematic evaluation of objective functions for predicting intracellular fluxes in Escherichia coli. Mol Syst Biol 3:119

Schuetz R et al (2012) Multidimensional optimality of microbial metabolism. Science 336 (6081):601–604

Smallbone K et al (2010) Towards a genome-scale kinetic model of cellular metabolism. BMC Syst Biol 4:6

ter Kuile BH, Westerhoff HV (2001) Transcriptome meets metabolome: hierarchical and metabolic regulation of the glycolytic pathway. FEBS Lett 500(3):169–171

Vogt AM et al (2002) Regulation of glycolytic flux in ischemic preconditioning. A study employing metabolic control analysis. J Biol Chem 277(27):24411–24419

Wallimann T et al (1992) Intracellular compartmentation, structure and function of creatine kinase isoenzymes in tissues with high and fluctuating energy demands: the 'phosphocreatine circuit' for cellular energy homeostasis. Biochem J 281(Pt 1):21–40

Ketogenic Diet and Cancer—a Perspective

Christopher Smyl

Abstract

Research of the last two decades showed that chronic low-grade inflammation, elevated blood glucose and insulin levels may play role in the onset of a number of non-communicable diseases such as type 2 diabetes and some forms of cancer. Regular exercise and fasting can ameliorate high blood glucose and insulin levels as well as increase the concentration of plasma ketone bodies. These, in consequence, may lead to reduction of inflammation. Exercise or severe restriction of caloric intake is not always advisable for patients, in particular those suffering from cancer. The ketogenic diet (KD), characterized by high fat, moderate protein and very low carbohydrate composition can evoke a physiological state similar to that triggered by exercise or fasting. These attributes of KD prompted its possible use in treatment of a number of metabolic diseases, including several types of malignancies. Although results from clinical studies employing KD in the treatment of cancer are still limited, the results obtained from animal models are encouraging and show that KD presents a viable option as an adjunct therapy for cancer.

Keywords

Ketogenic diet · Cancer therapy · Cancer · Ketone bodies · Metabolism

Research of the last two decades justifies to assert that low-grade inflammation, chronically elevated levels of glucose and insulin are outward manifestations of metabolic deregulations associated with a number of noncommunicable diseases, some of which reached epidemic proportions, e.g., diabetes and cancer (Onodera

C. Smyl (✉)
Department of Hepatology and Gastroenterology Charité—Universitätsmedizin Berlin,
Augustenburger Platz 1, 13353 Berlin, Germany
e-mail: christopher.smyl@charite.de

© Springer International Publishing Switzerland 2016
T. Cramer and C.A. Schmitt (eds.), *Metabolism in Cancer*,
Recent Results in Cancer Research 207, DOI 10.1007/978-3-319-42118-6_11

et al. 2014; Ferroni et al. 2015; Hojlund 2014; Chettouh et al. 2015). It is argued that diets rich in highly processed foods, in particular hydrogenated trans-fatty acids and refined carbohydrates, which are now consumed by the vast majority of the worlds population, significantly contribute to hyperglycemia, hyperinsulinemia, and chronic inflammation (Raiten et al. 2015; Bhattacharyya et al. 2015; Deer et al. 2015). It is widely accepted that health benefits associated with fasting—as discussed in this textbook by Brandhorst and Longo—and regular exercise can be, to a significant degree, attributed to the reduction of blood plasma glucose and insulin levels as well as elevation of ketone body production, which—in consequence—leads to reduced inflammation (Johnson et al. 2007; McGettrick and O'Neill 2013; Metsios et al. 2015; Lee et al. 2015). These physiological changes are regarded as key factors in reversing the decline in health associated with metabolic deregulation as well as providing a physiological milieu conducive in the treatment of metabolic diseases including cancer (Leitzmann et al. 2015; Sehdev and O'Neil 2015; Beulertz et al. 2015; Winters-Stone 2015). However, frequent fasting and/or exercise are not always applicable as an adjunct therapy for cancer as they can be too demanding or incompatible with the state of the patients. Against this background, it is interesting to note that the metabolic state evoked by fasting or exercise can also be induced by the ketogenic diet.

The ketogenic diet (KD) is a high-fat (65–90 % of total energy), adequate protein, low-carbohydrate (less than 10 % of total energy) diet, which originally was used to treat epilepsy (Nei et al. 2014; Li et al. 2013; Selter et al. 2015; Dressler et al. 2015). Ingestion of KD results in a switch from using glucose to fatty acids as the main energy source. As a consequence of the high fat content, in addition to lowering blood glucose, insulin, and insulin-like growth factor levels, KD results in significant increase of ketone body production and modulates concentration of hormones, neurotransmitters, and neuropeptides (Dang et al. 2015; Abdelwahab et al. 2012). KD has also been shown to have anti-inflammatory (Dupuis et al. 2015), anti-oxidative, and mitochondrial biogenesis-stimulating effects (Rhyu et al. 2014; Nazarewicz et al. 2007). These attributes of KD prompted for its application as a clinical dietary intervention in a number of disorders, e.g., obesity (Moreno et al. 2014; Yancy et al. 2004; Ruano et al. 2006), inborn disorders of glucose transport (Klepper et al. 2004; Lukyanova et al. 2015; Appavu et al. 2015), mitochondrial complex I deficiency syndrome (Veech 2004), amyotrophic lateral sclerosis (Zhao et al. 2006), diabetic nephropathy (Friedman et al. 2013), polycystic ovary syndrome (Mavropoulos et al. 2005), atherogenic dyslipidemia (Westman et al. 2006), type 2 diabetes mellitus (Westman et al. 2008), Alzheimer's and Parkinson's diseases (Baranano and Hartman 2008; Hertz et al. 2015; Newport et al. 2015), migraine (Di Lorenzo et al. 2015), depression (Murphy et al. 2004), and autism (Evangeliou et al. 2003). The effects of the KD on cancer have been investigated in animal models as well as in patients. Studies employing rodent models of colon (Tisdale et al. 1987; Beck and Tisdale 1989), prostate (Freedland et al. 2008; Mavropoulos et al. 2009), brain (Maurer et al. 2011; Stafford et al. 2010; Seyfried et al. 2014; Varshneya et al. 2015), skin (Magee et al. 1979), breast (Gluschnaider et al. 2014), lung (Allen et al. 2013), gastric (Otto et al. 2008), nonsmall cell lung and metastatic cancer (Poff et al. 2015) showed that KD is able to result in decreased tumor weight and improved overall survival.

To date, the investigations of clinical application of KD in treating cancer are sporadic and in essence limited to tumors of the brain. The reported trials show that the regime of KD is well tolerated by patients, without any adverse effects on their health or physiology improving overall patient's mood and fatigue (Schmidt et al. 2011). In addition to substantial reduction in blood glucose and elevation of ketone bodies, KD allows patients to maintain their body weight and appears to diminish physiological stress associated with chemotherapy treatment (Schmidt et al. 2011). KD intervention applied to two patients with malignant astrocytoma resulted in significant clinical improvement and arrested the progression of the disease in one patient (Nebeling et al. 1995). The study conducted by Seyfried and colleagues showed that two months of therapy with calorie-restricted KD led to a significant reduction of glioblastoma multiforme progression; however, tumor recurrence was found after 10 weeks of suspension of the dietary intervention (Zuccoli et al. 2010). Recently, trials in the Steinbach laboratory involving 20 glioblastoma patients led the authors to conclude that although KD can be safely applied as adjuvant agent to the glioblastoma patients, when used as a sole agent KD had no significant activity against cancer (Rieger et al. 2014). Clinical trials employing KD in the treatment of brain tumors are reviewed by Bozzetti and Zupec-Kania (2015) and Varshneya et al. (2015).

Although the clinical data as of today do not confirm KD as a valid, sole tool in the treatment of malignancies, the physiological state evoked by following the KD regime compels for in-depth research into KD impact on the onset and progression of proliferative cell growth. Research of recent years showed that ketone bodies, in particular β-hydroxybutyrate, play significant regulatory roles in inflammation. Preclinical studies showed that KD inhibited activity of the mammalian target of rapamycin (mTOR) pathway (McDaniel et al. 2011). mTOR inhibition is associated with diminished inflammatory responses and cancer growth (Thiem et al. 2013; Gomez-Pinillos and Ferrari 2012; Deng et al. 2010). NOD-like receptor (NLR) proteins are important mediators of inflammation, which control the release of a number of proinflammatory cytokines from macrophages (Martinon et al. 2007; Wen et al. 2013). Using in vitro and murine *in vivo* studies, Youm and coworkers showed that β-hydroxybutyrate suppressed activation of the NLRP3 complex, resulting in inhibition of inflammatory cytokine release and reduced severity of inflammation (Youm et al. 2015). Ketone bodies have also been shown to display protective effects against oxidative stress, most likely through the decrease of reactive oxygen species production by mitochondria (Kim et al. 2007). The high fat content of KD allows for manipulation of its fatty acids composition. The inclusion of highly ketogenic medium-chain fatty acids increases the circulating plasma β-hydrohybutyrate, whereas addition of monounsaturated (MUFA) and omega-3 polyunsaturated fatty acids (n-3 PUFA) augments anti-inflammatory properties of KD. It is well established that MUFA and n-3 PUFA suppress inflammatory responses (Mocellin et al. 2015; Sijben and Calder 2007; Thies et al. 2001; Harbige 2003; Roy et al. 2015; Sacchi et al.

2014) and proliferation of some cancers (Mansara et al. 2015; Rose et al. 1995; Calder 2015; Hu et al. 2015; Halder et al. 2015).

In conclusion, a thoughtfully designed and regularly controlled ketogenic combined with conventional medical care and physical exercise may present a useful addition to cancer treatment. More clinical trials are urgently needed to test safety, tolerability, and antiproliferative efficacy of the KD and other dietary modifications, e.g., intermittent fasting.

References

Abdelwahab MG, Fenton KE, Preul MC, Rho JM, Lynch A, Stafford P, S AC (2012) The ketogenic diet is an effective adjuvant to radiation therapy for the treatment of malignant glioma. PLoS ONE 7(5):36197. doi:10.1371/journal.pone.0036197

Allen BG, Bhatia SK, Buatti JM, Brandt KE, Lindholm KE, Button AM, Szweda LI, Smith BJ, Spitz DR, Fath MA (2013) Ketogenic diets enhance oxidative stress and radio-chemo-therapy responses in lung cancer xenografts. Clin Cancer Res 19(14):3905–3913. doi:10.1158/1078-0432.ccr-12-0287

Appavu B, Mangum T, Obeid M (2015) Glucose transporter 1 deficiency: a treatable cause of opsoclonus and epileptic myoclonus. Pediatr Neurol 53(4):364–366. doi:10.1016/j.pediatrneurol.2015.05.019

Baranano KW, Hartman AL (2008) The ketogenic diet: uses in epilepsy and other neurologic illnesses. Curr Treat Options Neurol 10(6):410–419

Beck SA, Tisdale MJ (1989) Effect of insulin on weight loss and tumour growth in a cachexia model. Br J Cancer 59(5):677–681

Beulertz J, Prokop A, Rustler V, Bloch W, Felsch M, Baumann FT (2015) Effects of a 6-month, group-based, therapeutic exercise program for childhood cancer outpatients on motor performance, level of activity, and quality of life. Pediatr Blood Cancer. doi:10.1002/pbc.25640

Bhattacharyya S, Feferman L, Unterman T, Tobacman JK (2015) Exposure to common food additive carrageenan alone leads to fasting hyperglycemia and in combination with high fat diet exacerbates glucose intolerance and hyperlipidemia without effect on weight. J Diabetes Res 2015:513429. doi:10.1155/2015/513429

Bozzetti F, Zupec-Kania B (2015) Toward a cancer-specific diet. Clin Nutr. doi:10.1016/j.clnu.2015.01.013

Calder PC (2015) Functional roles of fatty acids and their effects on human health. JPEN J Parenter Enteral Nutr. doi:10.1177/0148607115595980

Chettouh H, Lequoy M, Fartoux L, Vigouroux C, Desbois-Mouthon C (2015) Hyperinsulinaemia and insulin signalling in the pathogenesis and the clinical course of hepatocellular carcinoma. Liver Int. doi:10.1111/liv.12903

Dang MT, Wehrli S, Dang CV, Curran T (2015) The ketogenic diet does not affect growth of hedgehog pathway medulloblastoma in mice. PLoS ONE 10(7):0133633. doi:10.1371/journal.pone.0133633

Deer J, Koska J, Ozias M, Reaven P (2015) Dietary models of insulin resistance. Metabolism 64 (2):163–171. doi:10.1016/j.metabol.2014.08.013

Deng L, Zhou JF, Sellers RS, Li JF, Nguyen AV, Wang Y, Orlofsky A, Liu Q, Hume DA, Pollard JW, Augenlicht L, Lin EY (2010) A novel mouse model of inflammatory bowel disease links mammalian target of rapamycin-dependent hyperproliferation of colonic epithelium to inflammation-associated tumorigenesis. Am J Pathol 176(2):952–967. doi:10.2353/ajpath.2010.090622

Di Lorenzo C, Coppola G, Sirianni G, Di Lorenzo G, Bracaglia M, Di Lenola D, Siracusano A, Rossi P, Pierelli F (2015) Migraine improvement during short lasting ketogenesis: a proof-of-concept study. Eur J Neurol 22(1):170–177. doi:10.1111/ene.12550

Dressler A, Trimmel-Schwahofer P, Reithofer E, Groppel G, Muhlebner A, Samueli S, Grabner V, Abraham K, Benninger F, Feucht M (2015) The ketogenic diet in infants—Advantages of early use. Epilepsy Res 116:53–58. doi:10.1016/j.eplepsyres.2015.06.015

Dupuis N, Curatolo N, Benoist JF, Auvin S (2015) Ketogenic diet exhibits anti-inflammatory properties. Epilepsia 56(7):95–98. doi:10.1111/epi.13038

Evangeliou A, Vlachonikolis I, Mihailidou H, Spilioti M, Skarpalezou A, Makaronas N, Prokopiou A, Christodoulou P, Liapi-Adamidou G, Helidonis E, Sbyrakis S, Smeitink J (2003) Application of a ketogenic diet in children with autistic behavior: pilot study. J Child Neurol 18 (2):113–118

Ferroni P, Riondino S, Buonomo O, Palmirotta R, Guadagni F, Roselli M (2015) Type 2 diabetes and breast cancer: the interplay between impaired glucose metabolism and oxidant stress. Oxid Med Cell Longev 2015:183928. doi:10.1155/2015/183928

Freedland SJ, Mavropoulos J, Wang A, Darshan M, Demark-Wahnefried W, Aronson WJ, Cohen P, Hwang D, Peterson B, Fields T, Pizzo SV, Isaacs WB (2008) Carbohydrate restriction, prostate cancer growth, and the insulin-like growth factor axis. Prostate 68 (1):11–19. doi:10.1002/pros.20683

Friedman AN, Chambers M, Kamendulis LM, Temmerman J (2013) Short-term changes after a weight reduction intervention in advanced diabetic nephropathy. Clin J Am Soc Nephrol 8 (11):1892–1898. doi:10.2215/cjn.04010413

Gluschnaider U, Hertz R, Ohayon S, Smeir E, Smets M, Pikarsky E, Bar-Tana J (2014) Long-chain fatty acid analogues suppress breast tumorigenesis and progression. Cancer Res 74 (23):6991–7002. doi:10.1158/0008-5472.can-14-0385

Gomez-Pinillos A, Ferrari AC (2012) mTOR signaling pathway and mTOR inhibitors in cancer therapy. Hematol Oncol Clin North Am 26 (3):483–505, vii. doi:10.1016/j.hoc.2012.02.014

Halder RC, Almasi A, Sagong B, Leung J, Jewett A, Fiala M (2015) Curcuminoids and omega-3 fatty acids with anti-oxidants potentiate cytotoxicity of natural killer cells against pancreatic ductal adenocarcinoma cells and inhibit interferon gamma production. Front Physiol 6:129. doi:10.3389/fphys.2015.00129

Harbige LS (2003) Fatty acids, the immune response, and autoimmunity: a question of n-6 essentiality and the balance between n-6 and n-3. Lipids 38(4):323–341

Hertz L, Chen Y, Waagepetersen HS (2015) Effects of ketone bodies in Alzheimer's disease in relation to neural hypometabolism, beta-amyloid toxicity, and astrocyte function. J Neurochem 134(1):7–20. doi:10.1111/jnc.13107

Hojlund K (2014) Metabolism and insulin signaling in common metabolic disorders and inherited insulin resistance. Dan Med J 61(7):B4890

Hu Z, Qi H, Zhang R, Zhang K, Shi Z, Chang Y, Chen L, Esmaeili M, Baniahmad A, Hong W (2015) Docosahexaenoic acid inhibits the growth of hormone-dependent prostate cancer cells by promoting the degradation of the androgen receptor. Mol Med Rep 12(3):3769–3774. doi:10.3892/mmr.2015.3813

Johnson JB, Summer W, Cutler RG, Martin B, Hyun DH, Dixit VD, Pearson M, Nassar M, Telljohann R, Maudsley S, Carlson O, John S, Laub DR, Mattson MP (2007) Alternate day calorie restriction improves clinical findings and reduces markers of oxidative stress and inflammation in overweight adults with moderate asthma. Free Radic Biol Med 42(5):665–674. doi:10.1016/j.freeradbiomed.2006.12.005

Kim DY, Davis LM, Sullivan PG, Maalouf M, TA Simeone, van Brederode J, Rho JM (2007) Ketone bodies are protective against oxidative stress in neocortical neurons. J Neurochem 101 (5):1316–1326. doi:10.1111/j.1471-4159.2007.04483.x

Klepper J, Diefenbach S, Kohlschutter A, Voit T (2004) Effects of the ketogenic diet in the glucose transporter 1 deficiency syndrome. Prostaglandins Leukot Essent Fatty Acids 70(3):321–327. doi:10.1016/j.plefa.2003.07.004

Lee YY, Yang YP, Huang PI, Li WC, Huang MC, Kao CL, Chen YJ, Chen MT (2015) Exercise suppresses COX-2 pro-inflammatory pathway in vestibular migraine. Brain Res Bull 116:98–105. doi:10.1016/j.brainresbull.2015.06.005

Leitzmann M, Powers H, Anderson AS, Scoccianti C, Berrino F, Boutron-Ruault MC, Cecchini M, Espina C, Key TJ, Norat T, Wiseman M, Romieu I (2015) European code against cancer 4th edition: physical activity and cancer. Cancer Epidemiol. doi:10.1016/j.canep.2015.03.009

Li HF, Zou Y, Ding G (2013) Therapeutic success of the ketogenic diet as a treatment option for epilepsy: a meta-analysis. Iran J Pediatr 23(6):613–620

Lukyanova EG, Ayvazyan SO, Osipova KV, Pyreva EA, Sorvacheva TN (2015) Experience of using ketogenic diet in a patient with glucose transporter 1 deficiency syndrome (a case report). Zh Nevrol Psikhiatr In: Korsakova SS (ed) 115 (5 Part 2 Children's Neurology And Psychiatry):53–60

Magee BA, Potezny N, Rofe AM, Conyers RA (1979) The inhibition of malignant cell growth by ketone bodies. Aust J Exp Biol Med Sci 57(5):529–539

Mansara P, Ketkar M, Deshpande R, Chaudhary A, Shinde K, Kaul-Ghanekar R (2015) Improved antioxidant status by omega-3 fatty acid supplementation in breast cancer patients undergoing chemotherapy: a case series. J Med Case Rep 9:148. doi:10.1186/s13256-015-0619-3

Martinon F, Gaide O, Petrilli V, Mayor A, Tschopp J (2007) NALP inflammasomes: a central role in innate immunity. Semin Immunopathol 29(3):213–229. doi:10.1007/s00281-007-0079-y

Maurer GD, Brucker DP, Bahr O, Harter PN, Hattingen E, Walenta S, Mueller-Klieser W, Steinbach JP, Rieger J (2011) Differential utilization of ketone bodies by neurons and glioma cell lines: a rationale for ketogenic diet as experimental glioma therapy. BMC Cancer 11:315. doi:10.1186/1471-2407-11-315

Mavropoulos JC, Yancy WS, Hepburn J, Westman EC (2005) The effects of a low-carbohydrate, ketogenic diet on the polycystic ovary syndrome: a pilot study. Nutr Metab (Lond) 2:35. doi:10.1186/1743-7075-2-35

Mavropoulos JC, Buschemeyer WC 3rd, Tewari AK, Rokhfeld D, Pollak M, Zhao Y, Febbo PG, Cohen P, Hwang D, Devi G, Demark-Wahnefried W, Westman EC, Peterson BL, Pizzo SV, Freedland SJ (2009) The effects of varying dietary carbohydrate and fat content on survival in a murine LNCaP prostate cancer xenograft model. Cancer Prev Res (Phila) 2(6):557–565. doi:10.1158/1940-6207.capr-08-0188

McDaniel SS, Rensing NR, Thio LL, Yamada KA, Wong M (2011) The ketogenic diet inhibits the mammalian target of rapamycin (mTOR) pathway. Epilepsia 52(3):7–11. doi:10.1111/j.1528-1167.2011.02981.x

McGettrick AF, O'Neill LA (2013) How metabolism generates signals during innate immunity and inflammation. J Biol Chem 288(32):22893–22898. doi:10.1074/jbc.R113.486464

Metsios GS, Stavropoulos-Kalinoglou A, Kitas GD (2015) The role of exercise in the management of rheumatoid arthritis. Expert Rev Clin Immunol:1–10. doi:10.1586/1744666x.2015.1067606

Mocellin MC, Camargo CQ, Nunes EA, Fiates GM, Trindade EB (2015) A systematic review and meta-analysis of the n-3 polyunsaturated fatty acids effects on inflammatory markers in colorectal cancer. Clin Nutr. doi:10.1016/j.clnu.2015.04.013

Moreno B, Bellido D, Sajoux I, Goday A, Saavedra D, Crujeiras AB, Casanueva FF (2014) Comparison of a very low-calorie-ketogenic diet with a standard low-calorie diet in the treatment of obesity. Endocrine 47(3):793–805. doi:10.1007/s12020-014-0192-3

Murphy P, Likhodii S, Nylen K, Burnham WM (2004) The antidepressant properties of the ketogenic diet. Biol Psychiatry 56(12):981–983. doi:10.1016/j.biopsych.2004.09.019

Nazarewicz RR, Ziolkowski W, Vaccaro PS, Ghafourifar P (2007) Effect of short-term ketogenic diet on redox status of human blood. Rejuvenation Res 10(4):435–440. doi:10.1089/rej.2007.0540

Nebeling LC, Miraldi F, Shurin SB, Lerner E (1995) Effects of a ketogenic diet on tumor metabolism and nutritional status in pediatric oncology patients: two case reports. J Am Coll Nutr 14(2):202–208

Nei M, Ngo L, Sirven JI, Sperling MR (2014) Ketogenic diet in adolescents and adults with epilepsy. Seizure 23(6):439–442. doi:10.1016/j.seizure.2014.02.015

Newport MT, VanItallie TB, Kashiwaya Y, King MT, Veech RL (2015) A new way to produce hyperketonemia: use of ketone ester in a case of Alzheimer's disease. Alzheimers Dement 11 (1):99–103. doi:10.1016/j.jalz.2014.01.006

Onodera Y, Nam JM, Bissell MJ (2014) Increased sugar uptake promotes oncogenesis via EPAC/RAP1 and O-GlcNAc pathways. J Clin Invest 124(1):367–384. doi:10.1172/jci63146

Otto C, Kaemmerer U, Illert B, Muehling B, Pfetzer N, Wittig R, Voelker HU, Thiede A, Coy JF (2008) Growth of human gastric cancer cells in nude mice is delayed by a ketogenic diet supplemented with omega-3 fatty acids and medium-chain triglycerides. BMC Cancer 8:122. doi:10.1186/1471-2407-8-122

Poff AM, Ward N, Seyfried TN, Arnold P, D'Agostino DP (2015) Non-toxic metabolic management of metastatic cancer in vm mice: novel combination of ketogenic diet, ketone supplementation, and hyperbaric oxygen therapy. PLoS ONE 10(6):0127407. doi:10.1371/journal.pone.0127407

Raiten DJ, Sakr Ashour FA, Ross AC, Meydani SN, Dawson HD, Stephensen CB, Brabin BJ, Suchdev PS, van Ommen B (2015) Inflammation and nutritional science for programs/policies and interpretation of research evidence (INSPIRE). J Nutr 145(5):1039S–1108S. doi:10.3945/jn.114.194571

Rhyu HS, Cho SY, Roh HT (2014) The effects of ketogenic diet on oxidative stress and antioxidative capacity markers of Taekwondo athletes. J Exerc Rehabil 10(6):362–366. doi:10.12965/jer.140178

Rieger J, Bahr O, Maurer GD, Hattingen E, Franz K, Brucker D, Walenta S, Kammerer U, Coy JF, Weller M, Steinbach JP (2014) ERGO: a pilot study of ketogenic diet in recurrent glioblastoma. Int J Oncol 44(6):1843–1852. doi:10.3892/ijo.2014.2382

Rose DP, Connolly JM, Rayburn J, Coleman M (1995) Influence of diets containing eicosapentaenoic or docosahexaenoic acid on growth and metastasis of breast cancer cells in nude mice. J Natl Cancer Inst 87(8):587–592

Roy S, Brasky TM, Belury MA, Krishnan S, Cole RM, Marian C, Yee LD, Llanos AA, Freudenheim JL, Shields PG (2015) Associations of erythrocyte omega-3 fatty acids with biomarkers of omega-3 fatty acids and inflammation in breast tissue. Int J Cancer. doi:10.1002/ijc.29675

Ruano G, Windemuth A, Kocherla M, Holford T, Fernandez ML, Forsythe CE, Wood RJ, Kraemer WJ, Volek JS (2006) Physiogenomic analysis of weight loss induced by dietary carbohydrate restriction. Nutr Metab (Lond) 3:20. doi:10.1186/1743-7075-3-20

Sacchi R, Paduano A, Savarese M, Vitaglione P, Fogliano V (2014) Extra virgin olive oil: from composition to molecular gastronomy. Cancer Treat Res 159:325–338. doi:10.1007/978-3-642-38007-5_19

Schmidt M, Pfetzer N, Schwab M, Strauss I, Kammerer U (2011) Effects of a ketogenic diet on the quality of life in 16 patients with advanced cancer: A pilot trial. Nutr Metab (Lond) 8(1):54. doi:10.1186/1743-7075-8-54

Sehdev A, O'Neil BH (2015) The role of aspirin, vitamin d, exercise, diet, statins, and metformin in the prevention and treatment of colorectal cancer. Curr Treat Options Oncol 16(9):359. doi:10.1007/s11864-015-0359-z

Selter JH, Turner Z, Doerrer SC, Kossoff EH (2015) Dietary and medication adjustments to improve seizure control in patients treated with the ketogenic diet. J Child Neurol 30(1):53–57. doi:10.1177/0883073814535498

Seyfried TN, Flores R, Poff AM, D'Agostino DP, Mukherjee P (2014) Metabolic therapy: A new paradigm for managing malignant brain cancer. Cancer Lett. doi:10.1016/j.canlet.2014.07.015

Sijben JW, Calder PC (2007) Differential immunomodulation with long-chain n-3 PUFA in health and chronic disease. Proc Nutr Soc 66(2):237–259. doi:10.1017/s0029665107005472

Stafford P, Abdelwahab MG, Kim DY, Preul MC, Rho JM, Scheck AC (2010) The ketogenic diet reverses gene expression patterns and reduces reactive oxygen species levels when used as an adjuvant therapy for glioma. Nutr Metab (Lond) 7:74. doi:10.1186/1743-7075-7-74

Thiem S, Pierce TP, Palmieri M, Putoczki TL, Buchert M, Preaudet A, Farid RO, Love C, Catimel B, Lei Z, Rozen S, Gopalakrishnan V, Schaper F, Hallek M, Boussioutas A, Tan P, Jarnicki A, Ernst M (2013) mTORC1 inhibition restricts inflammation-associated gastrointestinal tumorigenesis in mice. J Clin Invest. doi:10.1172/jci65086

Thies F, Miles EA, Nebe-von-Caron G, Powell JR, Hurst TL, Newsholme EA, Calder PC (2001) Influence of dietary supplementation with long-chain n-3 or n-6 polyunsaturated fatty acids on blood inflammatory cell populations and functions and on plasma soluble adhesion molecules in healthy adults. Lipids 36(11):1183–1193

Tisdale MJ, Brennan RA, Fearon KC (1987) Reduction of weight loss and tumour size in a cachexia model by a high fat diet. Br J Cancer 56(1):39–43

Varshneya K, Carico C, Ortega A, Patil CG (2015) The efficacy of ketogenic diet and associated hypoglycemia as an adjuvant therapy for high-grade gliomas: a review of the literature. Cureus 7(2):251. doi:10.7759/cureus.251

Veech RL (2004) The therapeutic implications of ketone bodies: the effects of ketone bodies in pathological conditions: ketosis, ketogenic diet, redox states, insulin resistance, and mitochondrial metabolism. Prostaglandins Leukot Essent Fatty Acids 70(3):309–319. doi:10.1016/j.plefa.2003.09.007

Wen H, Miao EA, Ting JP (2013) Mechanisms of NOD-like receptor-associated inflammasome activation. Immunity 39(3):432–441. doi:10.1016/j.immuni.2013.08.037

Westman EC, Volek JS, Feinman RD (2006) Carbohydrate restriction is effective in improving atherogenic dyslipidemia even in the absence of weight loss. Am J Clin Nutr 84 (6):1549; author reply 1550

Westman EC, Yancy WS Jr, Mavropoulos JC, Marquart M, McDuffie JR (2008) The effect of a low-carbohydrate, ketogenic diet versus a low-glycemic index diet on glycemic control in type 2 diabetes mellitus. Nutr Metab (Lond) 5:36. doi:10.1186/1743-7075-5-36

Winters-Stone K (2015) Exercise and cancer risk-how much is enough? JAMA Oncol. doi:10.1001/jamaoncol.2015.2267

Yancy WS Jr, Olsen MK, Guyton JR, Bakst RP, Westman EC (2004) A low-carbohydrate, ketogenic diet versus a low-fat diet to treat obesity and hyperlipidemia: a randomized, controlled trial. Ann Intern Med 140(10):769–777

Youm YH, Nguyen KY, Grant RW, Goldberg EL, Bodogai M, Kim D, D'Agostino D, Planavsky N, Lupfer C, Kanneganti TD, Kang S, Horvath TL, Fahmy TM, Crawford PA, Biragyn A, Alnemri E, Dixit VD (2015) The ketone metabolite beta-hydroxybutyrate blocks NLRP3 inflammasome-mediated inflammatory disease. Nat Med. doi:10.1038/nm.3804

Zhao Z, Lange DJ, Voustianiouk A, MacGrogan D, Ho L, Suh J, Humala N, Thiyagarajan M, Wang J, Pasinetti GM (2006) A ketogenic diet as a potential novel therapeutic intervention in amyotrophic lateral sclerosis. BMC Neurosci 7:29. doi:10.1186/1471-2202-7-29

Zuccoli G, Marcello N, Pisanello A, Servadei F, Vaccaro S, Mukherjee P, Seyfried TN (2010) Metabolic management of glioblastoma multiforme using standard therapy together with a restricted ketogenic diet: Case Report. Nutr Metab (Lond) 7:33. doi:10.1186/1743-7075-7-33

Fasting and Caloric Restriction in Cancer Prevention and Treatment

Sebastian Brandhorst and Valter D. Longo

Abstract

Cancer is the second leading cause of death in the USA and among the leading major diseases in the world. It is anticipated to continue to increase because of the growth of the aging population and prevalence of risk factors such as obesity, smoking, and/or poor dietary habits. Cancer treatment has remained relatively similar during the past 30 years with chemotherapy and/or radiotherapy in combination with surgery remaining the standard therapies although novel therapies are slowly replacing or complementing the standard ones. According to the American Cancer Society, the dietary recommendation for cancer patients receiving chemotherapy is to increase calorie and protein intake. In addition, there are no clear guidelines on the type of nutrition that could have a major impact on cancer incidence. Yet, various forms of reduced caloric intake such as calorie restriction (CR) or fasting demonstrate a wide range of beneficial effects able to help prevent malignancies and increase the efficacy of cancer therapies. Whereas chronic CR provides both beneficial and detrimental effects as well as major compliance challenges, periodic fasting (PF), fasting-mimicking diets (FMDs), and dietary restriction (DR) without a reduction in calories are emerging as interventions with the potential to be widely used to prevent and treat cancer. Here, we review preclinical and preliminary clinical studies on

S. Brandhorst · V.D. Longo
Department of Biological Sciences, School of Gerontology, Longevity Institute, University of Southern California, Los Angeles, CA, USA

V.D. Longo (✉)
IFOM, FIRC Institute of Molecular Oncology, Milan, Italy
e-mail: vlongo@usc.edu

© Springer International Publishing Switzerland 2016
T. Cramer and C.A. Schmitt (eds.), *Metabolism in Cancer*,
Recent Results in Cancer Research 207, DOI 10.1007/978-3-319-42118-6_12

dietary restriction and fasting and their role in inducing cellular protection and chemotherapy resistance.

Keywords

Cancer · Caloric restriction · Dietary restriction · Fasting · Chemotherapy · Stress resistance · Sensitization

1 Introduction

In 2012, an estimated 14.1 million new cases of cancer occurred and approximately 8.2 million people died of cancer worldwide (Torre et al. 2015), although the exact number can only be estimated because incidence rates and treatment modalities for some parts of the world are not fully established. In the USA, one out of every four deaths was estimated to be due to cancer in 2012 (Siegel et al. 2012). The worldwide number of deaths is projected to increase to 13.2 million by 2030 due to the expected increase in the elderly population, as well as the adoption to cancer-causing behaviors (e.g., cigarette smoking) (Brawley 2011). Notably up to 35 % of all cancer deaths worldwide have been reported to be avoidable through adjustments to lifestyle and environmental factors, such as physical activity and dietary habits (Danaei et al. 2005). However, this estimate does not take into account the more recent advances in dietary interventions to affect aging and age-related diseases, which are likely to cause a major increase in the percentage of preventable cancers (Levine et al. 2014).

Compared to the improvements in the prevention and treatment of heart disease, cancer treatment and mortality have remained relatively unchanged over the past 30 years (Jemal et al. 2008). Chemotherapy and/or radiotherapy in combination with surgical removal of the tumor mass remain the standard therapies. Chemotherapy improves the survival rates of cancer patients but causes damage to normal tissues and leads to significant side effects such as emotional distress, myelo-suppression, fatigue, vomiting, diarrhea, and even death (Love et al. 1989; Partridge et al. 2001). Despite newly designed drugs that target specific cancer markers, cytotoxic drugs will have accompanying side effects unless novel and complimentary interventions or strategies are adopted. In the following sections, we discuss preclinical and preliminary clinical results on how specific forms of dietary restriction (DR) and periodic fasting (PF; 24 h or more) can have significant impact on the prevention and treatment of cancer while minimizing the adverse effects and limitations associated with chronic calorie restriction (CR).

The effects of CR on cancer prevention were first described more than 100 years ago (Moreschi 1909) and was followed by a large body of work demonstrating its tumor preventive effects in various animal models. However, recent studies in both mice and monkeys indicate that the effects of CR on longevity are either limited or absent and, in mice, CR can even reduce longevity (Colman et al. 2009; Mattison et al. 2012; Goodrick et al. 1990; Liao et al. 2010). The role of CR in cancer treatment is even more problematic considering that it would be extremely difficult to chronically restrict calorie intake in cancer patients. PF instead has been demonstrated to be effective in reducing pre-neoplastic lesions (Grasl-Kraupp et al. 1994) and in both cancer prevention and treatment, particularly in combination with cytotoxic drugs (Grasl-Kraupp et al. 1994; Brandhorst et al. 2015; Lee et al. 2012). PF protects normal cells/tissues from chemotherapy-induced toxicity (Lee et al. 2010; Raffaghello et al. 2008) while simultaneously increasing its therapeutic efficacy on a wide variety of malignant cells (Lee et al. 2012; Safdie et al. 2012). Newly designed dietary compositions aimed at inducing fasting-like effects that allow nourishment are slowly emerging as potential therapies to delay aging-associated diseases such as cancer (Brandhorst et al. 2015). These fasting-mimicking diets (FMD) and PF also promote multisystem regeneration and rejuvenation by promoting the self-renewal of stem cells and the generation of white blood cells (Brandhorst et al. 2015; Cheng et al. 2014). Thus, novel and periodic forms of extreme DR as well as targeted reductions in specific macronutrients, and a combination of both, are likely to replace the original balanced and chronic restriction of all calorie sources.

2 Conserved Role of Nutrient-Sensing Signaling Pathways in Life span and Stress Resistance

Fluctuations in food availability force most organisms into periods of drastically reduced caloric intake or even starvation. Whereas a sufficient nutrient supply enables growth and proliferation, periods of very low food availability, or the lack of specific macromolecules, activate alternative metabolic modes allowing organisms to remain protected against damage that could negatively affect fitness (Madia et al. 2009; Harrison and Archer 1989). High cellular growth rates and high stress protection rarely coexist, indicating that cells invest energy in either one or the other. In fact, when nutrients are abundant, the activation of nutrient-sensing signaling cascades promotes cellular growth while when they are scarce the down-regulation of these signaling pathways blocks cellular proliferation and activates stress resistance transcription factors which negatively regulate these pro-aging pathways (Fontana et al. 2010; Longo 1999; Guarente and Kenyon 2000; Kenyon 2001; Longo and Finch 2003). Over the last 20 years, diet-based approaches have been combined with genetic approaches to identify the genes and pathways that mediate nutrient-dependent effects on longevity and health span.

The most studied form of diet-based antiaging interventions is CR, which commonly describes a 20–40 % reduction in calorie intake. DR instead can also refer to restrictions of particular macromolecular components of the diet, without affecting daily calorie intake (e.g., low protein compensated with high fat). Alternate-day fasting (ADF), also often described as intermittent fasting (IF), is another well-studied and beneficial form of DR discussed in detail elsewhere. Finally, PF is distinct from ADF/IF as meal frequency is interrupted and caloric intake drastically reduced for periods generally ranging from 48 h to 3 weeks. In addition to CR, some forms of IF and PF have been shown to promote stress resistance and longevity in model organisms ranging from unicellular yeast to mammals, indicating that the molecular mechanisms responsible for the protective effects of CR have likely evolved billions of years ago and are partially conserved in many species (Fontana et al. 2010; Longo and Mattson 2014).

In the prokaryote *E. coli*, lack of glucose or nitrogen (comparable to protein restriction in mammals) increases the resistance to high levels of hydrogen peroxide (Jenkins et al. 1988). The complete lack of nutrients also extends the longevity of bacteria, which can be reversed by adding to the medium various nutrients except for acetate, a carbon source associated with starvation conditions (Gonidakis et al. 2010). In the yeast *S. cerevisiae*, cells switched from standard growth medium to water display a twofold increase in chronological life span and a major increase in resistance against oxidative insults and heat stress (Lee et al. 2012; Fabrizio and Longo 2003). Reducing the concentration of glucose from 2 to 0.5 % in the growth media can have similar, although less efficient effects on stress resistance (Longo et al. 1997; Wei et al. 2008). PF in yeast, which is implemented by switching yeast cells back and forth from nutrient-rich medium to water every 48 h, extends both medium and maximum longevity and increases the number of yeast cells surviving hydrogen peroxide treatment by more than 100-fold (Brandhorst et al. 2015). Although the precise mechanisms of DR-dependent life span extension in yeast are not yet fully understood, they include the down-regulation of the amino acid response Tor-S6K (Sch9) pathway and of the glucose-responsive Ras/adenylate cyclase(AC)/PKA pathway, and the activation of the serine threonine kinase Rim15 which increases stress resistance against oxidants, genotoxins, and heat shock through genes including superoxide dismutases and heat-shock proteins controlled by the transcription factors Msn2, Msn4, and Gis1 (Madia et al. 2009; Longo and Finch 2003; Fabrizio et al. 2003; Fabrizio et al. 2001). Deletion of these transcription factors reverses the protective effects of glucose restriction, demonstrating their requirement for maximum protection (Wei et al. 2008). Surprisingly, the deletion of Rim15, or of its downstream stress response transcription factors Msn2/4 and Gis1, does not prevent the life span effects of PF (Brandhorst et al. 2015). These results indicate that PF can protect simple organisms from toxins and aging by mechanisms that are in part independent of conserved pro-longevity transcription factors.

The protective effects of food restriction/starvation on the unicellular *E. coli* and *S. cerevisiae* are also observed in multicellular organisms. In the fruit fly *D. melanogaster*, dilution or reduction of food have been shown to extend life span, although intermittent food deprivation does not (Grandison et al. 2009; Piper and Partridge 2007). The fasting-induced protection against oxidative stress in flies is mediated by the repression of translation (consistent with energy diversion from cellular growth to protection) through increased expression of d4E-BP downstream of the PI3K/Akt/dFOXO3 pathway (Tettweiler et al. 2005; Villa-Cuesta et al. 2010). Of note is that moderate DR does not protect flies against the oxidative damage caused by re-oxygenation injury whereas a more stringent DR does (Vigne and Frelin 2007). Mutations in the insulin receptor substrate *chico* extend life span in *D. melanogaster* although its role in stress resistance remains unclear (Clancy et al. 2001; Giannakou and Partridge 2007). In the nematode *C. elegans*, CR and fasting, achieved by feeding either reduced amounts or no bacteria, also increase life span (Smith et al. 2008). Life span extension in *C. elegans* requires AMPK, a regulator of cellular glucose uptake and β-oxidation of fatty acids, as well as the stress resistance transcription factor DAF-16, analogous to Msn2/4 and Gis1 in yeast and FOXOs in *D. melanogaster* and mammals (Greer et al. 2007). Fasting every other day also increases oxidative stress resistance in *C. elegans* and increases life span by up to 56 % via modulation of the RHEB-1 and TOR signaling pathway, which are linked to DAF-16 (Weinkove et al. 2006; Honjoh et al. 2009). Conversely, excessive glucose shortens the life span of *C. elegans*, in part by decreasing DAF-16 activity (Lee et al. 2009). In adult *C. elegans*, mutations in age-1 (PI3K homolog) and daf-2 (insulin/IGF-1 receptor homolog) result in a 65–100 % life span extension by decreasing AKT-1/AKT-2 signaling, and by activating DAF-16, which is also associated with resistance to oxidative and ER stress (Johnson 1990; Paradis et al. 1999; Hsu et al. 2003; Henis-Korenblit et al. 2010).

Orthologues of the genes regulating life span and stress resistance in *S. cerevisiae* and *C. elegans* downstream of the growth hormone (GH)/insulin-like growth factor 1 (IGF-1) axis regulate stress resistance and/or life span in mice (Coschigano et al. 2003; Holzenberger et al. 2003; Bonkowski et al. 2009; Selman et al. 2009; Brown-Borg 2009). In laboratory mice, mutations in the insulin-GH/IGF-1 axis increase life span by up to 150 % and conversely, mice that overexpress GH have a shortened life span (Brown-Borg et al. 1996; Murakami 2006; Bartke et al. 2002). In agreement with the findings in lower eukaryotes, cultured cells derived from long-lived mice with deficiencies in the GH/IGF-1 axis have higher resistance against H_2O_2-induced oxidative stress, UV, genotoxins, as well as heat- and cadmium-induced stress (Murakami 2006; Salmon et al. 2005). Vice versa, activities of the antioxidant enzymes superoxide dismutases and catalase decrease in murine hepatocytes exposed to GH or IGF-1 and in transgenic mice overexpressing GH (Brown-Borg and Rakoczy 2000; Brown-Borg et al. 2002). IGF-1 sensitizes primary neurons to oxidative stress through Ras/Erk-dependent mechanisms (Li et al. 2008), and experiments in rat primary glia and mouse fibroblasts suggest that IGF-1 exposure sensitizes these cell types against oxidative damage and chemotherapeutic drugs (Lee et al. 2010). In addition, IGF-1 attenuates the cellular stress response and the expression of stress response proteins HSP72 and heme-oxygenase in rats (Sharma et al. 2000).

In mammals, various forms of partial or complete food deprivation have been investigated that range from daily 20–40 % CR, intermittent fasting (IF, including alternate-day fasting, ADF), and PF. The multisystemic benefits of CR on rodent life span have been extensively studied and are beyond the scope of this review. Generally, CR increases rodent life span by up to 60 % and delays the occurrence of many chronic diseases including cancer and improves stress resistance (Koubova and Guarente 2003; Mattson 2005; Mattson and Wan 2005). However, there is variability in the effects of CR and some genetic backgrounds do not experience a life span extension and can even display reduced longevity (Liao et al. 2010; Sohal and Forster 2014), likely because of trade-offs between effects on aging and effects on systems which may benefit from a higher calorie intake and/or fat level (Rikke et al. 2010). In mice, the longevity effects of CR involve reduced activity of the GHR/IGF-1 pathways since CR does not further extend the life span of GH signaling-deficient mice (Bonkowski et al. 2009; Arum et al. 2009). CR is associated with reduced oxidative stress and cellular proliferation (Youngman 1993; Sohal and Weindruch 1996) while enhancing DNA repair processes and autophagy (Weraarchakul et al. 1989; Cuervo et al. 2005; Wohlgemuth et al. 2007). CR effectively reduces the levels of plasma insulin, cholesterol, triglycerides, growth factors such as IGF-1, and inflammatory cytokines (Mahoney et al. 2006; Matsuzaki et al. 2001). CR also elevates plasma high-density lipoprotein levels, resulting in a reduced risk for atherosclerosis, diabetes, and obesity (Paoletti et al. 2006; Larson-Meyer et al. 2008). In a mouse model of DR-induced stress resistance, restriction of sulfur amino acids increased hydrogen sulfide production and the protection from oxidative damage induced by hepatic reperfusion injury (Hine et al. 2015). Of note, a severe restriction of dietary protein can extend the life span of rodents by up to 20 %, independently of the caloric intake (Pamplona and Barja 2006). Reduced levels of serum IGF-1 in rats and mice fed with protein-restricted diets might explain the beneficial effects on longevity (Sonntag et al. 1999; Brandhorst et al. 2013). Reducing protein intake and IGF-1 signaling also significantly reduces the incidence and progression of breast cancer and melanoma in murine models of tumor progression (Levine et al. 2014).

In rodents, the most studied fasting method has been IF which promotes protection against multiple diseases and causes a major reduction in the incidence of lymphomas (Goodrick et al. 1990; Mattson and Wan 2005; Descamps et al. 2005; Mattson 2012). The effects of alternate-day fasting on longevity in rodents depend on the species and age at diet initiation and can range from negative effects to as much as an 80 % life span extension (Goodrick et al. 1990; Arum et al. 2009; Goodrick et al. 1982). Similar to CR, IF increases the survivorship and improves insulin sensitivity of male wild-type mice, but fails to affect either parameter in GHR-KO mice (Arum et al. 2009). The major differences between IF and PF in mice are the duration and frequency of the fast. IF cycles usually last 24 h and are separated by one day of normal food intake (alternate-day fasting), whereas PF cycles last 2 or more days and are at least 1 week apart to allow for regaining of normal weight (Longo and Mattson 2014). Additional differences include the molecular changes caused by these fasting regimes on a variety of growth factors and metabolic markers, since IF causes more frequent but less pronounced changes than PF. Although less studied than IF, PF has been shown to have potent effects on cellular protection. PF for 48–60 h prior to etoposide induced lethal oxidative stress causes a significant increase in survival in three different mouse strains (Raffaghello et al. 2008). Similarly, 72 h of food deprivation before exposure to lethal doses of the chemotherapeutic drug doxorubicin, which causes oxidative stress-induced cardiotoxicity, protects CD-1 mice (Lee et al. 2010). Following ischemic reperfusion, which is associated with the production of reactive oxygen species in the affected tissue, fasting demonstrates protective effects in the mouse kidney and liver, rat brain as well as human liver against reperfusion injury (Mitchell et al. 2009; van Ginhoven et al. 2009; Verweij et al. 2011). Traumatic brain injury followed by fasting has neuroprotective effects, reduces oxidative damage, and improves cognitive function (Davis et al. 2008).

Despite its advantages, the extreme nature of prolonged water-only fasting could cause adverse effects, such as the exacerbation of previous malnourishments and dysfunctions. These concerns are now being addressed through the implementation of newly designed dietary interventions aimed to induce PF-like effects while minimizing the risk of adverse effects and the burden of complete food restriction. In C57Bl/6 mice, a 4-day-long diet that mimics the effects of fasting (fasting-mimicking diet, FMD) has been shown to lower blood glucose levels by 40 % and IGF-1 by 45 %, while increasing ketone bodies ninefold and IGFBP-1, which inhibits IGF-1, eightfold by the end of the FMD. The FMD extends health span and longevity, promotes hippocampal neurogenesis, lowers visceral fat, reduces skin lesions, rejuvenates the immune system, and retards bone mineral density loss in old mice (Brandhorst et al. 2015). The FMD started at middle age reduces tumor incidence, delays cancer onset, and causes a major reduction in the number of lesions, which may reflect a general switch from malignant to benign tumors.

Together, results from the range of studies and organisms described above indicate that enhanced stress resistance is a highly conserved phenotype of starved and long-lived organisms, which is in part mediated by reduced signaling through the GH and IGF-1 axis.

3 The Physiological Response to Fasting in Mammals

Mammals undergo distinct metabolic changes when deprived of food that are different from that of CR (Wang et al. 2006; Lee and Longo 2011). The post-absorptive phase, lasting for 10 or more hours after food ingestion involves the utilization of glycogen as the main stored energy source. Following the depletion of the hepatic glycogen storage, amino acids serve as gluconeogenic substrates while the brain consumes most of the remaining glucose. After about 72 h of fasting, serum glucose levels can reach 50–60 mg/dL in a healthy person, but return to normal levels within 30 min after oral administration of 100 g glucose (Unger et al. 1963). Glycerol and fatty acids are released from the adipose tissue and as a result, acetoacetate, β-hydroxybutyrate and acetone become the main carbon sources within a few days of fasting. During the last phase of prolonged food deprivation, the fat storage is eventually exhausted and muscles are being degraded to allow gluconeogenesis. In rats, this occurs after 4–5 days but does not result in a major increase in glucose levels (Wang et al. 2006). For humans, it is estimated that a 70-kg person can obtain basal caloric requirements from fat reserves for up to 2–3 months of fasting (Cahill and Owen 1968; Cahill et al. 1968), leading to an initially rapid weight loss that subsequently tapers off. During the first week of fasting, an average of 0.9 kg/day is lost, followed by a 0.3-kg/day weight loss by the third week of fasting, resulting in an approximately 20 % body weight loss after 30–35 days of fasting (Kerndt et al. 1982). Thus, PF is generally feasible and tolerable, but may be accompanied by side effects, such as headaches, light-headedness, nausea, weakness, edema, anemia, and amenorrhea (Bloom 1959; Runcie and Thomson 1970; Drenick and Smith 1964).

In accordance with the systemic effects of fasting, a plethora of extrinsic and intrinsic growth factors is affected, including mediators of evolutionarily conserved pathways that promote stress sensitivity and aging, such as IGF-1 (Fontana et al. 2010; Longo and Mattson 2014). IGF-1 levels decrease by up to 65 % following 120 h of PF in humans despite increased GH secretion (Ketelslegers et al. 1995; Underwood et al. 1994; Moller and Jorgensen 2009; Maccario et al. 2001) due to the increase in IGFBP-1, which decreases IGF-1 bioavailability and prevents the feedback inhibition of GH secretion by IGF-1 (Zapf et al. 1995). In humans, GH levels eventually decrease after 3–10 days of fasting and level off (Merimee and Fineberg 1974; Palmblad et al. 1977). In mice, PF (24–72 h) decreases IGF-1 level by 70 % and is associated with a 11-fold increase in IGFBP-1 (Lee et al. 2010; Tannenbaum et al. 1979; Frystyk et al. 1999), reduced protein synthesis, reduced AKT activity, reduced mTOR/S6K, increased 4E-BP1 activity, and increased FOXO-1, FOXO-3, and FOXO-4 (Sans et al. 2004; Imae et al. 2003).

Starving mice or mammalian cells triggers autophagy in various tissues, which is regulated by several genes that also regulate aging and stress resistance, such as AMPK, mTOR, and sirtuins (Kroemer et al. 2010; Marino et al. 2010; Morselli et al. 2009). Oncogenic signaling, e.g., through PI3K and Akt, has been shown to inhibit autophagy, whereas tumor suppressors such as PTEN and TSC2 can trigger autophagy (Morselli et al. 2009). Autophagy is elevated in many cancer cells and increases the resistance of cancer cells to chemotherapy (Morselli et al. 2009). Increased autophagy has been hypothesized to be mediated by ammonia as a by-product of glutaminolysis occurring in the mitochondria of malignant cells undergoing the Warburg effect to supply biosynthetic precursors required for their high proliferation rate (Vander Heiden et al. 2009). Therefore, autophagy triggered by fasting may be beneficial to normal cells but detrimental to malignant cells.

4 DR, Fasting and Cancer Treatment

Tumors arise over time from the combination of DNA damage and mutations associated with changes in the environmental niche surrounding precancerous cells (Hanahan and Weinberg 2011). Not surprisingly, aging is the major risk factor for the development of many cancer types (DePinho 2000; Campisi et al. 2011). Studies on caloric intake that delay organismal aging emphasize the preventive role of CR in cancer establishment. Over a century ago, the first direct relationship between chronic CR and the prevention of tumor transplantation in mice was demonstrated (Moreschi 1909). Subsequently, a multitude of studies has established that CR reduces the progression of spontaneous or induced tumors in various animal models (Tannenbaum 1996; Tannenbaum 1940; Berrigan et al. 2002;

Hursting et al. 1994). Further, CR has also been shown to reduce the cancer incidence in rodents. Roy Walford and Richard Weindruch reported that CR beginning either at the time of weaning or at 12 months of age increased life span, delayed immunologic aging in laboratory mice, and reduced spontaneous cancer incidence by more than 50 % (Weindruch and Walford 1982; Weindruch et al. 1986). In a multitude of auto- and xenograft tumor models, CR causes potent antigrowth effects, although some cell lines remain resistant. Resistance to CR has been associated with mutations that cause the constitutive activation of the phosphatidylinositol-3-kinase (PI3K) pathway; substitution of the mutant allele of PI3K with wild-type PI3K reestablishes a DR-sensitive cancer cell (Kalaany and Sabatini 2009).

Support for the potential application of CR in humans arose from a 20-year longitudinal adult-onset 30 % CR study in rhesus monkeys which decreased cancer incidence by 50 % and reduced mortality (Colman et al. 2009). These findings, however, have to be analyzed carefully as a CR regimen implemented in young and older age rhesus monkeys at the National Institute on Aging (NIA) has not improved survival outcomes (Mattison et al. 2012). Whether or not CR has the potential to reduce cancer incidence in humans remains largely unclear, although CR can reduce clinical markers associated with cancer if it also involves protein restriction (Longo and Fontana 2010). Despite data indicating benefits in cancer prevention at least in certain mouse genetic backgrounds, the use of CR in therapy is problematic in part because it would be extremely difficult to chronically restrict cancer patients even if a benefit was demonstrated. Further, chronic CR merely delays the progression of the disease (Mukherjee et al. 2004; Bonorden et al. 2009), and its effectiveness might be reduced to a subset of CR-responsive cancers (Kalaany and Sabatini 2009). In addition, in mice chronic CR is associated with weight loss, delayed wound healing, and impaired immune function, all of which may impose a significant risk to cancer patients receiving chemotherapy, surgery or immunity-based treatments, or who are at risk for losing weight and becoming frail and cachectic (Fontana et al. 2010; Kristan 2008; Reed et al. 1996; Kim and Demetri 1996).

Data suggest that PF (mice for 48–72 h, humans 4–5 days) may induce metabolic changes which delay the growth of damaged cells, without the negative side effects and limitations associated with chronic CR. In rats, fasting for 8 days or 40 % CR for 3 months, reduces pre-neoplastic liver masses by 20–30 %. Putative pre-neoplastic liver foci have significantly lowered DNA replications but increased apoptosis during PF, resulting in the reduced number and volume of putative pre-neoplastic liver foci by 85 % throughout the following 17 months (Grasl-Kraupp et al. 1994). However, in contrast to the protective effect of chronic CR on carcinogenesis (Tagliaferro et al. 1996), fasting followed by refeeding sustained tumor initiation in the rat liver enhances the growth of aberrant crypt foci in the colorectal mucosa and mammary tumors even by otherwise non-initiating carcinogen doses (Premoselli et al. 1998; Sesca et al. 1998; Tessitore et al. 1999). Given that the refeeding phase is associated with increased cellular proliferation in the liver and colorectal epithelium (Brandhorst et al. 2015; Cuervo et al. 2005), exposure to carcinogens during this period might explain these effects. Therefore, it is of importance to consider that normal food intake following PF should be initiated once the half-life of potential carcinogens indicates its expiration. Despite these caveats, PF has been shown to have potent protective effects in cancer treatment, in part due its impact on host metabolism. PF decreases blood glucose in mice by up to 75 % whereas long-term CR or IF cause a 15 % reduction (Longo and Mattson 2014; Lee and Longo 2011); this seems of particular importance given the glucose dependent phenotype of many cancer cells (Vander Heiden et al. 2009; Warburg 1956). PF reduces the pro-proliferative growth factor IGF-1 by up to 75 % (Lee et al. 2010; Underwood et al. 1994), while chronic CR causes a 25 % IGF-1 reduction in mice (Barger et al. 2008) and has no impact on IGF-1 levels in humans unless combined with protein restriction (Fontana et al. 2008). Furthermore, chronic DR with protein restriction only causes a 30 % reduction in IGF-1 for humans (Fontana et al. 2008).

5 Differential Stress Resistance by Fasting

As described above, PF inactivates partially conserved pro-proliferative pathways and increases resistance to oxidative and other stresses in lower eukaryotes, whereas the constitutive activation of analogous pathways play central roles in promoting cancer. Ras and Akt function in signal transduction pathways that are frequently found in a constitutively activated form in cancer cells (Kinzler and Vogelstein 1996; Hanahan and Weinberg 2000; Medema and Bos 1993). The connection between (de-regulated) cellular proliferation and stress resistance provides the theoretical foundation for a fasting-induced differential protection of normal cells from chemotoxicity (differential stress resistance, DSR). DSR is based on the hypothesis that in response to fasting, normal cells enter a stress-resistant state (Raffaghello et al. 2008; Longo et al. 1997) characterized by reduced/lack of cell divisions and the utilization of metabolites generated from the breakdown of fats, proteins and organelles. Self-sufficiency in growth signals enabled by gain-of-function mutations in oncogenes (e.g., Ras, Akt, mTOR) grants constitutive activation of proliferation pathways and together with loss-of-function mutations in tumor-suppressor genes (e.g., Rb, p53, PTEN) allows cancer cells to disregard antiproliferative signals such as those occurring during fasting, thereby prohibiting malignant cells to enter a nondividing and protected state (Hanahan and Weinberg 2000). The inability of cancer cells to respond to antigrowth signals or to grow in the absence of growth factors, a well-established hallmark of cancer, prevents the protective effect of fasting regardless of the type of cancer or oncogene mutations (Lee and Longo 2011). DSR therefore presents a method to preferentially target malignant cells with chemotherapy and other drugs (Raffaghello et al. 2008) while increasing the resistance against cytotoxic stressors through cell detoxification systems in normal dividing and/or nondividing cells. In fact, fasting induces the protection against lethal doses of the chemotherapeutic drug etoposide in A/J, CD-1, and athymic nude mice (Raffaghello et al. 2008). In a different set of experiments, CD-1 mice were protected from high-dose doxorubicin, known to cause cardiotoxicity (Lee et al. 2010). Analogously, transgenic mice with a conditional liver-specific IGF-1 gene deletion (LID), which results in a 70–80 % reduction in circulating IGF-1 levels, show enhanced protection to cytotoxic chemotherapy drugs including cyclophosphamide, doxorubicin, and 5-FU, although they are not protected against the topoisomerase inhibitor etoposide (Lee et al. 2010). Studies in mice demonstrate that restoring IGF-1 to normal levels during fasting reverses the protection against lethal doses of doxorubicin (Lee et al. 2010). Similarly, three days of fasting protects FabplCre;Apc$^{15lox/+}$ mice, which spontaneously develop intestinal tumors, against the side effects of a high dose of irinotecan whereas ad libitum-fed mice experience weight loss, reduced activity, ruffled coat, hunched-back posture, diarrhea, and leukopenia (Huisman et al. 2015). Fasting reduces the delayed-type chemotherapy-induced nausea and vomiting in cancer-bearing dogs receiving doxorubicin (Withers et al. 2014). The protective effect of fasting can promote potent changes even to the stem cell population: PF represses the immunosuppression and mortality caused by cyclophosphamide through signal transduction changes in long-term hematopoietic stem cells and niche cells that promote stress resistance, self-renewal, and lineage-balanced regeneration (Cheng et al. 2014), in agreement with preliminary data on the protection of lymphocytes from chemotoxicity in fasting patients (Safdie et al. 2009). Fasting has been shown to also promote regenerative effects in both the blood and other systems including the nervous system, muscle and liver indicating that, in addition to protecting against the toxicity of cancer drugs, it can stimulate the generation of healthy cells and tissue damaged by the therapy (Brandhorst et al. 2015; Cheng et al. 2014).

6 Differential Stress Sensitization of Cancer Cells by Fasting

Malignant cells are generated in a high nourishment environment and, therefore, thrive in environments resembling those in which they have evolved. Cancer cells generally prefer high levels of glucose to rely on glycolysis more than on oxidative phosphorylation (Warburg effect), which provides energy and biosynthetic precursors essential for proliferation (Vander Heiden et al. 2009). Glycolysis allows the metabolism of glucose-6-phosphate in the pentose shunt pathway, which provides substrates essential for nucleotide synthesis and NADPH production. The Warburg effect may restrict cytochrome c-mediated apoptosis, thereby favoring tumor cell survival and apoptosis evasion through decreased respiration (Vaughn and Deshmukh 2008; Ruckenstuhl et al. 2009). However, glucose alone does not supply all the building blocks required for cellular proliferation and cancer cells therefore depend on amino acids, particularly glutamine, as a nitrogen source (Vander Heiden et al. 2009). Glutamine, the circulating amino acid with the highest concentration in humans, plays a crucial part in the uptake of essential amino acids and can support NADPH production, making it necessary for lipid and nucleotide biosynthesis (Nicklin et al. 2009). Similarly to the effects of CR, protein/amino acid restriction have been demonstrated to increase longevity and to delay the onset of many aging-related diseases, including cancer (Mirzaei et al. 2014). The major reduction in circulating glucose and amino acids during fasting is a significant disadvantage to tumor cells that usually experience an almost unlimited supply through the bloodstream.

The same mutations that cause tumor cells to remain locked in a pro-growth mode render them also sensitive to alterations in the cellular environment. In evolutionary biology, it is well established that most acquired mutations become disadvantageous. The mutations generated in cancer cells are also mostly deleterious although such negative effects may not be observable until the cancer cells are in an environment that exposes that detrimental effect. For example, a mutation that requires high levels of glutamine would not be problematic to cancer cells until the concentration of glutamine becomes limited. In *S. cerevisiae*, expression of the oncogene-like RAS2val19 not only reverses the (water-) starvation induced protection against hydrogen peroxide and menadione, the constitutive activation of Ras sensitizes yeast cells compared to wild-type cells, a phenomenon called differential stress sensitization (DSS) (Lee et al. 2012).

In vitro models of fasting (by reducing glucose and/or serum availability in the growth medium) sensitize murine, rat and human glioblastoma multiforme cells, but not primary mixed glia, to temozolomide chemotherapy (Safdie et al. 2012). Similarly, fasting sensitizes 15 of 17 human and rodent cancer cell lines to the chemotherapeutic agents' doxorubicin and/or cyclophosphamide (Lee et al. 2012), while serum starvation alone is sufficient to induce sensitization to cisplatin in human mesothelioma and lung carcinoma cells (Shi et al. 2012). In a metastatic mouse neuroblastoma model, fasting for 48 h followed by a single administration of high-dose chemotherapy, but not either treatment alone, successfully improves survival limited by both drug toxicity and metastases and results in long-term cancer-free survival (Lee et al. 2012). In subcutaneous mouse models of melanoma and breast cancer, fasting cycles are as effective as chemotherapy alone while the combination of both treatment modalities significantly improves treatment efficacy (Lee et al. 2012). Similarly, fasting prior to gemcitabine injection significant decreases the progression of pancreatic cancer tumors by more than 40 %, in part through increased gemcitabine uptake of malignant cells (D'Aronzo et al. 2015). PF in combination with cisplatin for three weeks reduces the progression of mesothelioma by more than 60 % compared to the untreated control and a complete remission was observed in 60 % of the combination-treated mice (Shi et al. 2012). 48 h of PF also sensitizes both subcutaneous and intracranial glioma models to radio- and chemotherapy (Safdie et al. 2012), an effect not mimicked by dietary protein restriction alone (Brandhorst et al. 2013). 48 h of starvation, with or without oxaliplatin, reduces the progression of CT26 colorectal tumors through the down-regulation of aerobic glycolysis and glutaminolysis, while increasing oxidative phosphorylation in complex I and II of the mitochondrial electron transport chain, thereby resulting in reduced ATP production, increased oxidative stress, and apoptosis (Bianchi et al. 2015). In murine 4T1 breast cancer cells, PF increases the phosphorylation of the stress-sensitizing Akt and S6 kinases and is associated with increased oxidative stress, caspase-3 cleavage, DNA damage, and subsequent apoptosis (Lee et al. 2012), whereas the PF-induced activation of the ATM/Chk2/p53 signaling cascade is AMPK dependent and sensitizes mesothelioma cells to cisplatin (Shi et al. 2012). In breast cancer and melanoma cells fasting causes the sumoylation of the specialized DNA polymerase REV1 by SUMO2/3, resulting in the relief of REV1's inhibition of p53 and enhancing p53's effects on pro-apoptotic gene expression and apoptosis (Shim et al. 2015).

Many oncogenes are tyrosine kinases and thus provide a target for cancer treatment. PF potentiates the growth inhibiting efficacy of commonly administered tyrosine kinase inhibitors, such as erlotinib, gefitinib, lapatinib, crizotinib, and regorafenib, in in vitro and xenograft models by inhibiting the MAPK signaling and E2F-dependent inhibition of transcription (Caffa et al. 2015). In a non-small cell lung cancer xenograft model, subcutaneous tumor growth was effectively reduced by crizotinib or fasting whereas the combination of fasting with crizotinib was the most efficient treatment option. Similar results have been demonstrated in a colorectal cancer xenograft model for the combination of fasting with regorafenib (Caffa et al. 2015). These results demonstrate the dependence of malignant cells on steady levels of glucose, amino acids, and growth factors and indicate that reducing these factors may protect the organism while reducing tumor progression, particularly in combination with chemotherapy, by generating a challenging environment for the survival of the cancer cell.

Different molecule classes, including antisense RNA, monoclonal antibodies, and dominant negative IGF-1R gene variants have been employed toward this aim (Bahr and Groner 2004), and at least 12 different IGF-1R targeting compounds, including small antagonistic molecules and antibodies, have entered clinical trials (Gualberto and Pollak 2009). These agents have been able to reverse the transformed phenotype in several rodent and human cancer cell lines and, in analogy to fasting, sensitized cancer cells to conventional chemotherapeutic treatment and irradiation (Bahr and Groner 2004). However, IGF-1R blockade did not yield positive results in human trials, possibly because it also interferes with the positive function of IGF-1 in the proliferation of normal cells including stem and immune cells. In contrast, fasting promotes reduced levels of glucose and IGF-1 which return to the normal range after refeeding therefore generating an ideal environment to negatively affect the growth and survival of cancer cells but not that of normal cells.

7 Clinical Efficacy of Fasting

Several studies indicate that fasting has the potential to prevent and treat diseases and promote health in humans. For example, a water-only fast lasting 10–14 days followed by a low-fat, low-sodium vegan-based refeeding period (approximately 6–7 days on average) reduces systolic blood pressure points more than twofold compared to a combined vegan low-fat, low-salt diet and exercise (Goldhamer et al.

2001; Goldhamer 2002). Consuming a very-low-calorie diet of 350 kcal/day (an almost fasting-like approach) has been considered safe in a large cohort with over 2000 participants with chronic diseases, providing evidence that the very-low-calorie diet may protect against several diseases (Michalsen et al. 2005). Additionally, in a randomized clinical trial to evaluate the effects of 2 days of IF (500 kcal/day) a week on weight loss and metabolic disease risk markers in young overweight women found that both chronic CR and IF are equally efficient in reducing biomarkers associated with disease risk (Harvie et al. 2011). Fasting has also been proposed to protect patients from ischemic reperfusion damage following surgery, in which oxidative stress is largely responsible for the damage (Mitchell et al. 2009; van Ginhoven et al. 2009).

Although no randomized clinical trials are (yet) available to evaluate the effect of CR, IF or PF in cancer prevention, preclinical and clinical research supports the potential application in prevention and treatment of human cancers. PF, or fasting-like dietary regimen, can have pronounced effects on IGF-1, insulin, glucose, IGFBP1, and ketone body levels, thereby generating a protective environment for normal cells while creating a metabolic environment that does not favor precancerous and/or cancer cells (Barger et al. 2008; Mercken et al. 2013). In a pilot clinical trial, three monthly FMD cycles decreased risk factors/biomarkers for aging, diabetes, cardiovascular disease, and cancer without major adverse effects, providing support for the use of FMDs to promote health span (Brandhorst et al. 2015). Although not significant, the percentage of mesenchymal stem and progenitor cells in the peripheral blood mono-nucleated cell population showed a trend to increase at the end of FMD, in line with the results obtained in mice (Brandhorst et al. 2015). In these study participants, fasting blood glucose and IGF-1 levels were significantly reduced and remained lower than baseline levels even after resuming their normal diet following the final FMD cycle. Elevated circulating IGF-1 is associated with increased risk of developing certain malignancies (Giovannucci et al. 2003; Smith et al. 2000). In a population-based study of over 6000 American adults, respondents of 50–65 years that reported high protein intake had the highest circulating IGF-1 levels and experienced a 75 % increase in overall mortality and a fourfold increase in cancer death risk during the following 18 years compared to the low-protein/low-IGF-1 cohort (Levine et al. 2014). However, it is important to note that high protein intake was associated with reduced cancer and overall mortality in respondents over 65, indicating that high protein intake may be beneficial in older adults or at least that older adults reporting a low-protein diet are malnourished, frail, and sick (Levine et al. 2014; Dickinson et al. 2014). Although severe IGF-1 deficiency caused by growth hormone receptor deficiency (GHRD) known as Laron's syndrome leads to growth defects in humans, individuals with GHRD rarely develop cancer (Guevara-Aguirre et al. 2011; Steuerman et al. 2011).

In cancer treatment, preliminary studies of 10 patients with a variety of malignancies that voluntarily fasted for up to 180 h in combination with their prescribed chemotherapy indicate a reduction in common chemotherapy-associated side effects such as vomiting, diarrhea, fatigue and weakness, and in the cases where cancer progression could be followed, there was no evidence that fasting protected tumors

Table 1 Clinical trials using dietary restriction and fasting in patients with cancer

Identifier	Title	Purpose	Intervention	Location
NCT01175837	"Short-term fasting before chemotherapy in treating patients with cancer"	Assess the safety and feasibility of short-term fasting prior to administration of chemotherapy	Short-term fasting, 24–48 h	Mayo Clinic, Rochester, Minnesota, USA
NCT00936364	"Short-term fasting: impact on toxicity"	Reducing side effects in patients receiving gemcitabine hydrochloride and cisplatin for advanced solid tumors	Short-term fasting, 24–72 h	University of Southern California, Los Angeles, California, USA
NCT01304251	"Effects of short-term fasting on tolerance to chemotherapy"	Determine the effect of short-term fasting on tolerance to adjuvant chemotherapy in breast cancer patients	Short-term fasting, 24 h before and after chemotherapy	Leiden University Hospital, Leiden, Netherlands
NCT01954836	"Short-term fasting during chemotherapy in patients with gynecological cancer—a randomized controlled cross-over trial"	Assessments of adverse effects, quality of life and laboratory values	Short-term fasting, 60–72 h	Charite University, Berlin, Germany
NCT01802346	"Controlled low calorie diet in reducing side effects and increasing response to chemotherapy in patients with breast or prostate cancer"	Assessments of a low-calorie diet in reducing side effects and increasing response to chemotherapy in patients with breast or prostate cancer	Low-calorie diet 3 days prior to and 1 day after chemotherapy	University of Southern California, Los Angeles, California, USA
NCT02126449	"Dietary restriction as an adjunct to neoadjuvant chemotherapy for HER2 negative breast cancer"	Evaluate the impact of a FMD on tolerance to and efficacy of neoadjuvant chemotherapy in women with stage II or III breast cancer	FMD 3 days prior to and 1 day after chemotherapy	Leiden University Hospital, Leiden, Netherlands

or interfered with chemotherapy efficacy (Safdie et al. 2009). Analogous to the effects of fasting on the immune system in mice, the results from a phase I clinical trial indicate that 72 but not 24 h of PF in combination with chemotherapy were associated with normal lymphocyte counts and maintenance of a normal lineage balance in white blood cells (Cheng et al. 2014; Safdie et al. 2009). In a pilot study, 7 out of 13 women diagnosed with HER2-negative stage II/III breast cancer were randomized to fast 24 h before and 24 h after receiving neo-adjuvant (docetaxel/doxorubicin/cyclophosphamide) chemotherapy (de Groot et al. 2015). Fasting was well tolerated and protected from the chemotherapy-dependent reduction in erythrocyte and thrombocyte counts and possibly DNA damage in healthy cells compared to the non-fasted group. The study observed no changes in leukocyte or neutrophil counts which may be associated with a pegfilgrastim-induced production of white blood cells. Notably, the use of the antiemetic drug dexamethasone may explain the observed glucose increase despite 24 h of starvation and likely attenuated some metabolic benefits of starvation, e.g., a more prominent reduction in IGF-1 (de Groot et al. 2015). However, 48 h of fasting may not be sufficient to optimize the differential stress resistance or sensitization of normal and malignant cells. A number of clinical trials are now addressing the effects of fasting or fasting-mimicking diets on humans and the diet-induced protection of patients from the side effects of chemotherapy while sensitizing cancer cells to the treatment (Table 1).

8 Conclusions

CR has been known to have major health benefits in the vast majority of laboratory animal models. However, chronic CR is not feasible for the great majority of subjects and it often causes detrimental effects, possibly by negatively affecting the immune system, wound healing, and other important functions. PF but particularly higher calorie fasting-mimicking diets have the potential to promote the beneficial effects of CR, while reducing or eliminating adverse effects, and minimizing the burden of chronic restriction or of diets requiring drastic changes several times a week or every other day. The constitutive activation of nutrient signaling pathways which promotes the growth of cancerous cells might also be their Achilles' heel since they allow the use of fasting to promote the protection of normal cells and organs and sensitization of cancer cells, thus generating a wide-acting, consistent, and inexpensive strategy to increase therapeutic index.

Taken together, these results indicate that PF and FMD have the potential to play an important complementary role in medicine by promoting disease prevention, enhancing disease treatment, delaying the aging process, and stimulating stem cell-based regeneration.

References

Arum O, Bonkowski MS, Rocha JS, Bartke A (2009) The growth hormone receptor gene-disrupted mouse fails to respond to an intermittent fasting diet. Aging cell 8(6):756–760. doi:10.1111/j.1474-9726.2009.00520.x ACE520 [pii]

Bahr C, Groner B (2004) The insulin like growth factor-1 receptor (IGF-1R) as a drug target: novel approaches to cancer therapy. Growth Hormone IGF Res Off J Growth Hormone Res Soc Int IGF Res Soc 14(4):287–295. doi:10.1016/j.ghir.2004.02.004

Barger JL, Kayo T, Vann JM, Arias EB, Wang J, Hacker TA, Wang Y, Raederstorff D, Morrow JD, Leeuwenburgh C, Allison DB, Saupe KW, Cartee GD, Weindruch R, Prolla TA (2008) A low dose of dietary resveratrol partially mimics caloric restriction and retards aging parameters in mice. PLoS One 3(6):e2264. doi:10.1371/journal.pone.0002264

Bartke A, Chandrashekar V, Bailey B, Zaczek D, Turyn D (2002) Consequences of growth hormone (GH) overexpression and GH resistance. Neuropeptides 36(2–3):201–208. S0143417902908899 [pii]

Berrigan D, Perkins SN, Haines DC, Hursting SD (2002) Adult-onset calorie restriction and fasting delay spontaneous tumorigenesis in p53-deficient mice. Carcinogenesis 23(5):817–822

Bianchi G, Martella R, Ravera S, Marini C, Capitanio S, Orengo A, Emionite L, Lavarello C, Amaro A, Petretto A, Pfeffer U, Sambuceti G, Pistoia V, Raffaghello L, Longo VD (2015) Fasting induces anti-Warburg effect that increases respiration but reduces ATP-synthesis to promote apoptosis in colon cancer models. Oncotarget 6(14):11806–11819

Huisman SA, Bijman-Lagcher W, JN IJ, Smits R, de Bruin RW (2015) Fasting protects against the side effects of irinotecan but preserves its anti-tumor effect in Apc15lox mutant mice. Cell Cycle 14(14):2333–2339. doi:10.1080/15384101.2015.1044170

Bloom WL (1959) Fasting as an introduction to the treatment of obesity. Metabolism 8(3):214–220

Bonkowski MS, Dominici FP, Arum O, Rocha JS, Al Regaiey KA, Westbrook R, Spong A, Panici J, Masternak MM, Kopchick JJ, Bartke A (2009) Disruption of growth hormone receptor prevents calorie restriction from improving insulin action and longevity. PLoS One 4(2):e4567. doi:10.1371/journal.pone.0004567

Bonorden MJ, Rogozina OP, Kluczny CM, Grossmann ME, Grambsch PL, Grande JP, Perkins S, Lokshin A, Cleary MP (2009) Intermittent calorie restriction delays prostate tumor detection and increases survival time in TRAMP mice. Nutr Cancer 61(2):265–275 doi:10.1080/01635580802419798 908921174 [pii]

Brandhorst S, Wei M, Hwang S, Morgan TE, Longo VD (2013) Short-term calorie and protein restriction provide partial protection from chemotoxicity but do not delay glioma progression. Exp Gerontol doi:10.1016/j.exger.2013.02.016 S0531-5565(13)00047-8 [pii]

Brandhorst S, Choi IY, Wei M, Cheng CW, Sedrakyan S, Navarrete G, Dubeau L, Yap LP, Park R, Vinciguerra M, Di Biase S, Mirzaei H, Mirisola MG, Childress P, Ji L, Groshen S, Penna F, Odetti P, Perin L, Conti PS, Ikeno Y, Kennedy BK, Cohen P, Morgan TE, Dorff TB, Longo VD (2015) A periodic diet that mimics fasting promotes multi-system regeneration, enhanced cognitive performance, and healthspan. Cell Metab 22(1):86–99. doi:10.1016/j.cmet.2015.05.012

Brawley OW (2011) Avoidable cancer deaths globally. CA Cancer J Clin 61(2):67–68. caac.20108 [pii]

Brown-Borg HM (2009) Hormonal control of aging in rodents: the somatotropic axis. Mol Cell Endocrinol 299(1):64–71. doi:10.1016/j.mce.2008.07.001

Brown-Borg HM, Rakoczy SG (2000) Catalase expression in delayed and premature aging mouse models. Exp Gerontol 35(2):199–212. S0531-5565(00)00079-6 [pii]

Brown-Borg HM, Borg KE, Meliska CJ, Bartke A (1996) Dwarf mice and the ageing process. Nature 384(6604):33. doi:10.1038/384033a0

Brown-Borg HM, Rakoczy SG, Romanick MA, Kennedy MA (2002) Effects of growth hormone and insulin-like growth factor-1 on hepatocyte antioxidative enzymes. Exp Biol Med (Maywood) 227(2):94–104

Caffa I, D'Agostino V, Damonte P, Soncini D, Cea M, Monacelli F, Odetti P, Ballestrero A, Provenzani A, Longo VD, Nencioni A (2015) Fasting potentiates the anticancer activity of tyrosine kinase inhibitors by strengthening MAPK signaling inhibition. Oncotarget 6 (14):11820–11832

Cahill GF Jr, Owen OE (1968) Starvation and survival. Trans Am Clin Climatol Assoc 79:13–20

Cahill GJ Jr, Owen OE, Morgan AP (1968) The consumption of fuels during prolonged starvation. Adv Enzyme Regul 6:143–150

Campisi J, Andersen JK, Kapahi P, Melov S (2011) Cellular senescence: a link between cancer and age-related degenerative disease? Semin Cancer Biol 21(6):354–359. doi:10.1016/j. semcancer.2011.09.001 S1044-579X(11)00050-2 [pii]

Cheng C-W, Adams Gregor B, Perin L, Wei M, Zhou X, Lam Ben S, Da Sacco S, Mirisola M, Quinn David I, Dorff Tanya B, Kopchick John J, Longo Valter D (2014) Prolonged fasting reduces IGF-1/PKA to promote hematopoietic-stem-cell-based regeneration and reverse immunosuppression. Cell Stem Cell 14(6):810–823. doi:10.1016/j.stem.2014.04.014

Clancy DJ, Gems D, Harshman LG, Oldham S, Stocker H, Hafen E, Leevers SJ, Partridge L (2001) Extension of life-span by loss of CHICO, a Drosophila insulin receptor substrate protein. Science 292(5514):104–106. doi:10.1126/science.1057991 292/5514/104 [pii]

Colman RJ, Anderson RM, Johnson SC, Kastman EK, Kosmatka KJ, Beasley TM, Allison DB, Cruzen C, Simmons HA, Kemnitz JW, Weindruch R (2009) Caloric restriction delays disease onset and mortality in rhesus monkeys. Science 325(5937): 201–204. 325/5937/201 [pii]

Coschigano KT, Holland AN, Riders ME, List EO, Flyvbjerg A, Kopchick JJ (2003) Deletion, but not antagonism, of the mouse growth hormone receptor results in severely decreased body weights, insulin, and insulin-like growth factor I levels and increased life span. Endocrinology 144(9):3799–3810

Cuervo AM, Bergamini E, Brunk UT, Droge W, Ffrench M, Terman A (2005) Autophagy and aging: the importance of maintaining "clean" cells. Autophagy 1(3):131–140. 2017 [pii]

Danaei G, Vander Hoorn S, Lopez AD, Murray CJ, Ezzati M (2005) Causes of cancer in the world: comparative risk assessment of nine behavioural and environmental risk factors. Lancet 366(9499):1784–1793. S0140-6736(05)67725-2 [pii]

D'Aronzo M, Vinciguerra M, Mazza T, Panebianco C, Saracino C, Pereira SP, Graziano P, Pazienza V (2015) Fasting cycles potentiate the efficacy of gemcitabine treatment in in vitro and in vivo pancreatic cancer models. Oncotarget 6(21):18545–18557

Davis LM, Pauly JR, Readnower RD, Rho JM, Sullivan PG (2008) Fasting is neuroprotective following traumatic brain injury. J Neurosci Res 86(8):1812–1822. doi:10.1002/jnr.21628

de Groot S, Vreeswijk MP, Welters MJ, Gravesteijn G, Boei JJ, Jochems A, Houtsma D, Putter H, van der Hoeven JJ, Nortier JW, Pijl H, Kroep JR (2015) The effects of short-term fasting on tolerance to (neo) adjuvant chemotherapy in HER2-negative breast cancer patients: a randomized pilot study. BMC Cancer 15:652. doi:10.1186/s12885-015-1663-5

DePinho RA (2000) The age of cancer. Nature 408(6809):248–254. doi:10.1038/35041694

Descamps O, Riondel J, Ducros V, Roussel AM (2005) Mitochondrial production of reactive oxygen species and incidence of age-associated lymphoma in OF1 mice: effect of alternate-day fasting. Mech Ageing Dev 126(11):1185–1191. doi:10.1016/j.mad.2005.06.007 S0047-6374 (05)00147-8 [pii]

Dickinson JM, Gundermann DM, Walker DK, Reidy PT, Borack MS, Drummond MJ, Arora M, Volpi E, Rasmussen BB (2014) Leucine-enriched amino acid ingestion after resistance exercise prolongs myofibrillar protein synthesis and amino acid transporter expression in older men. J Nutr 144(11):1694–1702. doi:10.3945/jn.114.198671

Drenick EJ, Smith R (1964) Weight reduction by prolonged starvation. Practical management. Postgrad Med 36:A95–100

Fabrizio P, Longo VD (2003) The chronological life span of *Saccharomyces cerevisiae*. Aging Cell 2(2):73–81

Fabrizio P, Pozza F, Pletcher SD, Gendron CM, Longo VD (2001) Regulation of longevity and stress resistance by Sch9 in yeast. Science 292(5515):288–290. doi:10.1126/science.1059497 1059497 [pii]

Fabrizio P, Liou LL, Moy VN, Diaspro A, Valentine JS, Gralla EB, Longo VD (2003) SOD2 functions downstream of Sch9 to extend longevity in yeast. Genetics 163(1):35–46

Fontana L, Weiss EP, Villareal DT, Klein S, Holloszy JO (2008) Long-term effects of calorie or protein restriction on serum IGF-1 and IGFBP-3 concentration in humans. Aging Cell 7 (5):681–687

Fontana L, Partridge L, Longo VD (2010) Extending healthy life span–from yeast to humans. Science 328(5976):321–326. doi:10.1126/science.1172539 328/5976/321 [pii]

Frystyk J, Delhanty PJ, Skjaerbaek C, Baxter RC (1999) Changes in the circulating IGF system during short-term fasting and refeeding in rats. Am J Physiol 277(2 Pt 1):E245–E252

Giannakou ME, Partridge L (2007) Role of insulin-like signalling in Drosophila lifespan. Trends Biochem Sci 32(4):180–188 doi:10.1016/j.tibs.2007.02.007 S0968-0004(07)00064-3 [pii]

Giovannucci E, Pollak M, Liu Y, Platz EA, Majeed N, Rimm EB, Willett WC (2003) Nutritional predictors of insulin-like growth factor I and their relationships to cancer in men. Cancer Epidemiol Biomarkers Prev 12(2):84–89

Goldhamer AC (2002) Initial cost of care results in medically supervised water-only fasting for treating high blood pressure and diabetes. J Altern Complement Med 8(6):696–697. doi:10. 1089/10755530260511694

Goldhamer A, Lisle D, Parpia B, Anderson SV, Campbell TC (2001) Medically supervised water-only fasting in the treatment of hypertension. J Manipulative Physiol Ther 24(5):335– 339. doi:10.1067/mmt.2001.115263 S0161-4754(01)85575-5 [pii]

Gonidakis S, Finkel SE, Longo VD (2010) Genome-wide screen identifies *Escherichia coli* TCA-cycle-related mutants with extended chronological lifespan dependent on acetate metabolism and the hypoxia-inducible transcription factor ArcA. Aging Cell 9(5):868–881. doi:10.1111/j.1474-9726.2010.00618.x

Goodrick CL, Ingram DK, Reynolds MA, Freeman JR, Cider NL (1982) Effects of intermittent feeding upon growth and life span in rats. Gerontology 28(4):233–241

Goodrick CL, Ingram DK, Reynolds MA, Freeman JR, Cider N (1990) Effects of intermittent feeding upon body weight and lifespan in inbred mice: interaction of genotype and age. Mech Ageing Dev 55(1):69–87. 0047-6374(90)90107-Q [pii]

Grandison RC, Wong R, Bass TM, Partridge L, Piper MD (2009) Effect of a standardised dietary restriction protocol on multiple laboratory strains of *Drosophila melanogaster*. PLoS One 4(1): e4067. doi:10.1371/journal.pone.0004067

Grasl-Kraupp B, Bursch W, Ruttkay-Nedecky B, Wagner A, Lauer B, Schulte-Hermann R (1994) Food restriction eliminates preneoplastic cells through apoptosis and antagonizes carcinogenesis in rat liver. Proc Natl Acad Sci USA 91(21):9995–9999

Greer EL, Dowlatshahi D, Banko MR, Villen J, Hoang K, Blanchard D, Gygi SP, Brunet A (2007) An AMPK-FOXO pathway mediates longevity induced by a novel method of dietary restriction in *C. elegans*. Current Biol CB 17(19):1646–1656. doi:10.1016/j.cub.2007.08.047

Gualberto A, Pollak M (2009) Emerging role of insulin-like growth factor receptor inhibitors in oncology: early clinical trial results and future directions. Oncogene 28(34):3009–3021. doi:10. 1038/onc.2009.172

Guarente L, Kenyon C (2000) Genetic pathways that regulate ageing in model organisms. Nature 408(6809):255–262. doi:10.1038/35041700

Guevara-Aguirre J, Balasubramanian P, Guevara-Aguirre M, Wei M, Madia F, Cheng CW, Hwang D, Martin-Montalvo A, Saavedra J, Ingles S, de Cabo R, Cohen P, Longo VD (2011) Growth hormone receptor deficiency is associated with a major reduction in pro-aging signaling, cancer, and diabetes in humans. Sci Transl Med 3(70):70ra13 doi:10.1126/ scitranslmed.3001845 3/70/70ra13 [pii]

Hanahan D, Weinberg RA (2000) The hallmarks of cancer. Cell 100(1):57–70. S0092-8674(00)
   81683-9 [pii]
Hanahan D, Weinberg RA (2011) Hallmarks of cancer: the next generation. Cell 144(5):646–674.
   doi:10.1016/j.cell.2011.02.013 S0092-8674(11)00127-9 [pii]
Harrison DE, Archer JR (1989) Natural selection for extended longevity from food restriction.
   Growth Dev Aging GDA 53(1–2):3
Harvie MN, Pegington M, Mattson MP, Frystyk J, Dillon B, Evans G, Cuzick J, Jebb SA,
   Martin B, Cutler RG, Son TG, Maudsley S, Carlson OD, Egan JM, Flyvbjerg A, Howell A
   (2011) The effects of intermittent or continuous energy restriction on weight loss and metabolic
   disease risk markers: a randomized trial in young overweight women. Int J Obes 35(5):714–
   727. doi:10.1038/ijo.2010.171 ijo2010171 [pii]
Henis-Korenblit S, Zhang P, Hansen M, McCormick M, Lee SJ, Cary M, Kenyon C (2010)
   Insulin/IGF-1 signaling mutants reprogram ER stress response regulators to promote longevity.
   Proc Natl Acad Sci USA 107(21):9730–9735. doi:10.1073/pnas.1002575107 1002575107 [pii]
Hine C, Harputlugil E, Zhang Y, Ruckenstuhl C, Lee BC, Brace L, Longchamp A,
   Trevino-Villarreal JH, Mejia P, Ozaki CK, Wang R, Gladyshev VN, Madeo F, Mair WB,
   Mitchell JR (2015) Endogenous hydrogen sulfide production is essential for dietary restriction
   benefits. Cell 160(1–2):132–144. doi:10.1016/j.cell.2014.11.048
Holzenberger M, Dupont J, Ducos B, Leneuve P, Geloen A, Even PC, Cervera P, Le Bouc Y
   (2003) IGF-1 receptor regulates lifespan and resistance to oxidative stress in mice. Nature 421
   (6919):182–187. doi:10.1038/nature01298 nature01298 [pii]
Honjoh S, Yamamoto T, Uno M, Nishida E (2009) Signalling through RHEB-1 mediates
   intermittent fasting-induced longevity in C. elegans. Nature 457(7230):726–730. doi:10.1038/
   nature07583
Hsu AL, Murphy CT, Kenyon C (2003) Regulation of aging and age-related disease by DAF-16
   and heat-shock factor. Science 300(5622):1142–1145. doi:10.1126/science.1083701
   300/5622/1142 [pii]
Hursting SD, Perkins SN, Phang JM (1994) Calorie restriction delays spontaneous tumorigenesis
   in p53-knockout transgenic mice. Proc Natl Acad Sci USA 91(15):7036–7040
Imae M, Fu Z, Yoshida A, Noguchi T, Kato H (2003) Nutritional and hormonal factors control the
   gene expression of FoxOs, the mammalian homologues of DAF-16. J Mol Endocrinol 30
   (2):253–262
Jemal A, Siegel R, Ward E, Hao Y, Xu J, Murray T, Thun MJ (2008) Cancer statistics, 2008. CA
   Cancer J Clin 58 (2):71–96. CA.2007.0010 [pii]
Jenkins DE, Schultz JE, Matin A (1988) Starvation-induced cross protection against heat or H_2O_2
   challenge in Escherichia coli. J Bacteriol 170(9):3910–3914
Johnson TE (1990) Increased life-span of age-1 mutants in Caenorhabditis elegans and lower
   Gompertz rate of aging. Science 249(4971):908–912
Kalaany NY, Sabatini DM (2009) Tumours with PI3K activation are resistant to dietary restriction.
   Nature 458(7239):725–731. doi:10.1038/nature07782 nature07782 [pii]
Kenyon C (2001) A conserved regulatory system for aging. Cell 105(2):165–168. S0092-8674(01)
   00306-3 [pii]
Kerndt PR, Naughton JL, Driscoll CE, Loxterkamp DA (1982) Fasting: the history, pathophys-
   iology and complications. West J Med 137(5):379–399
Ketelslegers JM, Maiter D, Maes M, Underwood LE, Thissen JP (1995) Nutritional regulation of
   insulin-like growth factor-I. Metabolism 44(10 Suppl 4):50–57
Kim SK, Demetri GD (1996) Chemotherapy and neutropenia. Hematol Oncol Clin North Am 10
   (2):377–395
Kinzler KW, Vogelstein B (1996) Lessons from hereditary colorectal cancer. Cell 87(2):159–170.
   S0092-8674(00)81333-1 [pii]
Koubova J, Guarente L (2003) How does calorie restriction work? Genes Dev 17(3):313–321.
   doi:10.1101/gad.1052903

Kristan DM (2008) Calorie restriction and susceptibility to intact pathogens. Age (Dordr) 30(2–3):147–156. doi:10.1007/s11357-008-9056-1

Kroemer G, Marino G, Levine B (2010) Autophagy and the integrated stress response. Mol Cell 40 (2):280–293. doi:10.1016/j.molcel.2010.09.023 S1097-2765(10)00751-3 [pii]

Larson-Meyer DE, Newcomer BR, Heilbronn LK, Volaufova J, Smith SR, Alfonso AJ, Lefevre M, Rood JC, Williamson DA, Ravussin E (2008) Effect of 6-month calorie restriction and exercise on serum and liver lipids and markers of liver function. Obesity (Silver Spring) 16 (6):1355–1362. doi:10.1038/oby.2008.201 oby2008201 [pii]

Lee C, Longo VD (2011) Fasting vs dietary restriction in cellular protection and cancer treatment: from model organisms to patients. Oncogene 30(30):3305–3316. doi:10.1038/onc.2011.91 onc201191 [pii]

Lee SJ, Murphy CT, Kenyon C (2009) Glucose shortens the life span of *C. elegans* by downregulating DAF-16/FOXO activity and aquaporin gene expression. Cell Metab 10 (5):379–391. doi:10.1016/j.cmet.2009.10.003

Lee C, Safdie FM, Raffaghello L, Wei M, Madia F, Parrella E, Hwang D, Cohen P, Bianchi G, Longo VD (2010) Reduced levels of IGF-I mediate differential protection of normal and cancer cells in response to fasting and improve chemotherapeutic index. Cancer Res 70(4):1564–1572. doi:10.1158/0008-5472.CAN-09-3228 0008-5472.CAN-09-3228 [pii]

Lee C, Raffaghello L, Brandhorst S, Safdie FM, Bianchi G, Martin-Montalvo A, Pistoia V, Wei M, Hwang S, Merlino A, Emionite L, de Cabo R, Longo VD (2012) Fasting cycles retard growth of tumors and sensitize a range of cancer cell types to chemotherapy. Sci Transl Med. doi:10. 1126/scitranslmed.3003293 scitranslmed.3003293 [pii]

Levine ME, Suarez JA, Brandhorst S, Balasubramanian P, Cheng CW, Madia F, Fontana L, Mirisola MG, Guevara-Aguirre J, Wan J, Passarino G, Kennedy BK, Wei M, Cohen P, Crimmins EM, Longo VD (2014) Low protein intake is associated with a major reduction in IGF-1, cancer, and overall mortality in the 65 and younger but not older population. Cell Metab 19(3):407–417. doi:10.1016/j.cmet.2014.02.006

Li Y, Xu W, McBurney MW, Longo VD (2008) SirT1 inhibition reduces IGF-I/IRS-2/Ras/ERK1/2 signaling and protects neurons. Cell Metab 8(1):38–48. doi:10. 1016/j.cmet.2008.05.004 S1550-4131(08)00148-4 [pii]

Liao CY, Rikke BA, Johnson TE, Diaz V, Nelson JF (2010) Genetic variation in the murine lifespan response to dietary restriction: from life extension to life shortening. Aging Cell 9 (1):92–95. doi:10.1111/j.1474-9726.2009.00533.x

Longo VD (1999) Mutations in signal transduction proteins increase stress resistance and longevity in yeast, nematodes, fruit flies, and mammalian neuronal cells. Neurobiol Aging 20 (5):479–486. S0197-4580(99)00089-5 [pii]

Longo VD, Finch CE (2003) Evolutionary medicine: from dwarf model systems to healthy centenarians? Science 299(5611):1342–1346. doi:10.1126/science.1077991 299/5611/1342 [pii]

Longo VD, Fontana L (2010) Calorie restriction and cancer prevention: metabolic and molecular mechanisms. Trends Pharmacol Sci 31(2):89–98. doi:10.1016/j.tips.2009.11.004 S0165-6147 (09)00202-8 [pii]

Longo VD, Mattson MP (2014) Fasting: molecular mechanisms and clinical applications. Cell Metab 19(2):181–192. doi:10.1016/j.cmet.2013.12.008

Longo VD, Ellerby LM, Bredesen DE, Valentine JS, Gralla EB (1997) Human Bcl-2 reverses survival defects in yeast lacking superoxide dismutase and delays death of wild-type yeast. J Cell Biol 137(7):1581–1588

Love RR, Leventhal H, Easterling DV, Nerenz DR (1989) Side effects and emotional distress during cancer chemotherapy. Cancer 63(3):604–612

Maccario M, Aimaretti G, Grottoli S, Gauna C, Tassone F, Corneli G, Rossetto R, Wu Z, Strasburger CJ, Ghigo E (2001) Effects of 36 hour fasting on GH/IGF-I axis and metabolic parameters in patients with simple obesity. comparison with normal subjects and hypopituitary patients with severe GH deficiency. Int J Obes Relat Metab Disord 25(8):1233–1239. doi:10. 1038/sj.ijo.0801671

Madia F, Wei M, Yuan V, Hu J, Gattazzo C, Pham P, Goodman MF, Longo VD (2009) Oncogene homologue Sch9 promotes age-dependent mutations by a superoxide and Rev1/Polzeta-dependent mechanism. J Cell Biol 186(4):509–523. doi:10.1083/jcb. 200906011

Mahoney LB, Denny CA, Seyfried TN (2006) Caloric restriction in C57BL/6J mice mimics therapeutic fasting in humans. Lipids Health Dis 5:13. doi:10.1186/1476-511X-5-13 1476-511X-5-13 [pii]

Marino G, Madeo F, Kroemer G (2010) Autophagy for tissue homeostasis and neuroprotection. Curr Opin Cell Biol 23(2):198–206 doi:10.1016/j.ceb.2010.10.001 S0955-0674(10)00172-9 [pii]

Matsuzaki J, Kuwamura M, Yamaji R, Inui H, Nakano Y (2001) Inflammatory responses to lipopolysaccharide are suppressed in 40 % energy-restricted mice. J Nutr 131(8):2139–2144

Mattison JA, Roth GS, Beasley TM, Tilmont EM, Handy AM, Herbert RL, Longo DL, Allison DB, Young JE, Bryant M, Barnard D, Ward WF, Qi W, Ingram DK, de Cabo R (2012) Impact of caloric restriction on health and survival in rhesus monkeys from the NIA study. Nature 489(7415):318–321. doi:10.1038/nature11432 nature11432 [pii]

Mattson MP (2005) Energy intake, meal frequency, and health: a neurobiological perspective. Annu Rev Nutr 25:237–260. doi:10.1146/annurev.nutr.25.050304.092526

Mattson MP (2012) Energy intake and exercise as determinants of brain health and vulnerability to injury and disease. Cell Metab 16(6):706–722. doi:10.1016/j.cmet.2012.08.012 S1550-4131 (12)00402-0 [pii]

Mattson MP, Wan R (2005) Beneficial effects of intermittent fasting and caloric restriction on the cardiovascular and cerebrovascular systems. J Nutr Biochem 16(3):129–137. doi:10.1016/j. jnutbio.2004.12.007 S0955-2863(04)00261-X [pii]

Medema RH, Bos JL (1993) The role of p21ras in receptor tyrosine kinase signaling. Crit Rev Oncog 4(6):615–661

Mercken EM, Crosby SD, Lamming DW, JeBailey L, Krzysik-Walker S, Villareal DT, Capri M, Franceschi C, Zhang Y, Becker K, Sabatini DM, de Cabo R, Fontana L (2013) Calorie restriction in humans inhibits the PI3 K/AKT pathway and induces a younger transcription profile. Aging Cell 12(4):645–651. doi:10.1111/acel.12088

Merimee TJ, Fineberg SE (1974) Growth hormone secretion in starvation: a reassessment. J Clin Endocrinol Metab 39(2):385–386. doi:10.1210/jcem-39-2-385

Michalsen A, Hoffmann B, Moebus S, Backer M, Langhorst J, Dobos GJ (2005) Incorporation of fasting therapy in an integrative medicine ward: evaluation of outcome, safety, and effects on lifestyle adherence in a large prospective cohort study. J Altern Complement Med 11(4):601–607. doi:10.1089/acm.2005.11.601

Mirzaei H, Suarez JA, Longo VD (2014) Protein and amino acid restriction, aging and disease: from yeast to humans. Trends Endocrinol Metab TEM 25(11):558–566. doi:10.1016/j.tem. 2014.07.002

Mitchell JR, Verweij M, Brand K, van de Ven M, Goemaere N, van den Engel S, Chu T, Forrer F, Muller C, de Jong M, van IW, Jn IJ, Hoeijmakers JH, de Bruin RW (2009) Short-term dietary restriction and fasting precondition against ischemia reperfusion injury in mice. Aging Cell 9 (1):40–53. doi:10.1111/j.1474-9726.2009.00532.x ACE532[pii]

Moller N, Jorgensen JO (2009) Effects of growth hormone on glucose, lipid, and protein metabolism in human subjects. Endocr Rev 30(2):152–177. doi:10.1210/er.2008-0027 er.2008-0027 [pii]

Moreschi (1909) Beziehungen zwischen Ernährung und Tumorwachstum. Z Immunitätsforsch Orig (2): 651–675

Morselli E, Galluzzi L, Kepp O, Criollo A, Maiuri MC, Tavernarakis N, Madeo F, Kroemer G (2009a) Autophagy mediates pharmacological lifespan extension by spermidine and resveratrol. Aging (Albany NY) 1(12):961–970

Morselli E, Galluzzi L, Kepp O, Vicencio JM, Criollo A, Maiuri MC, Kroemer G (2009) Anti- and pro-tumor functions of autophagy. Biochimica et Biophysica Acta 1793(9):1524–1532. doi:10.1016/j.bbamcr.2009.01.006 S0167-4889(09)00024-X [pii]

Mukherjee P, Abate LE, Seyfried TN (2004) Antiangiogenic and proapoptotic effects of dietary restriction on experimental mouse and human brain tumors. Clin Cancer Res 10(16):5622–5629. doi:10.1158/1078-0432.CCR-04-0308 10/16/5622 [pii]

Murakami S (2006) Stress resistance in long-lived mouse models. Exp Gerontol 41(10):1014–1019. doi:10.1016/j.exger.2006.06.061 S0531-5565(06)00236-1 [pii]

Nicklin P, Bergman P, Zhang B, Triantafellow E, Wang H, Nyfeler B, Yang H, Hild M, Kung C, Wilson C, Myer VE, MacKeigan JP, Porter JA, Wang YK, Cantley LC, Finan PM, Murphy LO (2009) Bidirectional transport of amino acids regulates mTOR and autophagy. Cell 136 (3):521–534. doi:10.1016/j.cell.2008.11.044

Palmblad J, Levi L, Burger A, Melander A, Westgren U, von Schenck H, Skude G (1977) Effects of total energy withdrawal (fasting) on thelevels of growth hormone, thyrotropin, cortisol, adrenaline, noradrenaline, T4, T3, and rT3 in healthy males. Acta Medica Scandinavica 201(1–2):15–22

Pamplona R, Barja G (2006) Mitochondrial oxidative stress, aging and caloric restriction: the protein and methionine connection. Biochimica et biophysica acta 1757(5–6):496–508. doi:10.1016/j.bbabio.2006.01.009 S0005-2728(06)00022-3 [pii]

Paoletti R, Bolego C, Poli A, Cignarella A (2006) Metabolic syndrome, inflammation and atherosclerosis. Vasc Health Risk Manag 2(2):145–152

Paradis S, Ailion M, Toker A, Thomas JH, Ruvkun G (1999) A PDK1 homolog is necessary and sufficient to transduce AGE-1 PI3 kinase signals that regulate diapause in *Caenorhabditis elegans*. Genes Dev 13(11):1438–1452

Partridge AH, Burstein HJ, Winer EP (2001) Side effects of chemotherapy and combined chemohormonal therapy in women with early-stage breast cancer. J Natl Cancer Inst Monogr 30:135–142

Piper MD, Partridge L (2007) Dietary restriction in Drosophila: delayed aging or experimental artefact? PLoS Genet 3(4):e57. doi:10.1371/journal.pgen.0030057

Premoselli F, Sesca E, Binasco V, Caderni G, Tessitore L (1998) Fasting/re-feeding before initiation enhances the growth of aberrant crypt foci induced by azoxymethane in rat colon and rectum. Int J Cancer 77(2):286–294. doi:10.1002/(SICI)1097-0215(19980717)77:2<286:AID-IJC19>3.0.CO;2-9

Raffaghello L, Lee C, Safdie FM, Wei M, Madia F, Bianchi G, Longo VD (2008) Starvation-dependent differential stress resistance protects normal but not cancer cells against high-dose chemotherapy. Proc Natl Acad Sci USA 105(24):8215–8220. doi:10.1073/pnas.0708100105 0708100105 [pii]

Reed MJ, Penn PE, Li Y, Birnbaum R, Vernon RB, Johnson TS, Pendergrass WR, Sage EH, Abrass IB, Wolf NS (1996) Enhanced cell proliferation and biosynthesis mediate improved wound repair in refed, caloric-restricted mice. Mech Ageing Dev 89(1):21–43. 004763749601737X [pii]

Rikke BA, Liao CY, McQueen MB, Nelson JF, Johnson TE (2010) Genetic dissection of dietary restriction in mice supports the metabolic efficiency model of life extension. Exp Gerontol 45 (9):691–701. doi:10.1016/j.exger.2010.04.008

Ruckenstuhl C, Buttner S, Carmona-Gutierrez D, Eisenberg T, Kroemer G, Sigrist SJ, Frohlich KU, Madeo F (2009) The Warburg effect suppresses oxidative stress induced apoptosis in a yeast model for cancer. PLoS One 4(2):e4592. doi:10.1371/journal.pone.0004592

Runcie J, Thomson TJ (1970) Prolonged starvation—a dangerous procedure? Br Med J 3 (5720):432–435

Safdie FM, Dorff T, Quinn D, Fontana L, Wei M, Lee C, Cohen P, Longo VD (2009) Fasting and cancer treatment in humans: a case series report. Aging (Albany NY) 1(12):988–1007

Safdie F, Brandhorst S, Wei M, Wang W, Lee C, Hwang S, Conti PS, Chen TC, Longo VD (2012) Fasting enhances the response of glioma to chemo- and radiotherapy. PloS One 7(9):e44603. doi:10.1371/journal.pone.0044603 PONE-D-12-00123 [pii]

Salmon AB, Murakami S, Bartke A, Kopchick J, Yasumura K, Miller RA (2005) Fibroblast cell lines from young adult mice of long-lived mutant strains are resistant to multiple forms of stress. Am J Physiol Endocrinol Metab 289(1):E23–E29. doi:10.1152/ajpendo.00575.2004 00575.2004 [pii]

Sans MD, Lee SH, D'Alecy LG, Williams JA (2004) Feeding activates protein synthesis in mouse pancreas at the translational level without increase in mRNA. Am J Physiol Gastrointest Liver Physiol 287(3):G667–G675. doi:10.1152/ajpgi.00505.2003

Selman C, Tullet JM, Wieser D, Irvine E, Lingard SJ, Choudhury AI, Claret M, Al-Qassab H, Carmignac D, Ramadani F, Woods A, Robinson IC, Schuster E, Batterham RL, Kozma SC, Thomas G, Carling D, Okkenhaug K, Thornton JM, Partridge L, Gems D, Withers DJ (2009) Ribosomal protein S6 kinase 1 signaling regulates mammalian life span. Science 326 (5949):140–144 doi:10.1126/science.1177221 326/5949/140 [pii]

Sesca E, Premoselli F, Binasco V, Bollito E, Tessitore L (1998) Fasting-refeeding stimulates the development of mammary tumors induced by 7,12-dimethylbenz[a]anthracene. Nutr Cancer 30 (1):25–30. doi:10.1080/01635589809514636

Sharma HS, Nyberg F, Gordh T, Alm P, Westman J (2000) Neurotrophic factors influence upregulation of constitutive isoform of heme oxygenase and cellular stress response in the spinal cord following trauma. An experimental study using immunohistochemistry in the rat. Amino Acids 19(1):351–361

Shi Y, Felley-Bosco E, Marti TM, Orlowski K, Pruschy M, Stahel RA (2012) Starvation-induced activation of ATM/Chk2/p53 signaling sensitizes cancer cells to cisplatin. BMC Cancer 12:571 doi:10.1186/1471-2407-12-571 1471-2407-12-571 [pii]

Shim HS, Wei M, Brandhorst S, Longo VD (2015) Starvation promotes REV1 SUMOylation and p53-dependent sensitization of melanoma and breast cancer cells. Cancer Res 75(6):1056–1067. doi:10.1158/0008-5472.CAN-14-2249

Siegel R, Naishadham D, Jemal A (2012) Cancer statistics, 2012. CA Cancer J Clin 62(1):10–29. doi:10.3322/caac.20138

Smith GD, Gunnell D, Holly J (2000) Cancer and insulin-like growth factor-I. A potential mechanism linking the environment with cancer risk. Bmj 321(7265):847–848

Smith ED, Kaeberlein TL, Lydum BT, Sager J, Welton KL, Kennedy BK, Kaeberlein M (2008) Age- and calorie-independent life span extension from dietary restriction by bacterial deprivation in Caenorhabditis elegans. BMC Dev Biol 8:49. doi:10.1186/1471-213X-8-49

Sohal RS, Forster MJ (2014) Caloric restriction and the aging process: a critique. Free Radic Biol Med 73:366–382. doi:10.1016/j.freeradbiomed.2014.05.015

Sohal RS, Weindruch R (1996) Oxidative stress, caloric restriction, and aging. Science 273 (5271):59–63

Sonntag WE, Lynch CD, Cefalu WT, Ingram RL, Bennett SA, Thornton PL, Khan AS (1999) Pleiotropic effects of growth hormone and insulin-like growth factor (IGF)-1 on biological aging: inferences from moderate caloric-restricted animals. J Gerontol Ser A Biol Sci Med Sci 54(12):B521–B538

Steuerman R, Shevah O, Laron Z (2011) Congenital IGF1 deficiency tends to confer protection against post-natal development of malignancies. Eur J Endocrinol/Eur Fed Endocr Soc 164 (4):485–489. doi:10.1530/EJE-10-0859

Tagliaferro AR, Ronan AM, Meeker LD, Thompson HJ, Scott AL, Sinha D (1996) Cyclic food restriction alters substrate utilization and abolishes protection from mammary carcinogenesis female rats. J Nutr 126(5):1398–1405

Tannenbaum A (1940) The initiation and growth of tumours. Introduction. 1. Effects of underfeeding. Am J Cancer 38:335–350

Tannenbaum A (1996) The dependence of tumor formation on the composition of the calorie-restricted diet as well as on the degree of restriction. 1945. Nutrition 12(9):653–654. S0899900796001797 [pii]

Tannenbaum GS, Rorstad O, Brazeau P (1979) Effects of prolonged food deprivation on the ultradian growth hormone rhythm and immunoreactive somatostatin tissue levels in the rat. Endocrinology 104(6):1733–1738

Tessitore L, Tomasi C, Greco M (1999) Fasting-induced apoptosis in rat liver is blocked by cycloheximide. Eur J Cell Biol 78(8):573–579 doi:10.1016/S0171-9335(99)80023-5 S0171-9335(99)80023-5 [pii]

Tettweiler G, Miron M, Jenkins M, Sonenberg N, Lasko PF (2005) Starvation and oxidative stress resistance in Drosophila are mediated through the eIF4E-binding protein, d4E-BP. Genes Dev 19(16):1840–1843. doi:10.1101/gad.1311805

Torre LA, Bray F, Siegel RL, Ferlay J, Lortet-Tieulent J, Jemal A (2015) Global cancer statistics, 2012. CA Cancer J Clin 65(2):87–108. doi:10.3322/caac.21262

Underwood LE, Thissen JP, Lemozy S, Ketelslegers JM, Clemmons DR (1994) Hormonal and nutritional regulation of IGF-I and its binding proteins. Horm Res 42(4–5):145–151

Unger RH, Eisentraut AM, Madison LL (1963) The effects of total starvation upon the levels of circulating glucagon and insulin in man. J Clin Investig 42:1031–1039. doi:10.1172/JCI104788

van Ginhoven TM, Mitchell JR, Verweij M, Hoeijmakers JH, Ijzermans JN, de Bruin RW (2009) The use of preoperative nutritional interventions to protect against hepatic ischemia-reperfusion injury. Liver Transpl 15(10):1183–1191. doi:10.1002/lt.21871

Vander Heiden MG, Cantley LC, Thompson CB (2009) Understanding the Warburg effect: the metabolic requirements of cell proliferation. Science 324(5930):1029–1033. doi:10.1126/science.1160809 324/5930/1029 [pii]

Vaughn AE, Deshmukh M (2008) Glucose metabolism inhibits apoptosis in neurons and cancer cells by redox inactivation of cytochrome c. Nat Cell Biol 10(12):1477–1483. doi:10.1038/ncb1807

Verweij M, van Ginhoven TM, Mitchell JR, Sluiter W, van den Engel S, Roest HP, Torabi E, Ijzermans JN, Hoeijmakers JH, de Bruin RW (2011) Preoperative fasting protects mice against hepatic ischemia/reperfusion injury: mechanisms and effects on liver regeneration. Liver Transpl 17(6):695–704. doi:10.1002/lt.22243

Vigne P, Frelin C (2007) Diet dependent longevity and hypoxic tolerance of adult Drosophila melanogaster. Mech Ageing Dev 128(5–6):401–406. doi:10.1016/j.mad.2007.05.008

Villa-Cuesta E, Sage BT, Tatar M (2010) A role for Drosophila dFoxO and dFoxO 5'UTR internal ribosomal entry sites during fasting. PLoS One 5(7):e11521. doi:10.1371/journal.pone.0011521

Wang T, Hung CC, Randall DJ (2006) The comparative physiology of food deprivation: from feast to famine. Annu Rev Physiol 68:223–251. doi:10.1146/annurev.physiol.68.040104.105739

Warburg O (1956) On respiratory impairment in cancer cells. Science 124(3215):269–270

Wei M, Fabrizio P, Hu J, Ge H, Cheng C, Li L, Longo VD (2008) Life span extension by calorie restriction depends on Rim15 and transcription factors downstream of Ras/PKA, Tor, and Sch9. PLoS Genet 4(1):e13 doi:10.1371/journal.pgen.0040013 07-PLGE-RA-0807 [pii]

Weindruch R, Walford RL (1982) Dietary restriction in mice beginning at 1 year of age: effect on life-span and spontaneous cancer incidence. Science 215(4538):1415–1418

Weindruch R, Walford RL, Fligiel S, Guthrie D (1986) The retardation of aging in mice by dietary restriction: longevity, cancer, immunity and lifetime energy intake. J Nutr 116(4):641–654

Weinkove D, Halstead JR, Gems D, Divecha N (2006) Long-term starvation and ageing induce AGE-1/PI 3-kinase-dependent translocation of DAF-16/FOXO to the cytoplasm. BMC Biol 4:1. doi:10.1186/1741-7007-4-1

Weraarchakul N, Strong R, Wood WG, Richardson A (1989) The effect of aging and dietary restriction on DNA repair. Exp Cell Res 181(1):197–204. 0014-4827(89)90193-6 [pii]

Withers SS, Kass PH, Rodriguez CO Jr, Skorupski KA, O'Brien D, Guerrero TA, Sein KD, Rebhun RB (2014) Fasting reduces the incidence of delayed-type vomiting associated with doxorubicin treatment in dogs with lymphoma. Transl Oncol. doi:10.1016/j.tranon.2014.04.014

Wohlgemuth SE, Julian D, Akin DE, Fried J, Toscano K, Leeuwenburgh C, Dunn WA Jr (2007) Autophagy in the heart and liver during normal aging and calorie restriction. Rejuvenation Res 10(3):281–292. doi:10.1089/rej.2006.0535

Youngman LD (1993) Protein restriction (PR) and caloric restriction (CR) compared: effects on DNA damage, carcinogenesis, and oxidative damage. Mutat Res 295(4–6):165–179

Zapf J, Hauri C, Futo E, Hussain M, Rutishauser J, Maack CA, Froesch ER (1995) Intravenously injected insulin-like growth factor (IGF) I/IGF binding protein-3 complex exerts insulin-like effects in hypophysectomized, but not in normal rats. J Clin Investig 95(1):179–186. doi:10.1172/JCI117636

Printed in the United States
By Bookmasters